INTRODUCTORY ORGANIC CHEMISTRY
AND
HYDROCARBONS
A Physical Chemistry Approach

Caio Lima Firme
Organic Chemistry Professor
Chemistry Institute
Federal University of Rio Grande do Norte
Natal, Rio Grande do Norte
Brazil

CRC Press
Taylor & Francis Group
Boca Raton London New York

CRC Press is an imprint of the
Taylor & Francis Group, an **informa** business
A SCIENCE PUBLISHERS BOOK

CRC Press
Taylor & Francis Group
6000 Broken Sound Parkway NW, Suite 300
Boca Raton, FL 33487-2742

First issued in paperback 2021

Version Date: 20190705

ISBN-13: 978-0-367-77671-8 (pbk)
ISBN-13: 978-0-8153-8357-4 (hbk)

Visit the Taylor & Francis Web site at
http://www.taylorandfrancis.com

and the CRC Press Web site at
http://www.crcpress.com

To
my beloved daughter
Ananda França Firme

Preface

This book has several differentiating features from most, if not all, chemistry books: much of the information and most figures were obtained from the author's own Quantum mechanical calculations (the vast majority published in papers or to be published shortly). They were obtained using several methods and theories which are not well known by the average audience, such as quantum chemistry calculations using quantum theory of atoms in molecules (QTAIM), spin-coupled valence bond (SCVB), generalized valence bond (GVB), non-covalent interaction (NCI), intrinsic reaction coordinate (IRC), and molecular dynamics (MD).

The book begins with a brief description of the wave function and the antisymmetric wave function, which is the starting point to differentiate the molecular orbital (MO) and valence bond (VB) theories. The latter is widely used in this book (in its modern version: GVB and SCVB) to describe the electronic nature of several types of chemical bonds. Another important theory (based on the electronic density – the square of the wave function) is presented- the QTAIM, which is very important to describe the intermolecular interactions and chemical bonds.

In addition, a very important model used in this book is the electrostatic force and its relation to energy. Both concepts (force and energy) are used to understand the bond strength and relative stability of localized and delocalized systems besides all types of intermolecular interactions with the help of QTAIM analysis as well.

The concept of energy is expanded with the presentation of the electronic energy and the thermodynamic properties of enthalpy, internal energy, and Gibbs free energy (all four types of molecular energy). They are essential to discuss the stability of molecules by means of specific reactions. We also show that it is possible to obtain their absolute values theoretically and to compare to experimental values of the corresponding reaction by means of the concepts of statistical thermodynamics which are also explored in this book. From the theoretical data of the statistical thermodynamics, it is possible to understand the concepts of internal energy and entropy microscopically in each molecular entity, which is not possible in the classical thermodynamics.

The second part of the book deals with the introductory organic chemistry, where firstly the concepts of atomic radius and electronegative are presented as key points to understand the bond length and bond/molecular polarity/atomic charge, respectively. Afterwards, the resonance theory for delocalized systems is discussed with the help of electrostatic force model and its relation to energy to rationalize the stability of these systems with respect to the localized systems. The MO theory is also used to understand the relation between the increasing delocalization and decreasing HOMO-LUMO gap.

The historical relation between matrix mechanics and valence bond theory, plus its consequent onset of the concepts of the chemical bond and hybridization, are established and constructed in four chapters. Then, a comprehensive view of the concepts associated with the chemical bond is presented in a historical and mathematical approach.

Hereafter, the second part of the book deals with the geometric parameters of a molecule and the practical procedure of its optimization and the importance of this process for obtaining all theoretical properties of the molecule of interest. In addition, it advances to a thorough understanding of the transition state as a critical point of the potential energy surface. From this point on, the mechanistic aspects of a reaction and its relation to the potential energy surface, PES, are discussed. In a subsequent chapter, a comprehensive analysis of the transition state theory (from classical and statistical standpoints) is done as a key point to understand the kinetics of a chemical reaction, which is also important to understand the mechanism of the reaction.

Also, in the second part of the book, the models for representing the organic molecules and their specific applications are also presented as important tools to interpret the molecule in different perspectives.

In the third part of the book, a thorough analysis of nearly all types of intermolecular interactions and carbocations is done by means of QTAIM and NCI, besides the electrostatic force model as important auxiliary tools for rationalizing their geometric parameters, chemical bonds, interaction/bond strength, and stability. The third part of the book also deals with stereoisomerism (molecular symmetry, enantiomerism, diastereomerism, meso isomer, nomenclature, etc.) and its physical properties.

In the first three parts of the book, all prerequisites to a comprehensive understanding of the organic chemistry in a more profound perspective, supported by quantum chemistry, classical/statistical thermodynamics, and kinetics, are presented in an easy-to-understand mathematical/historical approach.

The fourth part of the book deals with the hydrocarbon chemistry itself in a physical chemistry approach using quantum chemistry to obtain: (i) optimized geometries; (ii) electronic nature of the chemical bonds and intermolecular interactions; (iii) the stability trend; (iv) the reactivity; (v) the regioselectivity; (vi) the potential energy surface and the structures of its critical points; (vii) deep insights on the mechanisms; (viii) thermodynamic data, etc.

The main target audience is made up of the undergraduate students from Chemistry, Chemical Engineering, and other related courses, plus graduate students of organic chemistry and physical chemistry.

Caio Lima Firme

Advice for Students

Students should bear in mind that an appropriate learning of organic chemistry depends on the basic concepts of general chemistry (for instance, electronegativity, polarizability, dipole moment, inter/intra-molecular interactions, nucleophilicity, acid-base reactions, formal charge, chemical bond, and hybridization) and some basic equations (see below).

Some general chemistry formulas to bear in mind:

$$FC_i = Z_i - Ne_i$$

$$\mu = Q.\mathbf{r}$$

Some basic physical chemistry formulas to bear in mind:

$$\Delta G = \Delta H - T\Delta S$$

$$\Delta G = -RT \ln K$$

$$K = \frac{\prod [product]^k}{\prod [reagent]^n}$$

$$\mathbf{F}_{elect} = K \frac{Q.q}{\mathbf{r}^2}$$

$$rate = k \prod [reagent]_{RdS}^v$$

All these formulas will be properly discussed in due time.

Note About the Next Volume

The title of the second book that succeeds this one is: "Introductory Organic Chemistry Continued and Beyond Hydrocarbons – a Physical Chemistry Approach". In this second book there are topics such as acidity/basicity, solubility, nucleophicility/electrophilicity, leaving groups, oxidation and reduction reactions, organometallic compounds, stereoselectivity, acid/base catalysis, properties and reactions of alcohols, amines, ethers and carbonyl compounds, and so on…following the same methodology of the present book.

Note About the Illustrations and Calculations

All illustrations of this book were done by the author. Drawings not derived from quantum chemistry calculations were mostly done using Accelrys Draw software an older version of Biovia Draw (Dessault Systèmes BIOVIA). All other illustrations were obtained from quantum chemistry calculations that were graphically generated by ChemCraft (Zhurko and Zhurko), AIM2000 (Biegler-König et al. 2002), or VMD (Humphrey et al. 1996), or Gausview v.5. Geometry optimization, frequency calculations, along with thermodynamic data used in this book were done in Gaussian09 (Frisch et al. 2009). Intrinsic reaction coordinate, IRC, calculations were based on HPC algorithm (Hratchian and Schlegel 2005). Subsequent calculations of the optimized molecules for QTAIM, NCI (Contreras-García et al. 2011), and GVB/SCVB calculations were done using AIM2000, MultiWFN (Lu and Chen2012), and VB2000 (Li et al. 2009), respectively.

REFERENCES CITED

Biegler-König, F., Schönbohm, J. and Bayles, D. 2001. AIM2000 - A program to analyze and visualize atoms in molecules. J. Comp. Chem. 22: 545-559

ChemCraft: graphical software for visualization of quantum chemistry computations. http://www.chemcraftprog.com

Contreras-Garcia, J., Johnson, E.R., Keinan, S., Chaudret, R., Piquemal, J.-P., Beratan, D.N. and Yang, W. 2011. NCIPLOT: A program for plotting non-covalent interactions. J. Chem. Theory Comput. 7: 625-632.

Dassault Systèmes BIOVIA, Biovia Draw, San Diego: Dassault Systèmes. Older version used: Accelrys Draw: Accelrys Draw 4.1 - Accelrys Inc.

Hratchian, N.H.P. and Schlegel, H.B. 2005. Using hessian updating to increase the efficiency of a hessian based predictor-corrector reaction path following method. J. Chem. Theory Comput. 1: 61–69.

Humphrey, W., Dalke, A. and Schulten, K. 1996. VMD: visual molecular dynamics. J. Mol. Graph 14: 33-38.

Li, J., Duke, B. and McWeeny, R. 2009. VB2000 v.2.1. SciNet Technologies, San Diego, CA.

Lu, T. and Chen, F. 2012. Quantitative analysis of molecular surface based on improved Marching Tetrahedra algorithm. J. Mol. Graph. Model. 38: 314-323.

Frisch, M.J., Trucks, G.W., Schlegel, H.B., Scuseria, G.E., Robb, M.A., Cheeseman, J.R., Scalmani, G., Barone, V., Mennucci, B., Petersson, G.A., Nakatsuji, H., Caricato, M., Li, X., Hratchian, H.P., Izmaylov, A.F., Bloino, J., Zheng, G., Sonnenberg, J.L., Hada, M., Ehara, M., Toyota, K., Fukuda, R., Hasegawa, J., Ishida, M., Nakajima, T., Honda, Y., Kitao, O., Nakai, H., Vreven, T., Montgomery, J.A., Jr., Peralta, J.E., Ogliaro, F., Bearpark, M., Heyd, J.J., Brothers, E., Kudin, K.N., Staroverov, V.N., Kobayashi, R., Normand, J., Raghavachari, K., Rendell, A., Burant, J.C., Iyengar, S.S., Tomasi, J., Cossi, M., Rega, N., Millam, J.M., Klene, M., Knox, J.E., Cross, J.B., Bakken, V., Adamo, C., Jaramillo, J., Gomperts, R., Stratmann, R.E., Yazyev, O., Austin, A.J., Cammi, R., Pomelli, C., Ochterski, J.W., Martin, R.L., Morokuma, K.,

Zakrzewski, V.G., Voth, G.A., Salvador, P., Dannenberg, J.J., Dapprich, S., Daniels, A.D., Farkas, Ö., Foresman, J.B., Ortiz, J.V., Cioslowski, J., Fox, D.J. 2009. Gaussian 09. Revision B.01. Gaussian, Inc., Wallingford CT.

Zhurko, G.A. and Zhurko, D.A., Chemcraft. Version 1.8 (Build 538). www.chemcraftprogram. com

Note about the ωB97X-D Functional used in this Book

In nearly all calculations we have used ωB97X-D functional (Chai and Head-Gordon 2008). In the assessment of the performance of DFT and DFT-D functionals for hydrogen bond interactions, ωB97X-D showed the best results (Thanthiriwatte et al. 2011). In another highly cited work on performance assessment of DFT functionals for intermolecular interactions in methane hydrates, ωB97X-D showed one of the best results (Liu et al. 2013). In another performance assessment (Forni et al. 2014), ωB97X-D was one of the best methods for the study of halogen bonds with benzene.

REFERENCES CITED

Chai, J. and Head-Gordon, M. 2008. Long-range corrected hybrid density functionals with damped atom-atom. Phys. Chem. Chem. Phys. 10: 6615-6620.

Forni, A., Pieraccini, S., Rendine, S., Sironi, M. 2014. Halogen bonds with benzene: An assessment of DFT functionals. J. Comp. Chem. 35: 386-394.

Liu, Y., Zhao, J., Li, F., Chen, Z. 2013. Appropriate description of intermolecular interactions in the methane hydrates: an assessment of DFT methods. J. Comp. Chem. 34: 121-131.

Thanthiriwatte, K.S., Hohenstein, E.G., Burns, L.A., Sherrill, C.D. 2011. Assessment of the performance of DFT and DFT-D methods for describing distance dependence of hydrogen-bonded interactions. J. Chem. Theory Comp. 7: 88-96.

Contents

Appendices 419

Index 433

Color Plate Section 437

Chapter One

Notions of Quantum Mechanics and Wave Function

NOTIONS OF OLD QUANTUM MECHANICS

Black-body radiation—an ideal material that absorbs all light and radiates electromagnetic energy according to its temperature — originated from Kirchhoff's law of thermal radiation in 1860. Quantum mechanics began in very early twentieth century when Max Planck found the expression for black body thermal radiation in which the emitted light was not a continuum as postulated by classical physics. He developed Planck's constant, h, to ensure that his expression matched experimental values. Planck's theory was based on statistical mechanics and postulated the blackbody as a *collection of isotropic oscillators with specific vibrational frequency for each oscillator*. Later, Albert Einstein proved Planck's quantization theory by means of theory of the photoelectric effect (Pilar 1990).

Henceforth, Niels Bohr succeeded in interpreting mathematically the hydrogen spectral lines (a type of bar code for each element) obtained from a gas tube discharge (Bohr 1925). The Bohr model established the circular orbit movement of electrons with definite energies, discrete (orbital) angular momentum, L, of the electron in orbit, and the electron energy jump between two discrete energy levels due to absorption or emission radiation.

$$\Delta E = E_2 - E_1 = h\nu$$

Where ν is the frequency of electromagnetic radiation.

In classical physics, angular momentum is given by the product of moment of inertia, I, (needed torque to yield angular acceleration) and angular velocity, ω:

$$L = I\omega \therefore I = mr^2 \therefore \omega = v/r \therefore L = mvr$$

Bohr also stated that the angular momentum of an electron in an atom is constrained to discrete values according to the quantum number, n.

$$L = \frac{nh}{2\pi}$$

For one electron, e, in a circular orbit around one nucleus with Z charge, the centripetal force equals the electrostatic force.

$$F_{centripetal} = F_{electrostatic(e-Z)}$$

$$F_{centripetal} = \frac{2T}{r} \therefore F_{electrostatic} = k\frac{(Zq_e)q_e}{r^2}$$

$$k\frac{Zq_e^2}{r^2} = \frac{m_e v^2}{r} \therefore r = \frac{n^2}{Z}a_B$$

Where a_B is Bohr radius. The total energy is given by:

$$E_T = \frac{m_e v^2}{2} + k\frac{Zq_e}{r}$$

Which gives the expression:

$$E_n = -\frac{Z^2}{n^2}E_0 \therefore (n = 1,2,3...)$$

Where E_0 is the ground-state energy ($n = 1$) of hydrogen atom which is 13.6 eV.

Another important experiment to prove the space quantization was realized by Pieter Zeeman. Curiously, his experiment was carried out before the birth of quantum physics. Initially, it confirmed that negatively charged particles (later discovered as electrons by Thomson) were the source of light from a determined substance, and that the emitted light was polarized under a magnetic field (Zeeman 1897). Zeeman's experiment was an important chapter in the history of spectroscopy initiated by Kirchhoff and Bunsen in 1860 (Kirchhoff and Bunsen 1860). However, the Zeeman effect was also important to prove the quantization of particles because of the splitting of the spectral lines under the magnetic field, B. The splitting occurs by the torque of B on magnetic dipole, $\mu_{orbital}$, which is associated with an orbital angular momentum, L.

$$\mu_{orbital} = -\frac{e}{2m_e}L$$

Where m_e is the electron mass.

When considering singlet substances, the normal Zeeman effect occurs, providing the discrete values of the orbital angular momentum (Fig. 1.1). In singlet atoms and molecules, all electrons are in parallel (or spin-paired). They are closed-shell substances. On the other hand, in an open-shell substance there is, at least, one electron anti-parallel (not spin-paired). In open-shell substances there occurs the "anomalous" Zeeman effect which was important for the discovery of spin (see the discussion in the subsequent section).

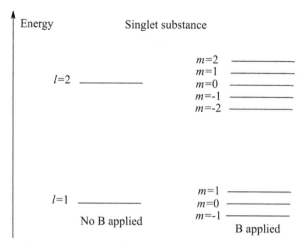

Figure 1.1   Schematic representation of normal Zeeman effect on spectral lines.

QUANTUM WAVE MECHANICS – GENERAL OVERVIEW

While Planck proposed that emission of light occurred in discrete values of energy, Einstein extended this concept to that in which light has a particle component (photon) and it propagates in discrete values of energy. Einsten's concept unified two classical definitions: (1) that of light as waves of electromagnetic fields; and (2) that of matter as localized particles. This work influenced De Broglie to propose the same duality (matter-wave) for the electron, which was confirmed three years later in an electron diffraction experiment (De Broglie 1925). De Broglie's work influenced Erwin Schrödinger to find a wave equation for matter. Schrödinger himself summarized his works about quantum matter-wave theory (Schrödinger 1926).

$$i\hbar \frac{\partial \psi}{\partial t} = -\frac{\hbar^2}{2m} \nabla^2 \psi + V \psi \therefore i\hbar \frac{\partial \psi}{\partial t} = H \psi$$

Where H is the Hamiltonian operator (sum of kinetic and potential operators) and ∇^2 is the Laplacian operator (see more discussion in the next chapter). For example, for cartesian coordinates, Laplacian operator is given by:

$$\nabla^2 = \frac{\partial^2}{\partial x^2} + \frac{\partial^2}{\partial y^2} + \frac{\partial^2}{\partial z^2}$$

The Hamiltonian for many-electron atom also involves a kinetic operator for the ith electron, potential operator for the interaction between nucleus Z and ith electron, and a second potential operator for the interaction between electrons.

$$H = \sum_i \left[-\frac{\hbar}{2m} \nabla_i^2 - \frac{Ze^2}{4\pi\varepsilon_0} \frac{1}{r_i} \right] + \sum_{i<j} \frac{e^2}{4\pi\varepsilon_0} \frac{1}{|r_i - r_j|}$$

This Hamiltonian regards that the nucleus motion is too slow with respect to the electron motion, and it is called adiabatic approximation. When one needs to include the nucleus motion in the Hamiltonian, the resultant wave function is called nonadiabatic (Kolos and Wolniewicz 1963).

Schrödinger applied his equation to solve the quantum harmonic oscillator, the quantum rigid rotor, and a hydrogen-like atom. The Schrödinger equation is an eigenequation where the Hamiltonian (an eigenvector) operates in a wave-matter function, ψ (an eigenfunction) in a linear transformation, yielding a parallel eigenfunction $E\psi$, where E is a scalar number (an eigenvalue), represented as a shortened form of the time-independent Schrödinger equation.

$$H\psi = E\psi$$

Pertaining to the solution of the Schrödinger equation to hydrogen atoms (which are also applicable to many-electron atoms) are the quantum numbers (n, l, m). They are called "quantum" because they vary in discrete integers or half-integers instead of a continuum range, as in classical physics. There is one quantum number associated with each quantum operator (Hamiltonian, total angular momentum, L^2, and total angular momentum projection, L_z). Since L^2, L_z, and H have commutative properties (i.e., the Hamiltonian operator commutes with total angular momentum and total angular momentum projection operators), they have simultaneous eigenfunctions:

$$H\psi\left(r,\theta,\varphi\right) = E\psi\left(r,\theta,\varphi\right)$$

$$L^2\psi\left(r,\theta,\varphi\right) = l(l+1)\hbar^2\psi\left(r,\theta,\varphi\right) \therefore l = 0,1,2...$$

$$L_z\psi\left(r,\theta,\varphi\right) = m\hbar\psi\left(r,\theta,\varphi\right) \therefore m = -l,-l+1,...,l-1,l$$

Since L^2 and L_z do not involve r variable, any spherical harmonic can be multiplied by any radial function containing r variable, and this product function is also an eigenfunction. Then, the wave function, ψ, is written as a product of the radial function and spherical harmonic function.

$$\psi\left(r,\theta,\varphi\right) = R(r)Y_{lm}\left(\theta,\varphi\right)$$

The general radial function, $R(r)$ and its solution (the total electron energy) is given by:

$$R_{nl}\left(r\right) = -\left(\frac{2}{na_0}\right)^{3/2}\sqrt{\frac{(n-l-1)!}{2n\left[(n+l)!\right]^3}}\left(\frac{2r}{na_0}\right)^l e^{-r/na_0} L_{n+l}^{2l+1}\left(\frac{2r}{na_0}\right)$$

$$E_n = -\frac{e^2}{2a_0}\frac{1}{n^2}$$

Where a_0 is Bohr radius, e is the electron charge, L_{n+l}^{2l+1} is the associated Laguerre polynomials, and E_n is the total energy of the electron belonging in the nth atomic shell of the hydrogen atom. Then, n ($n = 1,2,3...$) determines the total energy of the electron without any external magnetic field.

The general formula for the spherical harmonic function is:

$$Y_{lm}(\theta,\varphi) = \sqrt{(-1)^{m+|m|}}\sqrt{\frac{2l+1}{4\pi}}\sqrt{\frac{(l-|m|)!}{(l+|m|)!}}P^{|m|}(\cos\theta)e^{im\varphi}$$

Where $P^{|m|}$ is associated Legendre polynomials.

The atomic orbitals come from the product of the radial function and spherical harmonic function which, in turn, both depend on the quantum numbers n, l, and m. For example, the expression for the $1s$ orbital, $|1s\rangle$, is:

$$|1s\rangle = |100\rangle = R_{10}(r)Y_{00}(\theta,\varphi) = \frac{1}{\pi^{1/2}}\left(\frac{Z}{a_0}\right)^{3/2}e^{-Zr/a_0}$$

Henceforth, there appears the well-known geometries of s, p, d, and f atomic orbitals. Figure 1.2 shows a schematic representation of s and p orbitals.

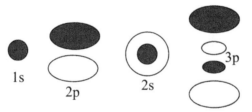

Figure 1.2   Schematic representation of s and p orbitals.

SPIN AND WAVE MECHANICS

In 1921, A. H. Compton was the first to establish the relation between electron and a tinny gyroscope. One year later, the existence of the magnetic quantum number was first observed by the Stern-Gerlach experiment. In that experiment, a gaseous silver beam (holding paramagnetic property) reached a non-uniform magnetic field, which caused deflection of the original beam into two new ones (Fig. 1.3). In 1925, Phipps and Taylor carried out nearly the same experiment using hydrogen beam, which definitely established the failure of the Schrödinger equation for not describing the electron quantum magnetic moment in hydrogen atom.

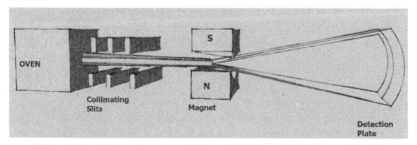

Figure 1.3   Schematic representation of the Stern–Gerlach experiment.

In an attempt to rationalize the anomalous Zeeman effect, Pauli proposed the fourth degree of freedom of the electron which he called "two-valuedness not describable classically" (Pauli 1925a, b). The anomalous Zeeman effect could be observed when the Zeeman apparatus was used with alkali metals. In this work, he established the exclusion principle where no electron can have equal values for all quantum numbers. Fifteen years later, Pauli extended this principle to all fermions (Pauli 1940).

Yet in 1925, one year before the birth of the Schrödinger equation, Uhlenbeck and Goudsmit discovered the spin from the fine structure of hydrogen-like spectra, and proposed that any particle should have magnetic angular moment with similar expression to the orbital angular moment (Uhlenbeck and Goudsmit 1925, 1926).

$$M_S = \frac{g_s \left| -e \right|}{2m_e} s = \frac{g_s \mu_B}{\hbar} s = \gamma s$$

Where the spin magnetic moment, M_S, is proportional to the spin operator (or spin angular momentum), s, and the gyromagnetic ratio, γ, which is a specific constant for electrons, protons, neutrons, and nuclei. Pauli himself said that his paper about exclusion principle was initially difficult to understand due to the lack of a model to explain the fourth degree of freedom of the electron, and that the work of Uhlenbeck and Goudsmit helped to connect the exclusion principle to the idea of spin.

One year after the Schrödinger equation was published, Pauli corrected it in order to describe the influence of an external magnetic field in the wave function as it occurs in the anomalous Zeeman effect and Stern–Gerlach experiment, i.e., Pauli intended to describe the interaction of the spin of a particle with an external magnetic field (Pauli 1927).

$$i\hbar \frac{\partial}{\partial t} |\psi_\pm\rangle = \left(\frac{(\mathbf{p} - q\mathbf{A})^2}{2m} + q\phi \right) \begin{pmatrix} 1 & 0 \\ 0 & 1 \end{pmatrix} |\psi\rangle - \frac{q\hbar}{2m} \boldsymbol{\sigma} \cdot \mathbf{B} |\psi\rangle$$

$$\mathbf{p} = \mathbf{P} - q\mathbf{A} \therefore \mathbf{B} = \nabla \times \mathbf{A} \therefore \mathbf{A} = \begin{pmatrix} A_x & A_y & A_z \end{pmatrix}$$

$$\boldsymbol{\sigma} = \begin{pmatrix} \sigma_x & \sigma_y & \sigma_z \end{pmatrix}$$

Where $\mathbf{B}$ is the external magnetic field, ϕ is the scalar electric potential, $\mathbf{A}$ is the magnetic vector potential, $\mathbf{P}$ is the canonical momentum, m and q are mass and charge of the particle, respectively, and $\boldsymbol{\sigma}$ are the Pauli matrices. It is important to emphasize that in this book matrices are represented by bold, non-italic, and upper-case characters. As one exception, $\mathbf{p}$ is also a matrix.

$$\sigma_x = \begin{pmatrix} 0 & 1 \\ 1 & 0 \end{pmatrix} \therefore \sigma_y = \begin{pmatrix} 0 & -i \\ i & 0 \end{pmatrix} \therefore \sigma_z = \begin{pmatrix} 1 & 0 \\ 0 & -1 \end{pmatrix}$$

Likewise orbital angular momentum, S^2, S_z, and H have commutative property, i.e., the Hamiltonian operator commutes with total spin and total spin projection operators. As a consequence, they have simultaneous eigenfunctions:

$$H\psi\left(r,\theta,\varphi,s\right) = E\psi\left(r,\theta,\varphi,s\right)$$

$$S^2\psi\left(r,\theta,\varphi,s\right) = s(s+1)\hbar^2\psi\left(r,\theta,\varphi,s\right) \therefore s = \frac{1}{2}$$

$$S_z\psi\left(r,\theta,\varphi,s\right) = m_s\hbar\psi\left(r,\theta,\varphi,s\right) \therefore m_s = \pm\frac{1}{2}$$

Then, it is also possible to separate spatial and spin functions.

$$\Psi_H = \phi_{spatial}\Phi_{spin}$$

$$\phi_{spatial} = R_{nl}(r)Y_{lm}(\theta,\varphi)$$

For fermions (electrons, protons, and neutrons), the corresponding wave functions follow Fermi-Dirac statistics, and are anti-symmetric (see more discussion in the next chapter).

Dirac himself in 1928 incorporated special relativity in the one-particle problem, where the spin appeared as a part of the solution of the so-called Dirac equation, unlike the Schrödinger equation (Dirac 1928a, b, Darwin 1928).

MATRIX MECHANICS – INSPIRATION FROM OLD QUANTUM MECHANICS

Heisenberg's work on the development of his quantum mechanics (matrix mechanics) in conjunction with Born and Jordan was enormously influenced by Bohr's lectures about Kramers's work on transition intensities over quadratic Stark-effect in hydrogen atom (Kramers 1920). Stark effect, an electric analog of Zeeman effect, leads to the shifting and splitting of spectral lines of atoms and molecules due to an electric field. Kramers established that transition between states occurred at multiples of the orbit frequencies and that the orbits in quantum system should be a Fourier expansion of harmonics at multiples of orbit frequency. Bohr school used Slater's idea of an atom having virtual harmonic oscillators in communication with distant atoms in order to explain electronic transitions. Heisenberg disagreed with Kramers's results and tried to arrive at the intensity formula for hydrogen transitions under an electric field using Fourier expansion for an anharmonic oscillator (Heisenberg 1925). Heisenberg argued that the electron orbit did not exist, and started formulating the new quantum theory by analyzing only spectral lines and observables instead of the atomic model. He found that the transition probabilities are proportional to the squares of the transition amplitudes. In Heisenberg's work, he used X_{nm} instead of X_n in the periodic motion of the electron, corresponding to the transition from n to m (i.e., $n-m$), following Born's idea. Heisenberg also established a numerical connection between $|X_{nm}|^2$ and the radiation emitted or absorbed in the transitions $n \rightarrow m$.

For large n and m but small $n-m$, the frequency of emission is

$$\frac{(E_n - E_m)}{h} = \frac{h(n-m)}{T},$$

where T is the period of the n orbit or m orbit, according to old quantum mechanics, i.e., the electron will emit radiation with integer multiples of orbital frequency. But for large $n\text{-}m$, such as in quadratic Stark effect, the frequencies are not integer multiples of any frequency. Then, Heisenberg realized that the equation of motion of the electron could not be a Fourier representation from harmonic oscillator:

$$X(t) = \sum_{n=-\infty}^{\infty} e^{2\pi i(nt)/T} \mathbf{X_n}$$

Where $\mathbf{X_n}$ are complex numbers.

Heisenberg derived another equation of motion for electrons, taking for granted the quantity $\mathbf{X_{nm}}$. He assumed $\mathbf{X_{nm}} = \mathbf{X_n}$ for large n and m but small $n-m$. His time-dependent equation of motion of electron became:

$$X_{nm}(t) = \sum_{n=-\infty}^{\infty} e^{2\pi i(E_n - E_m)t/h} \mathbf{X_{nm}(0)}$$

According to Heisenberg, a quantum Fourier series could not describe the motion of the electron from only one n state because each term in the series describes a transition process associated with two states, n and m.

When Heisenberg communicated his results to Born, the latter realized that the mathematics underlying the results of the former could be written in matrix formulation (Born and Jordan 1925). Both of them, along with Jordan, worked on a complete theory of atoms and their transitions (Born et al. 1926).

At that time, matrix algebra was poorly used by physicists, while differential equations (from quantum wave mechanics) were familiar to all of them. Then, matrix mechanics did not become as popular as wave quantum mechanics. Matrix mechanics was influenced by Niels Bohr's ideas about transitions from spectral lines and Heisenberg's own vision about observables. On the other hand, Schrödinger was influenced by wave-particle duality. Nonetheless, both quantum mechanical approaches agreed in their results, and both Schrödinger and Heisenberg tried to prove their theories using matrix mechanics and wave mechanics, respectively. As one example, see the section "QTAIM: extension of quantum mechanics for an open system" in chapter two.

SUMMING-UP

Ingold's overview of Schrödinger's wave mechanics: "What can be known about an electron, considered as a particle, is of a statistical nature, and is summarized in the behavior of a wave. The frequency of the wave measures the energy of the electron ($E = h \cdot v$), and the amplitude measures the probability that the electron is in a given elementary volume (prob. $= |\psi|^2 dv$). The fundamental wave-property in the theory of the atom is self-interference – the self-interference of a confined wave; this annihilates all frequencies except those that can form standing waves, which are perpetuated and describe the quantum states" (Ingold 1938). Birtwistle also added: "He (Schrödinger) assumed that the dynamics of an atom cannot be represented by a point moving through the coordinate space as in classical theory, but must be represented by a wave in that space, and obtained a differential equation which the wave function must satisfy" (Birtwistle 1928). As for Matrix Mechanics, Birtwistle wrote: "Heisenberg sought to develop a scheme of quantum kinematics by which the quantum formula would be obtained directly in terms of these experimentally observable magnitudes (...) without the intermediate use of orbital frequencies and amplitudes, which by their nature can never be probably observed. This meant that instead of representing a dynamical quantity, as in classical theory, by one-dimensional line of Fourier series (...) it should be, in the quantum theory, represented by a two-dimensional table of terms (a matrix)" (Birtwistle 1928).

REFERENCES CITED

Birtwistle, G. 1928. The New Quantum Mechanics. Cambridge University Press, New York.

Bohr, N. 1925. Atomic theory and mechanics. Nature 116: 845-852.

Born, M. and Jordan, P. 1925. Zur quantenmechanik. Z. Phys. 34: 858-888.

Born, M., Heisenberg, W. and Jordan, P. 1926. Zur quantenmechanik II. Z. Phys. 35: 557-615.

Dirac, P.A.M. 1928a. The quantum theory of the electron. Proc. Roy. Soc. Lond. A. 117: 610-624.

Dirac, P.A.M. 1928b. The quantum theory of the electron. Part II. Proc. Roy. Soc. Lond. A. 118: 351-361.

Darwin, C. 1928. The wave equations of the electron. Proc. Roy. Soc. Lond. A. 118: 654-680.

De Broglie, 1925. Recherches sur la théorie des quanta. Ann. De Phys. 10e série, t. III (Translated by A.F. Kracklauer, 2004).

Heisenberg, W. 1925. Über quantentheoretische umdeutung kinematischer und mechanischer beziehungen. Z. Phys. 33: 879-893.

Ingold, C.K. 1938. Resonance and mesomerism. Nature 141: 314-318.

Kirchhoff, G. and Bunsen, R. 1860, Chemical analysis by observation of spectra. Annalen der Physik und der Chemie 110: 161-189.

Kolos, W. and Wolniewicz, L. 1963. Nonadiabatic theory for diatomic molecules and its application to hydrogen molecule. Rev. Mod. Phys. 35: 473-483.

Kramers, H.A. 1920. Über den Einflub eines elektrischen feldes auf die feinstruktur der wasserstofflinien. Z. Phys. 3: 199-223.

Pauli, W. 1925a. Über den einflub der geschwindigkeitsabhängigkeit der elektronenmasse auf den Zeemaneffekt. Z. Phys. 31: 373-385.

Pauli, W. 1925b. Über den Zusammenhang des abschlusses der elektronengruppen im atom mit der komplexstruktur der spektren. Z. Phys. 31: 765-783.

Pauli, W. 1927. Zur quantenmechanik des magnetischen elektrons zeitschrift für physik. Z. Phys. 43: 601-623.

Pauli, W. 1940. The connection between spin and statistics. Phys. Rev. 58: 716-722.

Pilar, F.L. 1990. Elementary Quantum Chemistry. McGraw Hill Publishing Co, Singapore.

Schrödinger, E. 1926. An undulatory theory of the mechanics of atoms and molecules. Phys. Rev. 28: 1049-1070.

Uhlenbeck, G.E. and Goudsmit, S. 1925. Die Naturwissenschaften 47: 953-954.

Uhlenbeck, G.E. and Goudsmit, S. 1926. Spining electrons and the structure of spectra. Nature 117: 264-265.

Zeeman, P. 1897. The effect of magnetisation on the nature of light emitted by a substance. Nature 55: 347.

Chapter Two

Molecular Orbital, Valence Bond, Atoms in Molecules Theories, and Non-covalent Interaction Theories and their Applications in Organic Chemistry

ANTISYMMETRIC WAVE FUNCTION

Fermi developed the quantization of the motion of an ideal gas with the restriction of Pauli's exclusion principle one year after Pauli had published his exclusion principle (Fermi 1926), in which similar work was done by Dirac (Dirac 1926). This statistics of motion of identical (or indistinguishable) particles obeying Pauli's exclusion principle was termed Fermi-Dirac statistics. As a consequence of Fermi-Dirac statistics, identical fermions are represented by an antisymmetric wave function.

In an antisymmetric wave function, any permutation of two indistinguishable states shall result in the same function with a negative sign. Let's consider the following antisymmetric wave function for two electrons with states $|n_1\rangle$ and $|n_2\rangle$.

$$\psi = N\left(|n_1\rangle|n_2\rangle - |n_2\rangle|n_1\rangle\right)$$

Where N is the normalization constant.

When using the permutation operator, P_{12}, we have:

$$P_{12}\psi = N\left(|n_2\rangle|n_1\rangle - |n_1\rangle|n_2\rangle\right)$$

By multiplying $P_{12}\psi$ by -1, we have:

$$P_{12}\psi \times (-1) = -N\left(-|n_2\rangle|n_1\rangle + |n_1\rangle|n_2\rangle\right)$$
$$P_{12}\psi \times (-1) = -N\left(|n_1\rangle|n_2\rangle - |n_2\rangle|n_1\rangle\right)$$
$$P_{12}\psi \times (-1) = -\psi$$

Then one can write:

$$\psi(q_1, q_2) = -\psi(q_2, q_1) \therefore |n\rangle = q \therefore q = (n, l, m, s)$$

Let's see what happens when both q_1 and q_2 have the same four quantum numbers:

$$\psi(q_1, q_2) = -\psi(q_2, q_1)$$
$$\psi(q_1, q_2) + \psi(q_2, q_1) = 0$$
$$2\psi = 0$$

As a consequence, the probability to find two electrons with the same quantum numbers is null. Then, the antisymmetric wave function respects Pauli's exclusion principle.

Henceforth, any fermion wave function must be antisymmetric. Since the Schrödinger wave function is a product of spatial and spin functions, there are two simple ways to assure an antisymmetric wave function: either by constructing an antisymmetric spatial function and corresponding symmetric function or the opposite.

$$\Psi = \phi_{(spatial)} \cdot \Theta_{(spin)}$$
$$\Psi = \phi_{(antisymmetric)} \cdot \Theta_{(symmetric)} = \Psi_{(antisymmetric)}$$
$$\Psi = \phi_{(symmetric)} \cdot \Theta_{(antisymmetric)} = \Psi_{(antisymmetric)}$$

For many-electron wave function, there are two ways of constructing an antisymmetric fermion wave function: by means of molecular orbital theory or valence bond theory. The former uses Slater determinant and the latter uses group theory.

Atomic spin orbitals have a set of unique quantum numbers (at least, one different quantum number) for every electron. For a system with number of particles (nucleus plus electrons) higher than two, there is no analytic solution for that wave function. First approximation is to transform a many-electron wave function into a product of single electron wave functions.

$$\Psi_{n_1, n_2, \ldots, n_N}(q_1, q_2, \ldots, q_N) = \Psi_{n_1}(q_1)\Psi_{n_2}(q_2)\ldots\Psi_{n_N}(q_N) = \prod_{i=1}^{N} \Psi_{n_i}(q_i)$$

$$q = (n, l, m, s)$$

However, this function is not antisymmetric. The simplest way to solve this problem is to use Slater determinants.

MOLECULAR ORBITAL THEORY

The Slater determinant yields an antisymmetric wave function and it is a linear combination of one-electron functions (spin orbitals). Let's consider the lithium atom. It has six combination of spatial and spin functions (α, β).

$$1S(1)\alpha(1)1S(2)\beta(2)2S(3)\alpha(3)$$
$$1S(2)\alpha(2)1S(1)\beta(1)2S(3)\alpha(3)$$
$$1S(2)\alpha(2)1S(3)\beta(3)2S(1)\alpha(1)$$
$$1S(1)\alpha(1)1S(3)\beta(3)2S(2)\alpha(2)$$
$$1S(3)\alpha(3)1S(2)\beta(2)2S(1)\alpha(1)$$
$$1S(3)\alpha(3)1S(1)\beta(1)2S(2)\alpha(2)$$

The normalized and antisymmetric wave function for lithium atom is given by the following matrix:

$$\Psi = \frac{1}{\sqrt{6}} \begin{vmatrix} 1s(1)\alpha(1) & 1s(2)\alpha(2) & 1s(3)\alpha(3) \\ 1s(1)\beta(1) & 1s(2)\beta(2) & 1s(3)\beta(3) \\ 2s(1)\alpha(1) & 2s(2)\alpha(2) & 2s(3)\alpha(3) \end{vmatrix}$$

Which becomes the following expression (the determinant of the above matrix):

$$\Psi = \frac{1}{\sqrt{3!}} \begin{bmatrix} 1s(1)\alpha(1)1s(2)\beta(2)2s(3)\alpha(3) - 1s(2)\alpha(2)1s(1)\beta(1)2s(3)\alpha(3) \\ +1s(2)\alpha(2)1s(3)\beta(3)2s(1)\alpha(1) - 1s(1)\alpha(1)1s(3)\beta(3)2s(2)\alpha(2) \\ +1s(3)\alpha(3)1s(1)\beta(1)2s(2)\alpha(2) - 1s(3)\alpha(3)1s(2)\beta(2)2s(1)\alpha(1) \end{bmatrix}$$

The antisymmetric wave function for lithium atom can also be resolved in other way:

$$P_{12} \rightarrow -1s(2)\alpha(2)1s(1)\beta(1)2s(3)\alpha(3)$$
$$1s(1)\alpha(1)1s(2)\beta(2)2s(3)\alpha(3) \quad P_{23} \rightarrow -1s(1)\alpha(1)1s(3)\beta(3)2s(2)\alpha(2)$$
$$P_{13} \rightarrow -1s(3)\alpha(3)1s(2)\beta(2)2s(1)\alpha(1)$$
$$-1s(3)\alpha(3)1s(2)\beta(2)2s(1)\alpha(1) \quad P_{23} \rightarrow +1s(2)\alpha(2)1s(3)\beta(3)2s(1)\alpha(1)$$
$$-1s(2)\alpha(2)1s(1)\beta(1)2s(3)\alpha(3) \quad P_{23} \rightarrow +1s(3)\alpha(3)1s(1)\beta(1)2s(2)\alpha(2)$$

Whose sum of all distinguished products gives the non-normalized Slater determinant.

Slater determinant is a $\{N \times N\}$ or $\{2k \times 2k\}$ square matrix where N corresponds to the total number of electrons (using minimal basis set) and k is the total number basis functions in a chosen basis set. Every atomic spin orbital (in the construction of the Slater determinant) is doubly occupied. Every molecular orbital, ψ_i, comes from a linear combination of the product of expansion coefficient, $C_{\mu i}$, and atomic orbitals, ϕ_μ, from a specific basis set with k basis functions $\{\phi_\mu(r) | = 1, 2...k\}$.

$$\psi_i = \sum_{\mu=1}^{k} C_{\mu i} \phi_\mu$$

The expansion coefficients, $C_{\mu i}$ are found after the self-consistent field calculation, SCF, where all atomic orbitals become optimized molecular orbitals yielding doubly occupied molecular orbitals and virtual ones.

Figure 2.1(A) shows all the occupied molecular orbitals of ethene which has 16 electrons, i.e., it has 8 occupied molecular orbitals. The last occupied molecular orbital is called the highest occupied molecular orbital, HOMO. When using minimal basis set, the same number of virtual (unoccupied) and occupied molecular orbitals exists, i.e., there are also 8 virtual molecular orbitals for ethene. *As the basis set increases, the number of virtual molecular orbitals also increases.* From Fig. 2.1(A), one can see that, except for HOMO, besides HOMO-7 and HOMO-6 (which represent MOs for core electrons), no other occupied MO resembles Lewis picture of chemical bond – a consequence of delocalized nature of MO orbitals.

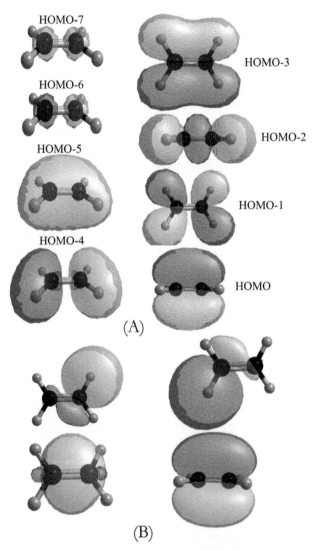

Figure 2.1   (A) Occupied molecular orbitals of ethene; (B) selected occupied NBOs of ethene.

Color version at the end of the book

Since the Slater determinant comes from a normal square matrix, there is not only one matrix which has the same determinant. By unitary similarity transformation, one can diagonalize the original matrix from Slater determinant using the characteristic equation {det($\mathbf{M}$-$\lambda_n\mathbf{I}_n$)=0, where $\mathbf{M}$ is the $n \times n$ square matrix, λ_n are eigenvalues, and $\mathbf{I}_n$ is the identity matrix}. Then, there are more than one matrix (set of molecular orbitals) for the Slater determinant and, as a consequence, there are more than one set of acceptable molecular orbitals for the same wave function.

Localized molecular orbitals are also acceptable matrices for the same Slater determinant. They are obtained from unitary transformations of canonical molecular orbitals.

Natural bonding orbital, NBO, is one of several ways to localize the molecular orbitals. Figure 2.1(B) shows four NBOs orbitals from ethene for comparison. Except for orbitals of C core electrons and π-bond, all other ethene NBOs are different from the canonical MOs, although they are acceptable solutions for the same molecular system. As a consequence, *molecular orbitals are not univocal and a set of different molecular orbitals is also acceptable for the same molecular system.*

From a historical perspective, just after the birth of Schrödinger equation and valence bond theory, the molecular orbital theory, MO, was developed by Mülliken with important contributions from Lennard-Jones, Slater, Hückel, Coulson, and mainly Hund.

In order to surpass some difficulties of Hund-Mulliken notation of electrons in molecules, Lennard-Jones introduced the linear combination of atomic orbitals method, LCAO, based on Heitler-London bonding and repulsive wave functions and resonance concept (Lennard-Jones 1929). A very important review of the LCAO method was done by Mülliken (Mülliken 1960).

Another important contribution to MO theory came from Hückel. By applying LCAO only to π bonds, Hückel developed a handy methodology (Hückel 1931, 1932) to predict, to some extent, the stability of conjugated systems (molecule with alternate double and single bonds) and aromatic systems in comparison with alkene analogs. Likewise, Coulson was another important proponent at the beginning of MO theory. He used MO and SCF methods to calculate the wave function of hydrogen molecule (Coulson 1938) since no exact analytic solution for three-particle (and higher) systems can be found from the "pure" Schrödinger equation (i.e., without any approximation).

Only two years after Pauli's paper about his exclusion principle, Slater developed the determinant (Slater determinant) to describe an antisymmetrical wave function (Slater 1929), which was incorporated in MO theory. Curiously, Slater is also known to contribute to valence bond theory (see the discussion in the next section).

Despite important contributions from Slater and Lennard-Jones, the origin of MO is credited to Hund and Mulliken. Based on Hund's early papers (Hund 1928), in which the rules of electron configuration for atoms were created (Hund's rules), Mülliken developed a theory to provide, at first, the electron configuration of diatomic molecules and their corresponding quantum states (Mülliken 1928). Later, Mülliken wrote about the main idea of MO: "A molecule is here regarded as a set of nuclei, around each of which is grouped an electron configuration closely similar to that of a free atom in an external field, except that the outer parts of the electron configuration surrounding each nucleus belong, in part, jointly to two or more nuclei" (Mülliken 1932a, b). Then, Mülliken constructed a molecular wave function from a distinguished strategy than that used in valence bond, VB (see the next section). Hund, in turn, showed that MO and VB are equivalent when regarding localized molecular orbitals for ordinary stable diatomic molecules (Hund 1929, 1931, 1932). In addition, Hund and Mülliken separately showed that the double and triple bonds can be described as $[\sigma]^2 [\pi]^2$ and $[\sigma]^2 [\pi]^4$, respectively, which is in accordance with the Lewis theory. Mülliken advocated that MO theory, unlike VB, can describe one-electron bonds, can describe bonding molecular orbitals from any degree of polarity or inequality of electron sharing, and can describe multicenter bonding (Mülliken 1932a, b). For example, in Fig. 2.2, the HOMO of bisnoradamantane (without multicenter bonding), bisnoradamantenyl cation (with $3c$-$2e$ multicenter bonding) and bisnoradamantenyl dication (with $4c$-$2e$ multicenter bonding) is shown, where one can see that MO theory was successfully used to rationalize the multicenter bondings (Firme et al. 2013). In addition, MO theory became an important tool for developing Hartree-Fock theory and Roothaan equations (Roothaan 1960) which, in conjunction, were the starting point for a new era of quantum chemistry.

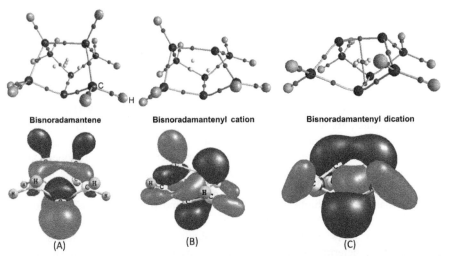

Figure 2.2   HOMO orbitals of (A) bisnoradamantane, (B) bisnoradamantenyl cation, and (C) bisnoradamantenyl dication, and their corresponding molecular graphs (see the next section). Courtesy of Springer-Verlag (see the Acknowledgment).

Color version at the end of the book

Woodward and Hoffmann had succeeded in deriving symmetry relations involving occupied and virtual molecular orbitals in unsaturated hydrocarbons to rationalize and predict the conditions for forbidden and allowed pericyclic reactions, known as Woodward-Hoffmann rules (Woodward and Hoffmann 1965), which recognized a former alternative methodology by Fukui et al., the so-called frontier molecular orbital theory (Fukui et al. 1952). *Although this (and others) are important triumph(s) of MO theory, one has to be aware of MO's limitations, since no theory (or model) is applicable to all situations (i.e., molecular systems and conditions) successfully.* For example, precautions have to be taken when using canonical orbitals for understanding the electronic nature of planar aromatic systems or even to rationalize chemical bond from canonical or certain types of localized MOs. Nonetheless, a recent method called Adaptive Natural Density Partitioning, AdNDP, has been successfully used for describing localized and delocalized chemical bonds in the same molecular system (Zubarev and Boldyrev 2008). The AdNDP orbitals (in which is generated only occupied orbitals) from ethene are depicted in Fig. 2.3.

Figure 2.3   AdNDP orbitals of C-C π-bond and σ-bond and C-H σ-bond from ethene.

As defined by Mülliken earlier, an N-electron molecular orbital is a product of singly N molecular orbitals yielding the following wave function for diatomic molecule (AB) which also included ionic terms $[a^2\phi_A(1)\phi_A(2)+b^2\phi_B(1)\phi_B(2)]$ and, as a consequence, all types of heteronuclear bonds.

$$\psi = \left[a\phi_A(1)+b\phi_B(1)\right]\left[a\phi_A(2)+b\phi_B(2)\right]$$
$$\psi = a^2\phi_A(1)\phi_A(2)+b^2\phi_B(1)\phi_B(2)+ab\left[\phi_A(1)\phi_B(2)+\phi_A(2)\phi_B(1)\right]$$

This first MO approach is called independent-particle model, IPM, and it does not take into account interaction between the electrons, which leads to an imprecise result.

Later, Hartree developed a method for solution of the wave equation for any atom from the approach of central non-Coulomb force field and, in a subsequent work, he developed the self-consistent field method, SCF, to find a field for each core electron in order to get the radial distribution of the charge for the core electrons and the corresponding wave function (Hartree 1928a, b). In Hartree's method, the momentary positions of the n-1 electrons with respect to the reference electron is changed into their averaged positions in order to simplify the force field that n-1 electrons exert on the reference electron and, as a consequence, the n-body problem reduces to n one-body problem by removing the position-dependent field of the other charges. His method found, for some cases, good agreement with X-ray experimental results. However, Hartree did not include exchange as part of an anti-symmetric wave function, which was later corrected by Fock by adding the exchange operator in the Schrödinger equation and deriving a system of linear differential equations where Hartree equation is the first approximation (Fock 1930).

Lennard-Jones also made an important contribution by improving the Hartree-Fock method (Lennard-Jones 1931). Henceforth, many-body MO antisymmetrical wave functions and their observables were solely solved by the Hartree-Fock method until the advent of its improvements, known as post Hartree-Fock methods.

CLASSICAL VALENCE BOND THEORY

Heitler and London wave function for H_2 was not based on IPM, but rather a distinguished approach where electrons from a diatomic molecule are assigned to atomic orbitals, called independent-atom approach.

Heitler and London's work was based on Heisenberg's idea of resonance (Heisenberg 1926). According to Bohr and Slater one can rationalize electron(s) in an atom having a set of virtual harmonic oscillators. Heisenberg, adept to Bohr's ideas, used oscillatory treatment to find the solution to spectrum of helium atom using matrix mechanics as well as the Schrödinger equation where he transformed a many-body problem into a system with two coupled oscillators by employing the idea of resonance. He stated that "resonance always occurs when the two systems were not originally in the same state". From the linear transformation of two oscillators in m and n states, he found eigenfunctions having a resonance character:

$$\frac{1}{\sqrt{2}}\left(\varphi_n^a \varphi_m^b \pm \varphi_m^a \varphi_n^b\right)$$

Heitler and London applied the idea of resonance from Heisenberg to describe the chemical bond in hydrogen molecule (Heitler and London 1927). They used the simple formula for H_2: $\psi = \phi_A(1)\phi_B(2)$ for electron 1 in atom A and electron 2 in atom B. Since electrons are identical, a second possibility should be taken into account: $\psi = \phi_A(2)\phi_B(1)$. Then, two wave functions from linear combination of $\psi = \phi_A(1)\phi_B(2)$ and $\psi = \phi_A(2)\phi_B(1)$ arise:

$$\alpha = a\left[\phi_A(1)\phi_B(2)\right] + b\left[\phi_A(2)\phi_B(1)\right]$$
$$\beta = c\left[\phi_A(1)\phi_B(2)\right] + d\left[\phi_A(2)\phi_B(1)\right]$$

After normalization and orthogonalization processes, both equations become:

$$\alpha = \frac{1}{\sqrt{2+2S}}\left[\phi_A(1)\phi_B(2) + \phi_A(2)\phi_B(1)\right]$$
$$\beta = \frac{1}{\sqrt{2-2S}}\left[\phi_A(1)\phi_B(2) - \phi_A(2)\phi_B(1)\right]$$

Where S is the overlap integral. One equation corresponds to the symmetric repulsive wave function, α, and the other to the antisymmetric attractive wave function, β.

Henceforth, they found the energies for each eigenfunction.

$$E_\alpha = \frac{1}{1+S}(E_{11} + E_{12})$$

$$E_\beta = \frac{1}{1-S}(E_{11} - E_{12})$$

Where E_{11} and E_{12} are energy components described in the Heitler-London paper, in which the terms $\phi_A(1)\phi_B(2)$ and $\phi_A(2)\phi_B(1)$ are included in both of them. As a consequence, the energy components take into account the resonance phenomenon.

The calculated value of H_2 bond energy from Heitler and London's work was 67% from experimental value (which can be improved using effective nuclear charge) and the calculated equilibrium distance was very close to the experimental value. Therefore, the wave function for a diatomic molecule has to take into account two situations: (i) electron 1 in atom A and electron 2 in atom B, and (ii) electron 1 in atom B and electron 2 in atom A, forming two resonance structures: $H_A(1)$ $H_B(2)$ and $H_A(2)H_B(1)$. *The chemical bond and the stabilization interaction through chemical bond are partly a consequence of this resonance phenomenon.*

Valence bond theory was formerly known as Heiltler-London-Slater-Pauling theory (HLSP), later known as classical valence bond theory. Slater has given a great contribution to quantum chemistry, not only for his determinantal wave function. Along with Bohr and Kramers, Slater developed a theory based on old quantum theory to understand the interaction between electromagnetic radiation and atoms, using virtual oscillators (Bohr et al. 1924). This theory was the initial step for Heisenberg and Born to create the matrix mechanics (see the last section in the previous chapter). He also mathematically developed the self-consistent field method (Slater 1928), developed the so-called Slater-type orbitals (Slater 1932), and introduced the virial theorem to the molecular system, which became the first scheme of energy partition of a molecule (Slater 1933). Simultaneously, Pauling and Slater created the concept of hybridization (see chapter six), and the latter helped to develop the valence bond theory (Slater 1931). Nonetheless, there is no doubt that Pauling was the most important proponent of VB theory.

A general expression for the HLSP wave function is:

$$\Psi_{HLSP} = NA\left(\varphi_{1,2}\varphi_{3,4}...\varphi_{2n-1,2n}\right)$$

Where N is the normalization constant, A is the anti-symmetrization operator, and $\varphi_{2i-1,2i}$ is the function corresponding to the covalent bond between atomic orbitals ϕ_{2i-1} and ϕ_{2i}. and n is the number of electron pairs (Mo et al. 2011).

$$\varphi_{2i-1,2i} = A\left(\phi_{2i-1}\phi_{2i}\right)\left[\alpha(i)\beta(j) - \beta(i)\alpha(j)\right]$$

In a series of papers entitled "The nature of chemical bond", Pauling used the idea of Heitler-London H_2 wave function (the concept of resonance) to develop the chemical bonds with larger molecules, where he derived the concept of hybridization as a linear combination of fundamental atomic orbitals (Pauling 1931a). Pauling established the rules for the electron-pair bond from quantum mechanics platform to support Lewis's rules of electron-pair bond (Lewis 1916) in which each atom with an unpaired electron forms a two-center two-electron, $2C$-$2e$, bond. Construction of a VB wave function was based upon the variational method (Eckart 1930), where $\psi_A = \Sigma_k C_{ik} \phi_{Ak}$ is the wave function for each isolated atom to form a chemical bond.

Pauling also derived a wave function having certain ionic character. Let us assume a polar diatomic MX molecule, then he stated that the corresponding wave function is a combination of pure covalent and pure ionic wave functions (Pauling 1931a):

$$\Psi_{MX} = \frac{\psi_M(1)\psi_X(2) + \psi_M(2)\psi_X(1)}{\sqrt{2 + 2S^2}} \therefore \text{Pure covalent character}$$

$$\Psi_{M^+X^-} = \psi_X(1)\psi_X(2) \therefore \text{Pure ionic character}$$

$$\Psi_+ = a\Psi_{MX} + \sqrt{1 - a^2}\,\Psi_{M^+X^-}$$

$$\Psi_- = -a\Psi_{M^+X^-} + \sqrt{1 - a^2}\,\Psi_{MX}$$

In this same paper, Pauling explained the principle which leads to orbital hybridization: "The type of bond formed by an atom is dependent on the ratio of bond energy to energy of penetration of the core (s-p separation). When this ratio is small, the bond eigenfunctions are p eigenfunctions, giving rise to bonds at right angles to one another; but when it is large, new eigenfunctions especially adapted to bond formation can be constructed".

Nonetheless, the same elegant mathematical approach of resonance could not be applied to one electron bond when there is a considerable energy difference between the two possible eigenfunctions, e.g., •H Li$^+$ and •Li H$^+$ (Pauling 1931b). Another difficulty arose with the representation of triplet oxygen with two three-electron bonds. In addition, resonance approach and electron-pair model were ineffective, from this original model, to rationalize multicenter bonding. As Mülliken formerly stated that these three difficulties from VB theory were easily solved in MO theory (see the previous section). Later, Shaik and Hiberty explained that these initial problems in classical VB theory were not actually failures (Shaik and Hiberty 2008). *Classical VB theory brought to chemistry the concepts of hybridization, resonance, and a very elegant mathematical approach to rationalize localized electron-pair bonds.*

Unfortunately, classical VB theory, or we can say more precisely, HLSP theory began to decline, from the viewpoint of most chemists, for a series of different reasons: (1) overwhelming success of MO theory in incorporating Slater determinants to construct antisymmetrical canonical wave-functions upon linear combination of atomic orbitals along the whole molecular system (and not only restricted to a set of electron-pair-bonded atoms as in VB theory), which enormously facilitated its implementation in program languages such as Fortran; (2) the handy method of Hückel to predict stability of aromatic and delocalized systems; (3) the great success of Fukui, Woodward, and Hoffmann in predicting the reaction conditions and stereoselectivity of pericyclic reactions; and, at last, (4) Pauling's abandonment of VB theory for many possible different reasons. Nonetheless, VB theory has never died. On the contrary, different research groups in Japan, Europe, and USA (during the rise of VB and in the period of its apparent decline) were continuously working on the improved versions of HLSP theory and their applications.

MODERN VALENCE BOND THEORY

Even before Slater determinants, other approaches to construct the antisymmetrical wave functions were elaborated. These approaches are based on group theory. Unlike MO wave function, construction of spin function for the antisymmetrical VB wave function uses group theory approach, since it is possible to separate the spatial and spin functions, as previously discussed (see the first chapter). Then, as stated by McWeeny: "the problem of constructing spin eigenfunctions can be isolated from any consideration of the orbital form of the wave function and also from the overall space-spin antisymmetry requirement. This is because the spin operators S^2 and S_z are symmetric in the particles" (McWeeny 1992). In another words, after one permutation of two spin variables the spin eigenfunction keeps the same corresponding eigenvalues (and sign). This can be accounted for by the fact that S^2 and S_z are sums of contributions from N electrons in the corresponding wave function, and then, they are independent of any permutation. Another reason for the separation of spin and spatial functions is the fact that the Hamiltonian does not operate on spin variables and similarly S^2 and S_z do not operate on spatial variables. Therefore, a wave function can be in the form:

$$\Psi(x_1, x_2, ...x_N) = \phi(r_1, r_2, ...r_N)\Phi_{s,m_s}^N(\mu_1, \mu_2, ..., \mu_N)$$
$$x = \{r, \theta, \varphi, s\} \therefore r = \{r, \theta, \varphi\} \therefore \mu = \{s, m_s\}$$

Henceforth, it is necessary to apply an antisymmetrizer operator, A, in both spin and spatial functions in order to construct an antisymmetric wave function.

$$A = \frac{1}{(n!)^{1/2}} \sum_p (-1)^p \mathbf{P}$$

Where summation runs over all $n!$ permutations of n electrons using permutation operator **P**.

For example, the singlet hydrogen VB wave function is:

$$\Psi_{H_2}^1 = N\hat{A}\left[1s_A(r_1)1s_B(r_2)\left\{\uparrow(1)\downarrow(2) - \downarrow(1)\uparrow(2)\right\}\right]$$

$$\Psi_{H_2}^1 = \frac{1}{\sqrt{2(1+S^2)}}\begin{bmatrix} 1s_A(r_1)\uparrow(1)1s_B(r_2)\downarrow(2) \\ -1s_A(r_1)\downarrow(1)1s_B(r_2)\uparrow(2) \\ +1s_B(r_1)\uparrow(1)1s_A(r_2)\downarrow(2) \\ -1s_B(r_1)\downarrow(1)1s_A(r_2)\uparrow(2) \end{bmatrix}$$

Alternatively, antisymmetric wave function can be a product of symmetric and antisymetric factors, one for a spatial function and other for the spin function, or vice-versa:

$$\Psi = \phi_{spatial} \Phi_{spin}$$

$$\Psi = \phi_{(antisym)} \cdot \Phi_{(sym)} = \Psi_{(antisym)}$$

$$\Psi = \phi_{(sym)} \cdot \Phi_{(antisym)} = \Psi_{(antisym)}$$

For example, the H$_2$ wave function from this approach is:

$$\Psi^1_{H_2} = \frac{1}{\sqrt{2(1+S^2)}} \Big[\{1s_A(r_1)1s_B(r_2)+1s_A(r_2)1s_B(r_1)\}\{\uparrow(1)\downarrow(2)-\downarrow(1)\uparrow(2)\} \Big]$$

The spatial function is symmetric and the spin function is antisymmetric.

$$\left[\underbrace{\{1s_A(r_1)1s_B(r_2)+1s_A(r_2)1s_B(r_1)\}}_{symmetric} \underbrace{\{\uparrow(1)\downarrow(2)-\downarrow(1)\uparrow(2)\}}_{antisymmetric} \right]$$

This hydrogen molecule wave function is different from that from Heitler-London, because the latter does not take into account the spin function.

CONSTRUCTING SPIN FUNCTION ON MODERN VALENCE BOND THEORY

For systems with three or more electrons there is a certain difficulty to construct all linearly independent spin functions for a given orbital configuration. It is important to note that the problem of constructing spin functions, $\Theta^{(S,M)}$, can be isolated from any consideration of the orbital form of the wave function. There are two ways for constructing spin functions: synthetic and analytic.

Serber considered that a specific configuration A with orthogonal orbits α_1, $\alpha_2...\alpha_n$ have n! wave functions when considering all the permutations of the orbits using permutation operator, P. Then, these wave functions are $P_a\psi^A, P_b\psi^A,...P_{n!}\psi^A$. However, not all n! states follow Pauli's exclusion principle, and then P has to be restricted to P^s which permutes the spins among themselves, whose formula is:

$$P^s_{ij} = \frac{1}{2}\left(1+4s_i \cdot s_j\right)$$

Serber proposed a method to obtain all $P^s_{ij}\psi^{s_i s_j}_{sm_s}$ wave functions for any possible multiplicity (Serber 1934a). Serber's method to obtain a set of wave functions that follow exclusion principle is based on the Dirac vector model, in which the latter found a correlation between spin alignment and exchange energy and the relation between spin-spin coupling and the Hamiltonian:

$$H = Q - \tfrac{1}{2}\sum_{j>i}\left(1+4s_i \cdot s_j\right)J_{ij}$$

$$J_{ij} = \int \varphi_i(1)\varphi_j(2)H\varphi_j(1)\varphi_i(2)d\tau$$

Where J_{ij} is the exchange integral, Q is the Coulomb energy, s_i and s_j are vector matrices of the electron in orbits i and j, respectively, and it considers the spins in singly-occupied orbitals (Dirac 1929).

Sherman and van Vleck also used the Serber building-up method for constructing spin function and found the analytical formula for a number of states, N, for a given total spin, S (being $S=0$, for singlet, $S=1$, for triplet, and so on) involving n electrons:

$$N = \binom{n}{k} - \binom{n}{k-1} \therefore \left(k = \tfrac{1}{2}n - S\right)$$

$$f_S^N = N = \frac{(2S+1)N!}{\left(\tfrac{1}{2}N+S+1\right)!\left(\tfrac{1}{2}N-S\right)!}$$

This formula restricts the number of states in accordance with Pauli exclusion principle, which is far lesser than $n!$ states. Alternatively, N can be found in a more intuitive methodology called branching diagram (van Vleck and Sherman 1935). Most authors use $f(N,S)$ instead of N, and it is named "number of independent spin states".

For example, for five electrons occupying five different orbitals with $S=1/2$, there are five linearly independent spin functions.

$$f_{1/2}^5 = \frac{\left(2\tfrac{1}{2}+1\right)5!}{\left(\tfrac{1}{2}5+\tfrac{1}{2}+1\right)!\left(\tfrac{1}{2}5-\tfrac{1}{2}\right)!} = 5$$

$$\Theta_k^{(1/2,1/2)} \Rightarrow \Theta_1^{(1/2,1/2)}, \Theta_2^{(1/2,1/2)}, \Theta_3^{(1/2,1/2)}, \Theta_4^{(1/2,1/2)}, \Theta_5^{(1/2,1/2)}$$

Sherman and van Vleck also derived the exchange energy, W, from the exchange integrals:

$$W = \sum_{i'>i} J_{ii'} - \tfrac{1}{2} \sum_{j>i(i\neq i')} J_{ij}$$

Where i and i' are paired electrons. Based on VB wave functions, they found very good agreement with experimental values for a series of compounds.

Serber himself developed his method to construct spin eigenfunctions similar to the genealogical construction by adding two-electron unit, a type of building-up method which starts from small systems (Serber 1934b, Pauncs 1979). The building-up method by adding one-electron unit to a small system has several approaches: diagonalization of S^2 matrix, construction of S^2 eigenfunctions by orthogonalization procedure, genealogical construction of spin eigenfunctions by Kotani, and branching diagram functions by Löwdin (Pauncs 1979).

Löwdin derived the spin function by means of a projection operator, where the idea of doubly-occupied orbitals is replaced by singly-occupied orbitals where antiparallel spins avoid each other (Löwdin 1955). In this method, it is not necessary to know the spin function from a small system. Instead, it uses trial function, S_z, and a projection operator on it, yielding S^2.

At last, one very intuitive method of constructing linearly independent spin functions is based on the Rumer diagram, where arrows unite atoms as points in a closed circuit that cannot cross themselves (Pauncs 1979). These set of spin functions are not orthogonal to each other. Rumer's method is called spin-paired spin eigenfunctions. For example, for a triplet system ($S=1$) with six electrons two spin pairs and two electrons with unpaired spins are needed. Below, the Rumer diagram for construction of two independent spin functions for six electrons with $S=1$ is shown.

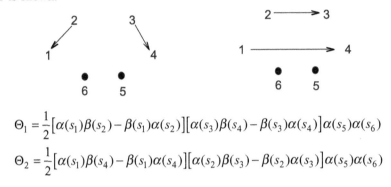

$$\Theta_1 = \frac{1}{2}[\alpha(s_1)\beta(s_2) - \beta(s_1)\alpha(s_2)][\alpha(s_3)\beta(s_4) - \beta(s_3)\alpha(s_4)]\alpha(s_5)\alpha(s_6)$$

$$\Theta_2 = \frac{1}{2}[\alpha(s_1)\beta(s_4) - \beta(s_1)\alpha(s_4)][\alpha(s_2)\beta(s_3) - \beta(s_2)\alpha(s_3)]\alpha(s_5)\alpha(s_6)$$

More information about the construction of Rumer spin basis can be viewed in Karadakov and team's work (Karadakov et al. 1995) and its implementation in the code called SPINS.

ANTISYMMETRIC WAVE FUNCTION ON MODERN VALENCE BOND

As previously stated, there are $f(N,S)$ wave functions that are a product of spatial function and spin function.

$$\psi_i = \phi(r_1, r_1...r_N)\Phi(N, S, M_s; i) \therefore i = 1, 2, ... f(N, S)$$

However, these wave functions have to become antisymmetric by the action of the antisymmetric operator A. It is important to emphasize that in this book matrices are represented by bold, non-italic, and upper-case characters.

$$\psi_i = \left[(N!)^{1/2} \sum_P (-1)^P \mathbf{P} \right] [\phi(r_1, r_1...r_N)\Phi(N, S, M_s; i)]$$

The summation on the equation above is over all permutations belonging to S_N (and not to $n!$). The antisymmetric wave function can also be separated into spatial and spin functions:

$$\psi_i = (N!)^{1/2} \sum_P (-1)^P \mathbf{P}^r \phi(r_1, r_1...r_N) \mathbf{P}^\sigma \Phi(N, S, M_s; i)$$

Where $\mathbf{P^r}$ operates on spatial variables and $\mathbf{P^\sigma}$ operates in spin variables. By knowing that:

$$\mathbf{P^\sigma}\Phi(N,S,M_s;i) = \sum_{i=1}^{f}\Phi(N,S,M_s;i)U(\mathbf{P})_{ij}^{S}$$

Where $U(\mathbf{P})_{ij}^{S}$ are expansion coefficients. Then, the antisymmetric wave function can be rearranged to:

$$\psi_i = f^{-1/2}\sum_{j=1}^{f}\Phi(N,M_s,S;j)\phi_{ij}^{S}$$

$$\phi_{ij}^{S} = \left(\frac{f}{N!}\right)^{1/2}\sum_{P}(-1)^P U(\mathbf{P})_{ij}^{S}\mathbf{P^r}\phi$$

Where ϕ and ϕ_{ij}^{S} are spatial functions, $\Phi(N,S,M_S;i)$ is one of the orthonormal spin eigenfunctions obtained by vector coupling the N spins in $f(N,S)$ (Goddard III 1967a). The Yamanouchi-Kotani method is based on the above equation (Kotani and Siga 1937, Yamanouchi 1937).

Therefore, ϕ_{ij}^{S} can be rationalized, in a simple way as a modified spatial function after all permutation operations in spatial function ϕ and its multiplication with $U(\mathbf{P})_{ij}^{S}$.

From the Yamanouchi-Kotani method, Goddard derived another method to generate antisymmetric VB wave functions (i.e., *without double occupation restriction and yet with the correct spin symmetry*), which was formerly named GF method and later generalized valence bond, GVB (Goddard III 1967a, b). As Goddard stated: "GF method consists of finding the best approximation to an eigenstate of the Hamiltonian by a wave function of the form:

$$\psi = G_f^\gamma[\phi\Phi]$$
$$\Phi = \alpha(1)\alpha(2)...\alpha(n)\beta(n+1)...\beta(N)$$
$$\phi = \phi_{1a}(1)\phi_{2a}(2)...\phi_{na}(n)\phi_{1b}(n+1)...\phi_{mb}(N)$$

And G_f^γ is an operator (...) involving permutations of the spatial and spin coordinates of the N electrons" (Goddard III 1968).

Like the Hartree-Fock method, the GF method uses the variational method, SCF equations, and basis set, but in the latter the orbitals and spin coupling are simultaneously optimized.

Goddard also explained the difference between the antisymmetrizer in Hartree-Fock wave function (Slater determinant) and the antisymmetrizer in GVB, G_f^γ: "In the Hartree-Fock method, the antisymmetrizer took care of the Pauli principle but not the spin symmetry (...). The group operator G_f^γ simultaneously takes care of both spin symmetry and the Pauli principle" (Goddard III and Ladner 1971).

In addition, by deriving H_2 wave function from the HF method, unrestricted HF, UHF, and GB methods (see the equations below) Goddard showed that the GVB wave function correctly describes the bond dissociation (Fig. 2.4).

$$\Psi_{HF(H_2)} = A[\phi(1)\alpha(1)\phi(2)\beta(2)] = \phi(1)\phi(2)[\alpha(1)\beta(2) - \beta(1)\alpha(2)]$$

$$\Psi_{UHF(H_2)} = A[\phi_a(1)\alpha(1)\phi_b(2)\beta(2)] = \phi_a\phi_b\alpha\beta - \phi_b\phi_a\beta\alpha$$

$$\Psi_{GVB(H_2)} = G_f^\gamma[\phi_a(1)\phi_b(2)\alpha(1)\beta(2)]$$

$$\Psi_{GVB(H_2)} = 1/4(\phi_a\phi_b + \phi_b\phi_a)(\alpha\beta - \beta\alpha)$$

On the other hand, *Hartree-Fock wave function does not describe a bond dissociation correctly* because it forces double occupation in one orbital leaving the other empty, which is incompatible in H_2 bond dissociation where each orbital must be singly occupied. As for the UHF method, its wave function correctly describes the bond dissociation when it includes both singlet and triplet states simultaneously, but it does not have the correct spin symmetry. For isolated singlet and triplet UHF wave functions, the description of bond dissociation also fails (Fig. 2.4). Other MO wave functions, e.g., MP2, and MO-based wave functions, e.g., DFT, have the same profile as HF wave function. *That is, only VB wave functions describe the bond dissociation correctly.*

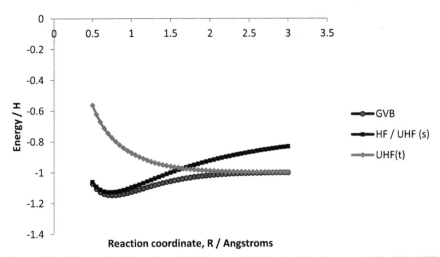

Figure 2.4 Energy (in Hartree) vs H-H interatomic distance (in Angstroms) for HF, UHF (singlet and triplet), and GVB wave functions using 6-311G basis set.

Goddard et al. have calculated GVB orbitals for many organic compounds (Hay et al. 1972, Goddard III et al. 1973). He also established the orbital phase continuity principle, OPCP, in which the selection rules for allowed and prohibited thermal additions depend only on the GVB orbitals without any symmetry analysis (Goddard III 1970).

Many other applications of GVB can be found elsewhere. More recently, this author and collaborators have used the GVB method to study the influence of oxygen lone pair from nitro substituent on the instability of the corresponding nitro-tetrahedrane by means of electronic repulsion between oxygen lone pairs and C-C bond from the cage. This analysis was done qualitatively (using GVB orbitals as indicated in Fig. 2.5) and quantitatively (by means of the overlap integrals between C-C bond and lone pair where the higher value indirectly indicated higher repulsion). Another indirect analysis of the electronic repulsion by the overlap integrals between proximate chemical bonds were used for rationalizing the instability of tetrahedrane with respect to more stable cubane and tetramethyl-tetrahedrane (Monteiro et al. 2014). Then, GVB became the first successful VB method after HLSP theory, although its implementation usually has two impositions: perfect-pairing (yielding only one VB structure) and strong orthogonality of GVB orbitals. These restrictions on GVB are known as GVB-SOPP.

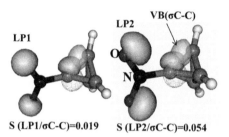

S (LP1/σC-C)=0.019   S (LP2/σC-C)=0.054

Figure 2.5   Selected GVB orbitals of both oxygen lone pairs (LP1 and LP2) and σC-C bond from the corresponding tetrahedrane cage in nitrotetrahedrane (see the Acknowledgment section). The GVB positive lobe (in non-bonding region) was omitted for simplicity.

Nearly at the same time GVB theory was first published, the spin coupled valence bond method, SCVB, was developed by Gerratt, who published his elegant paper in the early 1970's (Gerratt 1971). The SCVB method uses Young operators, ω_{lk}^{S}, which is quite similar to the Wigner projection operator, to build an antisymmetrical VB wave function with correct spin symmetry.

$$\omega_{lk}^{S} = \left(\frac{f(N,S)}{N!}\right)^{1/2} \sum_{P \in \ell_N} U_{lk}^{S}(P) \mathbf{P}^{\mathrm{r}} \therefore [l,k = 1,2,...f(N,S)]$$

Then, $\omega_{lk}^{S}\phi$ has the necessary permutation symmetry forming each k value as a basis for an irreducible representation of ℓ_N where ϕ is the spatial function. The approximate eigenfunction ψ_{SCVB}, is:

$$\psi_{SCVB} = \left(1/f(N,S)\right)^{1/2} \sum_{l=1}^{f(N,S)} \left(\omega_{lk}^{S}\phi\right)\Phi_{S,M_s,l}^{N}$$

The total electronic energy of SCVB (or SC) wave function can be expressed in the following energy partition:

$$E = E_c + E_{cv} + E_v$$

Where E_c is the core energy, E_{cv} is the core-valence interaction energy, and E_v is the valence energy (Karadakov et al. 1992).

The SCVB method also leans on the variational methodology to obtain its wave function, which can be divided into core and valence subsets. The core subset comprises of doubly occupied HF orbitals and the valence electrons are described by SCVB theory. The SCVB does not have the GVB impositions, that is, its orbitals are not orthogonal and its wave function comes from a combination of all modes of spin pairing. The SCVB is recommended for describing delocalized systems.

The GVB wave function can be related to only one resonance structure of a delocalized molecular system or as a result of all resonance structures (Li et al. 2002). Then, *it is important to add that the GVB wave function involving multiple structures (all resonance structures of the molecular system) is equivalent to the SCVB wave function.* For example, see benzene's GVB singly occupied p orbitals of one Kekule resonance structure, the GVB singly occupied p orbitals of resonance hybrid, and SCVB singly occupied p orbitals in Fig. 19.4 in chapter nineteen. One can note that the GVB wave function of all five resonance structures of benzene is equivalent to that of the SCVB wave function.

There are several applications of SCVB in the literature, mainly owing to Cooper and collaborators' works where we cite two of them: the mechanistic description of Diels-Alder reaction (Karadakov et al. 1998) and the SC orbitals for benzene and six-membered heteroaromatic compounds (Cooper et al. 1989). For further information about SC orbitals of benzene and other aromatic systems, please see the chapter on aromaticity.

Other VB methods appeared later, such as VBSCF (van Lenthe and Balint-Kurti 1983), CASVB (Hirao et al. 1996, Thorsteinsson and Cooper 1998), among other methods. The VBSCF applies a multiconfiguration SCF from non-orthogonal orbitals, resulting in a wave function from a linear combination of VB structures. As a consequence, it is possible to obtain weights for each configuration (or structure) from VBSCF of small molecules, i.e., to know the importance, for instance, of $H^- F^+$, $H^+ F^-$, and H-F in an HF molecule. See practical examples in Shaik and Hiberty's book (Shaik and Hiberty 2008). From the CASVB method, it is possible to obtain the corresponding orbitals for the excited states of the studied molecule.

Henceforth, some very useful VB programs appeared, for instance, VB2000 (Li and McWeeny 2002), TURTLE (van Lenthe and Balint-Kurti 1980, 1983), XMVB (Song et al. 2005), among others. All GVB and SCVB calculations from this author were done by means of VB2000.

THE QUANTUM THEORY OF ATOMS IN MOLECULES(QTAIM): BASIC CONCEPTS

All properties of matter are somewhat related to its corresponding density of electronic charge (or electron density). The electron density arises from a variational calculation (being HF-MO, post-HF-MO, HF-DFT, or VB) yielding an optimized wave function. From the electron density, the density matrix is obtained, which is used for further QTAIM calculation. The electron density in an equilibrium geometry is the distribution that minimizes the energy for that geometry. The quantum theory of atoms in molecules, QTAIM, is based on the electron density, $\rho(\chi_1)$, which is the probability of finding any electron from the set of N electrons in infinitesimal volume, $d\chi_1$,

$$\rho(\chi_1) = N \int \Psi(\chi_1, \chi_2, ..., \chi_N) \Psi^*(\chi_1, \chi_2, ..., \chi_N) d\chi_2 ... d\chi_N$$

The probability of finding any electron from the set of N electrons in infinitesimal volume, $d\chi_1$, regardless the spin value is:

$$P(r_1) = \int \rho(\chi_1) ds_1$$

This is the real electron density obtained by X-ray diffraction because $P(r_1)$ derives from the real variable r_1. Then, the total number of electrons is given by:

$$N = \int P(r_1) dr_1$$

Alternatively, electron density can be obtained from:

$$\rho(r) = \Sigma_i \eta_i |\varphi_i(r)|^2 = \Sigma_i \eta_i |\Sigma_l C_{l,i} \zeta_l(r)|^2$$

Where η_i is the occupation number of the atomic/molecular orbital i, φ_i, $C_{i,l}$ is the expansion coefficient (a matrix element from density matrix generated after SCF procedure) associated with the basis function ζ_l from a chosen basis set.

A simple way to understand electron density is its analogy with population density. Note the relation below:

$$\frac{\text{electron density}}{\text{molecule}} \equiv \frac{\text{population density}}{\text{city}}$$

Electron density is a real function that can be represented in 2-D plots with contour lines or 3-D plots (from the whole molecule or from one specific plane in the molecule, as shown in Fig. 2.6). Each contour line has the same density value, i.e., it is an isodensity surface. The outer surfaces are smaller in magnitude, while the inner surfaces are higher in magnitude.

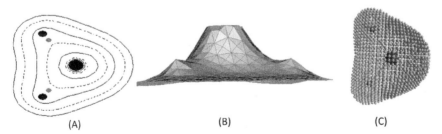

(A) (B) (C)

Figure 2.6 Electron density of water from (A) contour plot from H-O-H plane; (B) 3-D plot from H-O-H plane; (C) 3-D plot from the whole molecule.

Similar to any 2-D or 3-D function, electron density has critical points which are characterized by the first derivative and second derivatives of electron density (Fig. 2.7). A critical point is a point in the function where its tangent is zero (i.e., its first derivative is null).

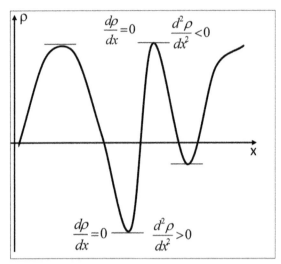

Figure 2.7   2-D representation of a hypothetical electron density depicting their critical points.

The concept of gradient is very important in QTAIM. A gradient, ∇ (del), is a vector function having unit vectors ($\mathbf{x}$, $\mathbf{y}$ and $\mathbf{z}$ in cartesian coordinates) as its basis and each corresponding coefficient given by the partial derivative of the scalar function, φ, in the corresponding coordinate. It is important to emphasize that in this book, vectors are represented by bold, non-italic, and lower-case characters.

$$\nabla\varphi = \mathbf{u_x}\frac{\partial\varphi}{\partial x} + \mathbf{u_y}\frac{\partial\varphi}{\partial y} + \mathbf{u_z}\frac{\partial\varphi}{\partial z}$$

The gradient describes the variation of the scalar function (e.g., electron density) in each point of this scalar function being represented by a gradient vector. The gradient vector in each point of the scalar function points towards the highest value of the scalar function. Let's suppose the function $\rho = -\left[\cos^2 x + \cos^2 y\right]^2$ as one electron density surface of a generic isomorphic four-membered ring. Figure 2.8 shows this scalar surface (in 3-D) and the corresponding schematic representation of the vector function of the gradient (in 2-D). The highest values of the scalar function are the hypothetical atoms in the generic isomorphic four-membered ring and the second highest values are the bond critical points, BCPs (see the discussion ahead). The BCP is a saddle point (lower point in a horse saddle). The gradient vectors point either towards each atom or towards the bond critical point (when in the middle point between a pair of homonuclear bonded atoms).

The set of gradient vectors pointing in the same direction make up the *gradient path*. Each gradient path begins or ends in maximum points, where $\nabla\varphi=0$ (in case of electron density: $\nabla\rho=0$). Two gradient paths from the *bond critical point, BCP* (the second highest maximum value of the electron density), go in opposite senses towards each corresponding bonding nucleus. These two special gradient paths are called *bond paths*. The bond path is an atomic interaction line linking BCP and one bonded atom along which the electron density is maximum with respect to its neighboring transverse (Bader 1998). According to Bader, *a bond path has to be mirrored by a virial path where the potential energy density is maximally stabilizing when the molecular system is in a stable electrostatic equilibrium* (Keith et al. 1996, Bader 2009). Bond path is associated with all types of chemical interactions, including not only chemical bonds but also van der Waals interactions, and *it is a universal indicator of bonded interactions*, i.e., all types of chemical interactions (Bader 1994), although it is possible for a few cases to exist in a chemical interaction without bond path, e.g., in some bent metallocenes (dos Santos et al. 2013).

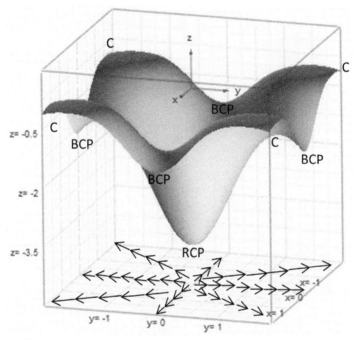

Figure 2.8   Representation of the electron density of a generic isomorphic four-membered ring and the corresponding schematic representation of the gradient function.

The nuclei, as well as BCPs, are critical points where $\nabla\rho=0$. For cyclic molecules, there is another critical point called ring critical point, RCP, in the middle of the ring (Fig. 2.9 and also the schematic representation in Fig. 2.8).

Another important characteristic from the gradient is the *zero flux surface*, ZFS. Any transverse surface that cuts the electron density has non-zero scalar product $\nabla\rho.\mathbf{n}$, where $\mathbf{n}$ is the normal vector to that surface except for the zero flux surface, where all $\nabla\rho.\mathbf{n}$ in it are zero. In addition, zero flux surface contains the bond critical point (Fig. 2.9).

From Fig 2.9, one can see that a set of zero-flux surfaces delimit the region of a specific atom which is called *atomic basin*. While in Fig. 2.9 a planar view of atomic basins is shown, in Fig. 2.10, one can see the 3-D view of oxygen and hydrogen atomic basins in the water molecule.

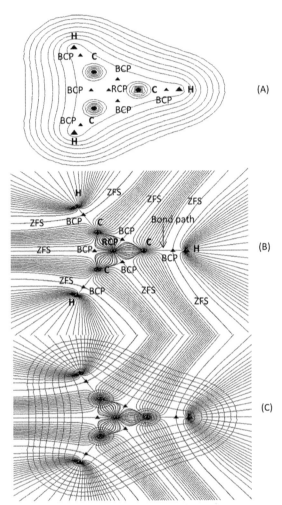

Figure 2.9 (A) Electron density contour plots of cyclopropenyl cation; (B) Gradient paths and zero flux surfaces of cyclopropenyl cation; (C) Mixture of (A) and (B).

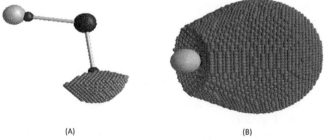

Figure 2.10 3-D atomic basins of oxygen and hydrogen from the water molecule.

The critical points of any function are characterized by two derivatives: first derivative and second derivative. The former determines the critical points (where the first derivative vanishes) and the latter determines whether the critical point is a maximum or minimum. Critical point is a very important property of a function because from them one can evaluate the function variations along the coordinates. For example, for the function $y = \sin(x)$, the solutions for its first derivative with zero value are:

$$\frac{d}{dx}(\sin(x)) = 0$$

$$x = \pi\{N_1\} + \frac{\pi}{2}$$

Its second derivative is $y'' = -\sin(x)$ which exemplifies that where $y = -1$, $y' = 0$ (first derivative) and $y'' = 1$ (minimum). On the other hand, where $y = 1$, $y' = 0$ and $y'' = -1$ (maximum). Then, second derivative function indicates the maximum ($y'' < 0$) or minimum ($y'' > 0$) of the corresponding scalar function.

While gradient is the del vector operator acting on a scalar function (or field) yielding a corresponding vector function (which points towards the higher magnitude of the scalar function), the divergence is the del vector operator acting on a vector function yielding a scalar function. For example, the divergence of space vector, **r**, is:

$$\nabla \cdot \mathbf{r} = \left(\mathbf{x}\frac{\partial}{\partial x} + \mathbf{y}\frac{\partial}{\partial y} + \mathbf{z}\frac{\partial}{\partial z} \right) \cdot (\mathbf{x}x + \mathbf{y}y + \mathbf{z}z)$$

$$\nabla \cdot \mathbf{r} = \frac{\partial x}{\partial x} + \frac{\partial y}{\partial y} + \frac{\partial z}{\partial z} = 3$$

The Laplacian is the divergence of the gradient (a vector function) of a scalar field (or function). The Laplacian of the electron density is the divergence of the electron density gradient.

$$\nabla \cdot \nabla\rho = \nabla^2\rho = \left(\mathbf{x}\frac{\partial}{\partial x} + \mathbf{y}\frac{\partial}{\partial y} + \mathbf{z}\frac{\partial}{\partial z} \right) \cdot \left(\mathbf{x}\frac{\partial\rho}{\partial x} + \mathbf{y}\frac{\partial\rho}{\partial y} + \mathbf{z}\frac{\partial\rho}{\partial z} \right)$$

$$\nabla^2\rho = \frac{\partial^2\rho}{\partial x^2} + \frac{\partial^2\rho}{\partial y^2} + \frac{\partial^2\rho}{\partial z^2}$$

In QTAIM, the starting point for ranking the critical points is the Hessian matrix instead of the Laplacian itself because the former contains all possible second partial derivatives. The Hessian of electron density is a 3×3 square matrix.

$$\nabla \cdot \nabla\rho = \begin{pmatrix} \dfrac{\partial^2\rho}{\partial x^2} & \dfrac{\partial^2\rho}{\partial x\partial y} & \dfrac{\partial^2\rho}{\partial x\partial z} \\[2mm] \dfrac{\partial^2\rho}{\partial y\partial x} & \dfrac{\partial^2\rho}{\partial y^2} & \dfrac{\partial^2\rho}{\partial y\partial z} \\[2mm] \dfrac{\partial^2\rho}{\partial z\partial x} & \dfrac{\partial^2\rho}{\partial z\partial y} & \dfrac{\partial^2\rho}{\partial z^2} \end{pmatrix}$$

The Laplacian function can be solved by diagonalizing the Hessian matrix. This diagonalization can be achieved by a proper rotation of the cartesian coordinates $(x \rightarrow x', y \rightarrow y', z \rightarrow z')$:

$$\nabla \cdot \nabla \rho = \begin{pmatrix} \dfrac{\partial^2 \rho}{\partial x'^2} & 0 & 0 \\ 0 & \dfrac{\partial^2 \rho}{\partial y'^2} & 0 \\ 0 & 0 & \dfrac{\partial^2 \rho}{\partial z'^2} \end{pmatrix}_{r'=r_{cc}}$$

Then, in QTAIM the Laplacian is obtained by the trace of the diagonalized Hessian matrix.

$$\nabla^2 \rho = \frac{\partial^2 \rho}{\partial x'^2} + \frac{\partial^2 \rho}{\partial y'^2} + \frac{\partial^2 \rho}{\partial z'^2}$$

For any point, r_1, in the electron density field, the corresponding Laplacian is given by the trace of the eigenvalues from the Hessian diagonalized matrix, λ, at this point.

$$\nabla \cdot \nabla \rho(r_1) = \begin{pmatrix} \lambda_1 & 0 & 0 \\ 0 & \lambda_2 & 0 \\ 0 & 0 & \lambda_3 \end{pmatrix} = \lambda_1 + \lambda_2 + \lambda_3$$

Then, firstly, QTAIM uses a specific algorithm (which depends on the QTAIM program) to find the critical points from the density matrix (a .wfn or .wfx file generated by one quantum chemistry package). For example, in AIM 2000, the algorithm for finding the critical points is based on the Newton-Raphson method (Biegler-König and Schönbohm 2002). Soon afterwards, by diagonalizing the Hessian matrix in these critical points, their corresponding eigenvalues are obtained, which are used for classifying the critical points according to Table 2.1. This is the first application of the Laplacian of the electron density (or charge density). When there is no degenerate critical point, all eigenvalues of Hessian matrix are non-zero values, ranging from minus to plus sign. In the critical point whose all three eigenvalues are negative, it is the absolute maximum of electron density. This is where one atom (or more correctly the atomic basin) is located. In the critical point whose two eigenvalues are negative and one positive, it is the relative maximum of electron density and it is called bond critical point. The BCP is maximum in two directions and minimum between the bonding nuclei, characterizing a saddle point. Figure 2.11 depicts the 3-D charge density of the fluorine molecule from the plane containing both atoms (where there is the saddle point) and transverse plane passing through the bond critical point. In this transverse plane the BCP is maximum in both directions.

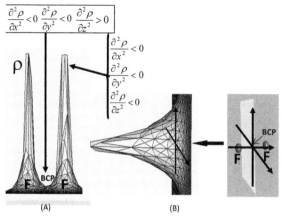

Figure 2.11   3-D charge density topology of fluorine molecule from (A) plane containing both atoms, and from (B) transverse plane passing through the bond critical point.

Then, a set of critical points and bond paths determine the *topology of the electron density* (or charge density) and it is represented by the *molecular graph*, as depicted in Fig. 2.10 and Fig. 2.2.

Table 2.1   Ranking of the critical points.

Name of the critical point	Acronym	Sign of			(ω,σ)
		λ_1	λ_2	λ_3	
Nuclear attractor	NA	−	−	−	(3,−3)
Bond critical point	BCP	−	−	+	(3,−1)
Ring critical point	RCP	−	+	+	(3,+1)
Cage critical point	CCP	+	+	+	(3,+3)

The bond critical point is the most important topological data from the charge density topology (or electron density topology). A lot of topological data from BCP gives information about the type (and strength) of different chemical bonds or chemical interactions, for example, metal-ligand interactions, covalent bonds, ionic bonds, weak interactions in noble gas, or hydrogen bonds (Table 2.2).

Table 2.2   Topological data of the bond critical point of CC bond in benzene, and ethane, and closed shell interactions in LiCl, NaF, and Ar_2 from MP2/aug-cc-pVTZ//CCSD(T)/6-311G++(d,p) wave function.

Level of theory: MP2/aug-cc-pVTZ//CCSD(T)/6-311G++(d,p)					
Chemical bond	ρ_b	$\nabla^2\rho_b$	ε	G_b/ρ_b	$|\lambda_1|/\lambda_3$
Ethane CC bond	0.2575	−0.7439	0.0000	0.2057	1.8101
Benzene CC bond	0.3321	−1.2003	0.2144	0.3004	4.1573
Ethene CC bond	0.3660	−1.3962	0.3897	0.3834	8.8927
LiCl	0.0430	0.2619	0.0000	1.4274	0.1666
NaF	0.0436	0.3851	0.0000	1.9287	0.1336
Ar_2	0.0029	0.0118	0.0000	0.7515	0.1158

There are basically two types of bonded interactions: (1) shared shell interaction from covalent bonds; and (2) closed shell interaction from ionic bonds and intermolecular/intramolecular interactions (Bader 1994). In Table 2.2, there are three molecules with shared shell interactions and three other molecules with closed shell interactions.

The charge density of the BCP, ρ_b, varies from 0.1 to 0.5 au. for shared shell interactions and below 0.07 au. for closed shell interaction. *The charge density of the BCP, ρ_b, can be used as an indirect measure of the strength of the bonded interaction.* The higher ρ_b the stronger is the bonded interaction.

As previously discussed, the Laplacian $\nabla^2 \rho_b$ is negative for charge accumulation or charge concentration (for shared shell interaction) and positive for charge dispersion or charge depletion (for closed shell interaction).

The ellipticity, ε, is the ratio between λ_1 and λ_2 minus one $[(\lambda_1/\lambda_2)-1]$. The λ_1 and λ_2 are eigenvalues of the Hessian matrix of the charge density. The λ_1 and λ_2 are eigenvalues of the corresponding eigenvectors $\vec{u}_1$ and $\vec{u}_2$, respectively. The eigenvectors $\vec{u}_1$ and $\vec{u}_2$ belong to a plane perpendicular to the bond path at the BCP between two bonding atoms. The BCP with $\varepsilon > 0$ means that the corresponding bonded interaction does not have elliptical symmetry. Elliptical symmetry is characteristic of single bond or triple bond (where $\varepsilon = 0$).

The other two ratios are G_b/ρ_b and $|\lambda_1|/\lambda_3$. The former is higher than 1 for ionic bonds, higher than 0.5 and smaller than 1 for intermolecular/intramolecular interactions (see ahead for more discussion about the types of intermolecular interactions according to QTAIM) and coordinate bonds, and smaller than 0.5 for covalent bonds. The latter is smaller than 1 (near 0.1) for closed shell interactions and higher than 1 for shared shell interactions.

This topological information can also determine the strength of the bonded interactions. For example, the magnitudes of all topological data for ethane (single bond), benzene (half double bond), and ethene (double bond) follow a linear relation with their corresponding bond orders (1, 1.5, and 2). Then, *QTAIM represents an important tool to classify chemical bonds or intermolecular/intramolecular interactions and also to quantify them as strong, moderate, or weak* (Bader and Essén 1984a).

In addition, QTAIM can also generate the *topology of the Laplacian of the charge density or topology of the Laplacian of the electron density*. This topology gives a lot of important information, e.g., (1) the type of chemical bond or intermolecular/intramolecular interactions; (2) the region with charge concentration or charge depletion in the molecule; and (3) it locates the lone pair from heteroatoms (Bader et al. 1984b). In Fig. 2.12, the partial Laplacian graph of water containing their atoms and only Laplacian $(3,-3)$ critical points are shown. In the topology of Laplacian of the charge density, the $(3,-3)$ critical point is the most important topological data because it indicates the *valence shell charge concentration*, VSCC, i.e., the highest concentration of the electron density in the molecule. The Laplacian $(3,-3)$ CPs are located in the atoms (of course!), but also in the lone pairs and one point between bonding atoms near the most electronegative atom for a heteronuclear bond. The Laplacian $(3,-3)$ CP in bonding region involving heteronuclear atoms does not coincide with the BCP from the topology of charge density. In Fig. 2.12(A), all of these (3,–3) CPs (atoms, two oxygen lone pairs and two between each O-H bond) are shown, except for one H atom, which does not appear at the chosen position of the molecule.

Figure 2.12(B) shows the 3-D Laplacian of the charge density of the water depicting the region of charge density depletion (near hydrogen atoms), and the region with charge density concentration (near the oxygen atoms). It also shows the partial Laplacian graph inside which the atoms and other (3,–3) CPs appear. Please, note the differences between 2-D/3-D electron density and corresponding 2-D/3-D Laplacian of the charge density by comparing Fig. 2.11 to Fig. 2.6(B) and (C) of the water molecule. One can realize that 2-D and 3-D Laplacian of the charge density graphs are much more informative than their corresponding 2-D and 3-D electron density.

Figure 2.12(C) shows the 3-D Laplacian of the charge density from H-O-H plane, where one can see both K and L shells of the oxygen atom. This graph goes up (highest concentration) and down (highest depletion), although the picture's arbitrary position does not permit us to see its lower part. This gives us another important information: *The Laplacian of an atom or atomic basin is zero because the sum of* $\nabla^2\rho(\Omega)<0$ *and* $\nabla^2\rho(\Omega)>0$ *is null*, where Ω represents the generic atomic basin. This is known as local zero condition where path integration of $\nabla\rho.\mathbf{n}$ over the whole ZFS is zero or the volume integration of $\nabla^2\rho$ over the atomic basin is also zero, as derived from Gauss theorem (see more discussion ahead).

$$\oint dS\nabla\rho\cdot\mathbf{n}=\int_A d\tau\nabla^2\rho=0$$

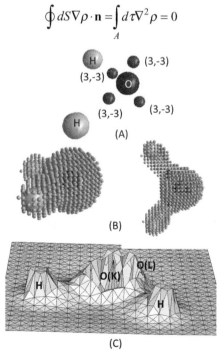

(A)

(B)

(C)

Figure 2.12 Laplacian of the charge density of water molecule from (A) partial Laplacian graph showing only (3,–3) CPs; (B) 3-D view of the whole molecule; and (C) 3-D view from plane passing through H-O-H plane.

The 2-D Laplacian of the charge density also gives important information with respect to the type of chemical bond or chemical interaction. The corresponding isodensity lines indicate: (1) covalent bond where the VSCC overlap appears in the bonding region of a covalent bond (Fig. 2.13(A)); (2) ionic bond where there is a polarization effect due to the strong electrostatic interaction between cation and anion (Fig. 2.13(B)); (3) weak interactions where there is an absence of any sharing of isodensity lines, such as in gas noble complexes (Fig. 2.13(C)). Here, in the 2-D Laplacian of the charge density, some isodensity lines are negative (darker lines) and others are positive (lighter lines).

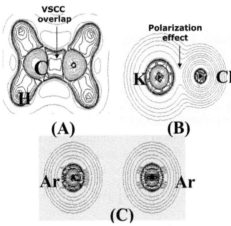

Figure 2.13   2-D Laplacian of the charge density of (A) ethene; (B) KCl; and (C) Ar_2 complex.

Besides the great amount of information regarding the topologies of electron density and its Laplacian, QTAIM has another outstanding differentiation with respect to orbital-based theories: it can provide a lot of atomic information from atoms in a molecule. Before the advent of QTAIM, quantum mechanics could not provide atomic information in a molecule. Quantum mechanics excluding QTAIM can only provide atomic information from isolated atoms! Bader has made a very huge contribution to science, both physics and chemistry, with his QTAIM theory.

QTAIM: Extension of Quantum Mechanics for an Open System

The QTAIM is an extension of quantum mechanics to an open system which is made up of atomic basins in a molecule and it can be developed by deriving Heisenberg's equations of motion from Schrödinger's equation, wherein any atomic property in a molecule can be calculated (Bader 2007). We will later advance in this definition, but firstly one should know that the Heisenberg equation of motion (HEM) is equivalent to the Schrödinger equation of motion (SEM) by only changing the time-dependency between operators (variable in HEM and constant in SEM) and vector states (constant in HEM and variable in SEM). Then, it is a matter of convenience to use either of them. Knowing that the time-dependent wave function is:

$$|\psi(t)\rangle = U(t)|\psi(0)\rangle \therefore U(t) = e^{-\frac{iHt}{\hbar}}$$

Where the Hamiltonian does not vary with time and $U(t)$ is the time-evolution operator. The observables for SEM and HEM are:

$$\textbf{SEM}: \langle \psi(t)|O|\psi(t)\rangle$$

$$\textbf{HEM}: \langle \psi(0)|U^{t}(t)O_{S}U(t)|\psi(0)\rangle$$

Then, for obtaining the observable of a given time-dependent operator, $A(t)$, one has to use the Heisenberg equation of motion.

$$\langle A\rangle_{t} = \langle \psi(0)|e^{\frac{iHt}{\hbar}} Ae^{-\frac{iHt}{\hbar}}|\psi(0)\rangle$$

The equation of motion for $A(t)$ is:

$$\frac{d}{dt}A(t) = \frac{i}{\hbar}He^{\frac{iHt}{\hbar}} Ae^{-\frac{iHt}{\hbar}} + e^{\frac{iHt}{\hbar}}\left(\frac{\partial A}{\partial t}\right)e^{-\frac{iHt}{\hbar}} + \frac{i}{\hbar}e^{\frac{iHt}{\hbar}} A(-H)e^{-\frac{iHt}{\hbar}}$$

Then, the Heisenberg equation of motion for $A(t)$ becomes:

$$\frac{d}{dt}A(t) = \frac{i}{\hbar}[H, A(t)] + e^{\frac{iHt}{\hbar}}\left(\frac{\partial A}{\partial t}\right)e^{-\frac{iHt}{\hbar}}$$

The corresponding eigenvalues or expectation values of $A(t)$, $\langle A\rangle_{t}$, from the HEM are obtained from the expression (Bader 1994):

$$\frac{d\langle A\rangle}{dt} = \frac{i}{\hbar}\langle \psi(0),[H, A(t)]\psi(0)\rangle + \langle \partial A/\partial t\rangle$$

For time-independent arbitrary linear operator, F, one finds the following expression (Hirschfelder 1960), which some authors consider to be the hypervirial theorem.

$$\frac{d}{dt}F = \frac{i}{\hbar}[H, F]$$

For time-independent arbitrary linear operators, F, the corresponding eigenvalues are time-invariant in stationary states. This leads to the hypervirial theorem (Hirschfelder 1960). This result is achieved by considering the Hermitian property of the Hamiltonian operator ($H\psi^{*}\psi = \psi^{*}H\psi$), the restriction on F that preserves its Hermiticity, and the use of the Schrödinger equation for stationary state ($H\psi = E\psi$ or $H\psi^{*} = E\psi^{*}$).

$$\langle \psi|[H, F]|\psi\rangle = \langle \psi|HF - FH|\psi\rangle$$

$$= \langle \psi|HF|\psi\rangle - \langle \psi|FH|\psi\rangle = \langle H\psi^{*}F\psi\rangle - \langle \psi^{*}FH\psi\rangle$$

$$= \langle E\psi^{*}F\psi\rangle - \langle \psi^{*}FE\psi\rangle = E\left[\langle \psi^{*}F\psi\rangle - \langle \psi^{*}F\psi\rangle\right] = 0$$

$$\frac{d}{dt}\langle F\rangle = \frac{i}{\hbar}\langle [H,F]\rangle = 0$$

The last equation above is the hypervirial theorem. The operator F is called the hypervirial operator. Chen defined it as: "a time-independent linear operator with an arbitrary functional structure expressed in terms of dynamical variables of the system under consideration .(...) In the energy representation, the diagonal matrix elements of the hypervirial operator are constant in time; this is known as the hypervirial theorem" (Chen 1964).

Bader used the above expression to subsystem (or open system), i.e., for an atomic basin, Ω, in a molecule in a stationary state, where the Hermiticity of H is lost, and by assuming a one-electron Hamiltonian, he arrived at the expression below (further details in Bader 1994):

$$\langle\psi|[H,F]|\psi\rangle_\Omega = \left(-\hbar^2/2m\right)\oint dS(r;\Omega)\mathbf{j}_F(r)\cdot\mathbf{n}(r)$$

$$\mathbf{j}_F(r) = \psi^*\nabla\left(F\psi\right) - \nabla\psi^*\left(F\psi\right)$$

Where $\mathbf{j}_F(r)$ is the vector current, giving the velocity of density of F at position r, and $\mathbf{j}_F(r)\bullet\mathbf{n}(r)$ is called *flux*.

While the Heisenberg equation of motion vanishes for a total system, for an open one-electron system, it is equal to the surface integral of the normal component of the current property of F. The same expression is found for an open N-electron system by multiplying with N. This is the great achievement of Bader's theory. As Bader stated: "this approach enables one to define all properties, including those that depend upon inter-particle coordinates, such as energy, in terms of real-space density distribution" (Bader 2007).

Bader used the "principle of least action" and the "principle of stationary action" to define the relation between the Lagrangian integral and zero surface condition (Bader et al. 1978, Srebrenik 1975, Srebrenik et al. 1978, Bader 1994). The "principle of least action" or the "principle of stationary action" is a variational principle (from the calculus of variations) applied to the action (S) of a physical system on the coordinates q_i along a path from time t_1 and t_2 where no change occurs at these two time end-points. For more details, see the Appendix 1 section.

The quantum theory of atoms in molecules depends on the zero flux surface of the gradient of charge density vector field for the application of Schwinger's principle of stationary action based on the formalism of variable domains in the calculus of variations in order to define the topological and quantum atom (Appendix section and Bader 1994). Local and net zero flux conditions must be equally satisfied for the interatomic zero flux surface (Nasertayoob and Shahbazian 2008). *The zero-flux condition is the determining point to obtain the atomic properties of each atom in a molecule.*

Therefore, the quantum condition of the subsystem states that the surface bounding the subsystem shall not be crossed by any gradient vectors of charge density. Since the gradient vector of charge density always points to the direction of greatest increase in electron density, it must always be perpendicular to lines of constant density.

QTAIM: Atomic Properties and Delocalization Index

Another important point is related to the virial theorem applied to QTAIM. Application of the hypervirial theorem (Hirschfelder 1960) for subsystems (atomic basins) and total system leads to distinguished results, while the atomic virial theorem (see the next chapter) warranties that the sum of the total energy of each subsystem yields the energy of the total system (Bader 1985, Bader and Beddall 1972). From virial partitioning, it is possible to obtain the total atomic energy for each atom in a molecule (see the next chapter). *The sum of all atomic energies in the molecule gives the total molecular energy.*

From the atomic virial theorem for a system in equilibrium geometry the electronic energy of a subsystem, $E_e(\Omega)$, and the total electronic energy, E_e, are given by:

$$E_e(\Omega) = -T(\Omega) = \frac{1}{2}V(\Omega) = \frac{1}{2}\int Tr\rho(r)d\tau$$

$$E_e = N\langle H_i \rangle = N\int d\tau \int d\tau' \left\{ \psi * \left(-\hbar^2/2m\right)\nabla_i^2 - r_i \cdot \nabla_i V \right\} \psi$$

$$E_e = \sum_\Omega E_e(\Omega)$$

The mode of integration for obtaining an atomic property is determined by the atomic variation principle (Bader 1980). The atomic average of an observable A is:

$$A(\Omega) = \langle A \rangle_\Omega = \int_\Omega d\tau \rho_A$$

$$\rho_A = (N/2)\int d\tau' \{\psi * A\psi + (A\psi) * \psi\}$$

For example, the atomic dipole moment, $M_1(\Omega)$, is a three-component vector, where r_Ω is centered on the atom and it measures the magnitude and direction of dipole polarization of the atom charge density. Its formula is:

$$M_1(\Omega) = -\int_\Omega d\tau \rho(r)r_\Omega$$

Another important atomic property is the QTAIM atomic charge, $q(\Omega)$, which is regarded as one of the most precise methods to calculate the atomic charge for molecules. The sum of all atomic charges gives the total molecular charge.

$$M_0(\Omega) = -\int_\Omega d\tau \rho(r) = -N(\Omega)$$

$$q(\Omega) = M_0(\Omega) + Z_\Omega$$

Not published in his seminal book in 1990 (and after in 1994), Bader later incorporated the concepts of delocalization index, DI, and localization index, LI, in QTAIM. However, in 1976, Bader had already launched the first mathematical proposition to obtain DI and LI from the total correlation contained in an atomic basin, $F(\Omega,\Omega)$, which measures the total Fermi and Coulomb correlations in an ideal wave function. Fermi and Coulomb correlation are the repulsive forces in a reference electron to avoid other electron with same spin (Fermi correlation) or other electron spin-independent (Coulomb correlation). Thereafter, the concepts of DI and LI were developed from the Fermi hole density (Bader et al. 1996a). The Fermi hole represents the surrounding region of a reference electron that cannot be occupied by another electron with the same spin (Lennard-Jones 1952). As Bader stated: "As an electron moves through space it carries with it a Fermi hole of ever changing shape, the density of the electron being spread out in the manner described by its Fermi Hole". The Fermi hole equation, $h^{\alpha}(r_1,r_2)$, for α-spin electron as reference is given by the sum of the ratio between exchange density, $\phi_i^{*}(r_1)\phi_i(r_2)\phi_j^{*}(r_2)\phi_j(r_1)$, and the density of α-spin electron, $\rho^{\alpha}(r_1)$.

$$h^{\alpha}(r_1,r_2) = \sum_j \sum_i \phi_i^{*}(r_1)\phi_i(r_2)\phi_j^{*}(r_2)\phi_j(r_1) \Big/ \rho^{\alpha}(r_1)$$

The integration of the Fermi hole of the reference electron over the whole space corresponds to removal of the electron with same spin and describes the spatial delocalization of the reference electron. Fermi hole, for example, accounts for the higher stability of triplet excited helium atom in comparison with singlet excited helium atom. In the former, the Fermi hole reduces the electric repulsion between the electrons.

Alternatively, the Fermi hole, $h^{\sigma1\sigma2}(r_1,r_2)$, can be obtained from the pair density, $\pi^{\sigma1\sigma2}(r_1,r_2)$, of completely independent electrons with spins σ_1 and σ_2, represented as $\rho^{\sigma1}(r_1)\,\rho^{\sigma2}(r_2)$ corrected by the exchange correlation density, $\Gamma^{\sigma1\sigma2}(r_1,r_2)$, where we know the position of the electron with spin σ_1 at r_1 and we wonder the probability of finding the electron with spin σ_2 at r_2, and the Lennard-Jones function, $\Omega^{\sigma1\sigma2}(r_1,r_2)$.

$$\pi^{\sigma1\sigma2}(r_1,r_2) = \rho^{\sigma1}(r_1)\rho^{\sigma2}(r_2) + \Gamma^{\sigma1\sigma2}(r_1,r_2)$$

$$\Omega^{\sigma1\sigma2}(r_1,r_2) = \frac{\pi^{\sigma1\sigma2}(r_1,r_2)}{\rho^{\sigma1}(r_1)} = \rho^{\sigma2}(r_2) + \frac{\Gamma^{\sigma1\sigma2}(r_1,r_2)}{\rho^{\sigma1}(r_1)}$$

$$h^{\sigma1\sigma2}(r_1,r_2) = \frac{\Gamma^{\sigma1\sigma2}(r_1,r_2)}{\rho^{\sigma1}(r_1)} = \Omega^{\sigma1\sigma2}(r_1,r_2) - \rho^{\sigma2}(r_2)$$

Henceforth, Bader stated that the localization index, the total number of electrons localized in an atomic basin, is given by the double integration of the product Fermi hole and the density of α-spin electron. In another words, LI or $\lambda(A)$ is the double integration over atomic basin A of the exchange density, P_{XC}, which can alternatively be thought of as the sum of squares of overlap integrals (Wang and Werstiuk 2003).

$$F^\alpha(\Omega,\Omega) = \int_\Omega dr_1 \int_\Omega dr_2 \left[\rho^\alpha(r_1) h^\alpha(r_1,r_2) \right] = -\sum_{i,j}^{N\alpha} S_{i,j}^2(\Omega)$$

$$F(\Omega,\Omega) = F^\alpha(\Omega,\Omega) + F^\beta(\Omega,\Omega)$$

$$\mathrm{LI} = \lambda(A) = F(\Omega,\Omega)$$

$$\mathrm{LI} = \lambda(A) = -\iint_\Omega P_{XC}(r_1,r_2) dr_1 dr_2$$

Similarly, the quantity of electrons in a basin A that is delocalized into another basin B, i.e., the delocalization index, DI or $\delta(A,B)$, is given by double integration of exchange correlation, each integration for each atom, or the sum of overlap integral in A and in B.

$$F^\alpha(\Omega,\Omega') = \int_\Omega dr_1 \int_{\Omega'} dr_2 \left[\rho^\alpha(r_1) h^\alpha(r_1,r_2) \right] = -\sum_{i,j}^{N\alpha} S_{i,j}(\Omega) S_{i,j}(\Omega')$$

$$F(\Omega,\Omega') = F^\alpha(\Omega,\Omega') + F^\beta(\Omega,\Omega')$$

$$\mathrm{DI} = \delta(A,B) = 2F(\Omega,\Omega')$$

$$\mathrm{DI} = \delta(A,B) = -2 \iint_{\Omega\Omega'} P_{XC}(r_1,r_2) dr_1 dr_2$$

Therefore, the total number of electrons in a molecule is

$$N_t = \sum_\Omega \mathrm{LI} + \frac{1}{2} \sum_\Omega \mathrm{DI}$$

The delocalization index is a measure of the number of electrons that are shared or exchanged between two atoms or basins. The delocalization of the same-spin density, DI, informs the interaction between distant atoms, bonded or not-bonded. It is *important to emphasize that the Lewis model of bonding electron pairs does not exist in the quantum theory of atoms in molecules as a consequence of the results from Fermi hole* (Firme et al. 2009). Bader and collaborators stated that the complete localization of electrons and consequent existence of electron pairs exist only in core electrons and in ionic systems (Bader et al. 1996b).

It is important to emphasize that the delocalization index of very small CC interactions lies between 10^{-2} and 10^{-3} order of magnitude. The DI value of 10^{-1} order of magnitude for CC bonds is characteristic of moderate CC interactions.

Although DI does not carry the restricting concept of electron pair in covalent bonding, the DI can be used to give the formal bond order (number of electron pairs in a bond). There is a direct relation between delocalization index and some different definitions of bond order. The relation between DI and formal bond is linear for CC, NN, GeGe, SiC, GeC, and CN bonds. The DI of single, double, and triple CC bonds are nearly one, two, and three electrons, respectively unlike the Lewis picture of two, four, and six electrons, respectively (Firme et al. 2009).

NON-COVALENT INTERACTION (NCI) THEORY

The Non-Covalent Interaction, NCI method, provides the surface associated with intra/intermolecular interactions, including the same QTAIM bond critical point for intra/intermolecular interactions. The NCI theory spans a broader region of intermolecular interactions than QTAIM does. The NCI function, so-called S function or reduced density gradient (RDG), is given by the equation below:

$$S = \frac{1}{2(3\pi^2)^{1/3}} \cdot \frac{|\nabla\rho|}{\rho^{4/3}}$$

There is an equivalence between NCI and QTAIM, since the formula of NCI includes the gradient of the charge density used for finding bond critical points of the molecular graph (Contreras-García et al. 2011).

From the NCI equation it is possible to obtain the NCI surface relative to intermolecular and/or intramolecular interactions from a determined isovalue, known as S isovalue surface. However, a more useful NCI surface is the RDG function isosurfaces with sign(λ_2)ρ coloring scheme, which might inform intra/intermolecular repulsion (in red), weaker van der Waals interactions (in light green), stronger van der Waals interactions (in dark green), and H-bonds (in light blue). The sign(λ_2)ρ, where sign(λ_2) is the sign of the second eigenvalue of the charge density Hessian matrix (λ_2), is the same Hessian matrix used in QTAIM. It can qualitatively discriminate between strong and weak non covalent bonds. Then it is possible to plot S isovalue isosurface as a function of sign(λ_2)ρ to provide isosurfaces in the regions containing interatomic interactions.

In the standard coloring scheme provided by sign(λ_2)ρ, the red region is related to van der Waals repulsion and green region is related to van der Waals attraction. Besides the S surface with sign(λ_2)ρ coloring scheme, the plot of S function versus sign(λ_2)ρ shows the regions of attractive interaction (negative sign of sign(λ_2)ρ axis at the right of this axis) and regions of repulsive interaction (positive sign of sign(λ_2)ρ axis at the left of this axis).

Then, NCI is a very useful method to obtain the whole surface of attractive and repulsive interatomic interaction, which also includes the corresponding bond critical points from QTAIM. See examples of application of NCI in organic chemistry in chapters eleven, fourteen, fifteen, and sixteen.

EXERCISES

1. Give the VB wave function for triplet hydrogen molecule

$$\Psi^3(x_1,x_2) = N\hat{A}\left[1s_A(r_1)1s_B(r_2)\alpha(\mu_1)\alpha(\mu_2)\right]$$

2. How many total spin representations exist for a system with 5 electrons for $S=3/2$, where each electron occupies a specific orbital?
3. For a system with $S=1/2$ and 5 electrons occupying specific orbitals, the five linearly independent spin eigenfunctions can be represented in the following diagram where only the unpaired spin-electron appears.

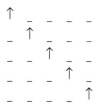

a) Use the Rummer diagram (Tip: points forming a pentagon) to represent these five linearly independent spin eigenfunctions;

b) Use same diagram as above and the Rummer diagram to represent all linearly independent spin eigenfunctions of a system with 5 electrons occupying different orbitals and $S = 3/2$.

Acknowledgment

Figure 2.5 was published in our own article (http://dx.doi.org/10.1039/C4NJ01271B), whose copyrights belong to Royal Society of Chemistry, and we appreciate the publisher's permission to print this figure in this book. Figure 2.2 was published in our own article, whose copyrights belong to Springer-Verlag and we appreciate the publisher's permission to print this figure in this book.

REFERENCES CITED

Bader, R.F.W. and Beddall, P.M. 1972. Virial field relationship for molecular charge distributions and the spatial partitioning of molecular properties. J. Chem. Phys. 56: 3320-3329.

Bader, R.F.W. Srebrenik, S., Nguyen-Dang, T.T. 1978. Subspace quantum dynamics and the quantum action principle. J. Chem. Phys. 68: 3680-3691.

Bader, R.F.W. 1980. Quantum topology of molecular charge distributions. III. The mechanics of an atom in a molecule. J. Chem. Phys. 73: 2871-2883.

Bader, R.F.W. and Essén, H. 1984a. The characterization of atomic interactions. J. Chem. Phys. 80: 1943-1960.

Bader, R.F.W., MacDougall, P.J. and Lau, C.D.H. 1984b. Bonded and nonbonded charge concentration and their relation to molecular geometry and reactivity. J. Am. Chem. Soc. 106: 1594-1605.

Bader, R.F.W. 1985. Atoms in molecules. Acc. Chem. Res. 18: 9-15.

Bader, R.F.W. 1994. Atoms in Molecules: A Quantum Theory. Oxford University Press, New York.

Bader, R.F.W, Streitwieser, A., Neuhaus, A., Laidig, K.E. and Speers, P. 1996a. Electron delocalization and the Fermi hole. J. Am. Chem. Soc. 118: 4959-4965.

Bader, R.F.W, Johnson, S. and Tang, T.-H. 1996b. The electron pair. J. Phys. Chem. 100: 15398-15415.

Bader, R.F.W. 1998. A bond path: a universal indicator of bonded interactions. J. Phys. Chem. A. 102: 7314-7323.

Bader, R.F.W. 2007. Everyman's derivation of the theory of atoms in molecules. J. Phys. Chem. 111: 7966-7972.

Bader, R.F.W. 2009. Bond paths are not chemical bonds. J. Phys. Chem. A 113: 10391-10396.

Biegler-König, F and Schönbohm, J. 2002. Update of the AIM2000 program for atoms in molecules. J. Comput. Chem. 23: 1489-1494.

Bohr, N., Kramers, H.A., Slater, J.C. 1924. The quantum theory of radiation. Phil. Mag. 47: 785-802.

Chen, J.C.Y. 1964. Off-diagonal hypervirial theorem and its applications. J. Chem. Phys. 40: 615-621.

Contreras-Garcia, J., Jonhson, E.R., Keinan, S., Chaudret, R., Piquemal, J.-P., Beratan, D.N. and Yang, W. 2011. NCIPLOT: A program for plotting non-covalent interactions. J. Chem. Theory Comput. 7: 625-632.

Cooper, D.L., Wright, S.C., Gerrat, J. and Raimondi, M. 1989. The electronic structure of heteroaromatic molecules. Part 1. Six-membered rings. J. Chem. Soc. Perkin Trans. 2. 255-261.

Coulson, C.A. 1938. Self-consistent field for molecular hydrogen. Math. Proc. Cambridge Phil. Soc. 34: 204-212.

Dirac, P.A.M. 1926. On the theory of quantum mechanics. Proc. R. Soc. Lond. A 112: 661-677.

Dirac, P.A.M. 1929. Quantum mechanics of many-electron system. Proc. R. Soc. Lond. A 123: 714-733.

Dos Santos, H.F., Pontes, D.L. and Firme, C.L. 2013. Relation between topology and stability of bent titanocenes. J. Mol. Model. 19: 2955-2964.

Eckart, C. 1930. The theory and calculation of screening constants. Phys. Rev. 36: 878-892.

Fermi, E. 1926. Zur quantelung des idealen einatomigen gases. Z. Phys. 36: 11-12.

Firme, C.L., Antunes, O.A.C. and Esteves, P.M. 2009. Relation between bond order and delocalization index of QTAIM. Chem. Phys. Lett. 468: 129-133.

Firme, C.L., Costa, T.F., Penha, E.T. and Esteves, P.M. 2013. Electronic structures of bisnoradamantenyl and bisnoradamantanyl dications and related species. J. Mol. Model. 19: 2485-2497.

Fock, V.A. 1930. Näherungsmethode zur lösung des quantenmechanischen mehrkörper-problems. Z. Phys. 61: 126-148.

Fukui, K., Yonezawa, T. and Shingu, H. 1952. A molecular orbital theory of reactivity in aromatic hydrocarbons. J. Chem. Phys. 20: 722-725.

Gerratt, J. 1971. General theory of spin-coupled wave functions for atoms and molecules. Adv. Atom. Mol. Phys. 7: 141-221.

Goddard III, W.A. 1967a. Improved quantum theory of many-electron systems. I. Construction of eigentunctions of S^2 which satisfy Pauli's principle. Phys. Rev. 157: 73-80.

Goddard III, W.A. 1967b. Improved quantum theory of many-electron systems. II. Basic method. Phys. Rev. 157: 81-93.

Goddard III, W.A. 1968. Improved quantum theory of many-electron systems. IV. Properties of GF wavefunctions. J. Chem. Phys. 48: 5337-5347.

Goddard III, W.A. 1970. The orbital phase continuity principle and selection rules for concerted reactions. J. Am. Chem. Soc. 92: 7520-7521.

Goddard III, W.A. and Ladner, R.C. 1971. A generalized orbital description of the reactions of small molecules. J. Am. Chem. Soc. 93: 6750-6756.

Goddard III, W.A., Dunning, Jr. T.H., Hunt, W.J. and Hay, P.J. 1973. Generalized valence bond description of bonding in low-lying states of molecules. Acc. Chem. Res. 6: 368-376.

Hartree, D.R. 1928a. The wave mechanics of an atom with a non-Coulomb central field. Part I. Theory and methods. Math. Proc. Camb. Phil. Soc. 24: 89-110.

Hartree, D.R. 1928b. The wave mechanics of an atom with a non-Coulomb central field. Part II. Some results and discussion. Math. Proc. Camb. Phil. Soc. 24: 111-132.

Hay, P.J., Hunt, W.J. and Goddard III, W.A. 1972. Generalized valence bond description of simple alkanes, ethylene, and acetylene. J. Am. Chem. Soc. 94: 8293-8301.

Heitler, W. and London, F. 1927. Wechselwirkung neutraler atome und homoopolare bindung nach der quantenmechanik. Z. Phys. 44: 455-472.

Heisenberg, W. 1926. Mehrkörperproblem und resonanz in der quantenmechanik. Z. Phys. 38: 411-426.

Hirao, K., Nakano, H., Nakayama, K. and Dupuis, M. 1996. A complete active space valence bond (CASVB) method. J. Chem. Phys. 105: 9227-9239.

Hirschfelder, J.O. 1960. Classical and quantum mechanical hypervirial theorem. J. Chem. Phys. 33: 1462-1466

Hückel, E. 1931 Quantentheoretische beiträge zum benzolproblem. I. Die elektronenkonfiguration des benzols und verbindungen. Z. Phys. 70: 204-286.

Hückel, E. 1932. Quantentheoretische beiträge zum problem der aromatischen und ungesättigten verbindungen III. Z. Phys. 76: 628-648.

Hund, F. 1928. Zur deutung der molekelspektren. IV. Z. Phys. 51: 759-795.

Hund, F. 1929. Chemical binding. Trans. Faraday Soc. 2: 646-648.

Hund, F. 1931. Zur frage der chemischem bindung. Z. Phys. 73: 1-30

Hund, F. 1932. Zur frage der chemischem bindung. II. Z. Phys. 73: 565-577.

Li, J., Duke, B. and McWeeny, R. 2002. VB2000: Pushing valence bond theory to new limits. Int. J. Quantum Chem. 89: 208-216.

Karadakov, P.B., Gerrat, J., Cooper, D.L., and Raimondi, M. 1992. Core-valence separation in the spin-coupled wave function: a fully variational treatment based on a second-order constrained optimization procedure. J. Chem. Phys. 97: 7637-7655.

Karadakov, P.B., Gerratt, J., Cooper, D.L. and Raimondi, M. 1995. SPINS: A collection of algorithms for symbolic generation and transformation of many-electron spin eigenfunctions. Theor. Chim. Acta 90: 51-73.

Karadakov, P.B., Cooper, D.L. and Gerratt, J. 1998. Modern valence-bond description of chemical reaction mechanisms: Diels-Alder reaction. J. Am. Chem. Soc. 120: 3975-3981.

Keith, T.A., Bader, R.F.W. and Aray, Y. 1996. Structural homeomorphism between the electron density and the virial field. Int. J. Quantum Chem. 183-198.

Kotani, M. and Siga, M. 1937. On the valence theory of the methane molecule I. Proc. Phys. Math. Soc. Jap. 19: 471-486.

Li, J., Duke, B. and McWeeny, R. 2009. VB2000 v.2.1. SciNet Technologies, San Diego, CA.

Li, J. and McWeeny, R. 2002. VB2000: Pushing valence bond theory to new limits. Int. J. Quantum Chem 89: 208-216.

Lennard-Jones, J.E. 1929. The electronic structure of some diatomic molecules. Trans. Faraday Soc. 25: 668-686.

Lennard-Jones, J.E. 1931. Wave functions of many-electron atoms. Math. Proc. Cambridge Phil. Soc. 27:469-480.

Lennard-Jones, J.E. 1952. The spatial correlation of electrons in molecules. J. Chem. Phys. 20: 1024-1029.

Lewis, G.N. 1916. The atom and the molecule. J. Am Chem. Soc. 38: 762-785.

Löwdin, P-O. 1955. Quantum theory of many-particle systems. III. Extension of the Hartree-Fock scheme to include degenerate systems and correlation effects. Phys. Rev. 97: 1509-1520.

McWeeny, R. 1992. Methods of Molecular Quantum Mechanics. Academic Press. San Diego.

Mo, Y., Bao, P. and Gao, J. 2011. Energy decomposition analysis based on a block-localized wave function and multistate density functional theory. Phys. Chem. Chem. Phys. 13: 6760-6775.

Monteiro, N.K.V., de Oliveira, J.F. and Firme, C.L. 2014. Stability and electronic structures of substituted tetrahedranes, silicon and germanium parents - a DFT, ADMP, QTAIM and GVB study. New J. Chem. 38: 5892-5904.

Mülliken, R.S. 1928. The assignment of quantum numbers for electrons in molecules. I. Phys. Rev. 32: 186-222.

Mülliken, R.S. 1932a. Electronic structures of polyatomic molecules and valence. Phys. Rev. 40: 55-62.

Mülliken, R.S. 1932b. Electronic structures of polyatomic molecules and valence.II. General considerations. Phys. Rev. 41: 49-71.

Mülliken, R.S. 1960. Self-consistent field atomic and molecular orbitals and their approximations as linear combinations of Slater-tyoe orbitals. Rev. Mod. Phys. 32: 232-238.

Nasertayoob, P. and Shahbazian, S. 2008. Revisiting the foundations of quantum theory of atoms in molecules (QTAIM): the variational procedure and the zero-flux conditions. Int. J. Quantum Chem. 108: 1477-1484.

Pauling, L. 1931a. The nature of the chemical bond. Application of results obtained from the quantum mechanics and from a theory of paramagnetic susceptibility to the structure of molecules. J. Am. Chem. Soc. 53: 1367-1400.

Pauling, L. 1931b. The nature of the chemical bond. II. The one-electron bond and the three-electron bond. J. Am. Chem. Soc. 53: 3225-3237.

Pauncs, R. 1979. Spin Eigenfunctions. Construction and Use. Plenun Press. New York and London.

Roothaan, C.C.J. 1960. Self-consistent field theory for open shells of electronic systems. Rev. Mod. Phys. 32: 179-185.

Serber, R. 1934a. Extension of the Dirac vector model to include several configurations. Phys. Rev. 45: 461-467.

Serber, R. 1934b. The solution of problems involving permutation degeneracy. J. Chem. Phys. 2: 697-710.

Shaik, S.S. and Hiberty, P.C. 2008. A Chemist's Guide to Valence Bond Theory. John Wiley and Sons. Hoboken.

Slater, J.C. 1928. The self-consistent field and the structure of atoms. Phys. Rev. 32: 339-348.

Slater, J.C. 1929. The theory of complex spectra. Phys. Rev. 34: 1293-1322.

Slater, J.C. 1931. Directed valence in polyatomic molecules. Phys. Rev. 37: 481-489.

Slater, J.C. 1932. Analytic atomic wave functions. Phys. Rev. 42: 33-43.

Slater, J.C. 1933. The virial theorem and molecular structure. J. Chem. Phys. 1: 687-691.

Song, L., Mo, Y., Zhang, Q. and Wu, W. 2005. XMVB: a program for ab initio nonorthogonal valence bond computations. J. Comput. Chem. 26: 514-521.

Srebrenik, S. 1975. Rayleigh principle for a subspace of a quantum system. Int. J. Quantum Chem. 9: 375-383.

Srebrenik, S., Bader, R.F.W. and Nguyen-Dang, T. 1978. Subspace quantum mechanics and the variational principle. J. Chem. Phys. 68: 3667-3679.

Thorsteinsson, T. and Cooper, D.L. 1998. Modern valence bond descriptions of molecular excited states: an application of CASSVB. Int. J. Quantum Chem. 70: 637-650.

van Vleck, J.H. and Sherman, A. 1935. The quantum theory of valence. Rev. Mod. Phys. 7: 167-228.

van Lenthe, J.H. and Balint-Kurti, G.G. 1983. The valence-bond SCF (VB SCF) method. Synopsis of theory and test calculation of OH potential energy curve. Chem. Phys. Lett. 76: 138-142.

van Lenthe, J.H. and Balint-Kurti, G.G. 1980. The valence-bond self-consistent field method (VB-SCF). Theory and test calculations. J. Chem. Phys. 78: 5699-5713.

Wang, Y-G. and Werstiuk, N.H. 2003. A pratical and efficient method to calculate AIM localization and delocalization indices at post-HF levels of theory. J. Comp. Chem. 24: 379-385.

Woodward, R.B. and Hoffmann, R. 1965. Stereochemistry of electrocyclic reactions. J. Am. Chem. Soc. 87: 395-397.

Woodward, R.B. and Hoffmann, R. 1965. Selection rules for concerted cycloaddition reactions. J. Am. Chem. Soc. 87: 2046-2048.

Yamanouchi, T. 1937. On the construction of unitary irreducible representations of the symmetric group. Proc. Phys. Math. Soc. Jap. 19: 436-450.

Zubarev, D.Y. and Boldyrev, A.I. 2008. Developing paradigms of chemical bonding: adaptive natural density partitioning. Phys. Chem. Chem. Phys. 10: 5207-5217.

Chapter Three

Quantum Mechanics and Electrostatic Force in Molecules

FUNDAMENTAL FORCES

The fundamental forces of nature are the set of forces that can explain any type of interaction between particles in nature. They are of four types: strong, weak, gravity, and electromagnetic. The strong force is the strongest force of nature, which holds the nucleus together, even though there is a high repulsive force between protons. The weak force is responsible for beta decay through interaction between intermediate vector bosoms, and it is necessary for building up a heavy nucleus. The gravity is the smallest force, being 10^{38} times weaker than the strong force. On the other hand, electromagnetic force is only 137 times weaker than strong force. Both gravity and electromagnetic forces obey the "inverse square law". Electric and magnetic phenomena were unified by Maxwell equations, for example, in generation of electric energy by means of mechanic energy. However, electrostatic phenomena or magnetostatic phenomena can exist in isolation by Coulomb law and Biot-Savart law, respectively.

Carrier particles are responsible for transmitting a specific force between the involved bodies. Each type of force has a specific carrier particle (Table 3.1).

Table 3.1 Relative strength, range, and particle carrier for each type of force.

Force	Relative strength	Range	Carrier particle
Gravity	10^{-38}	infinite	graviton
Electromagnetic	10^{-2}	infinite	photon
Weak force	10^{-13}	$<10^{-18}$ m	W^+, W^-, Z^0
Strong force	1	$<10^{-15}$ m	Gluon

The electromagnetic forces are responsible for keeping the atoms and molecules together. The electrostatic force is the dominant force in atoms and molecules, and magnetic force is only observable as small corrections. Then, *all bonded interactions (all chemical bonds, intramolecular, and intermolecular interactions) are under electrostatic (or electromagnetic) law, i.e., the nature of all bonded interactions are electrostatic.*

As Bader stated: "The electrostatic force, the only force operative in a field-free molecule, determines the potential energy operators in the Hamiltonian. The resulting wave function enables one to determine the average electrostatic forces that act on the electrons and on the nuclei. The force acting on the electrons (on the entire atom in a molecule), the Ehrenfest force, is obtained from the equation of motion for the electronic momentum operator, p, while the force acting on a nucleus, the Feynman force, is obtained from a corresponding equation for the nucleus gradient operator. (...) Thus, through the Ehrenfest and Feynman theorems, one has the tools that are needed to describe the forces acting in a molecule, and through the virial theorem to relate these forces to the molecule's energy and its kinetic and potential contributions" (Bader and Fang 2005). Then, *there are two types of electrostatic forces operating on a molecule: the Ehrenfest force (operating on the electrons) and the Feynman force (operating on the nuclei).*

It is very common to find the terms London force and van der Waals force associated with intermolecular interactions. But, as mentioned above, there are only four fundamental forces in nature. Then, these terms are not correct. See more discussion in chapter eleven.

HELLMANN-FEYNMAN THEOREM

From the Schrödinger equation, the energy of stationary state, E_n, is determined by the electronic Hamiltonian, which depends on the nuclear coordinates, X_α.

$$\hat{H}\psi_n = E_n\psi_n$$

$$E_n = \int \psi_n^* \hat{H}\psi_n d\tau$$

By deriving the above equation in relation to the parameter λ, one gets the general Hellmann-Feynman theorem (Levine 1999).

$$\frac{\partial E_n}{\partial \lambda} = \frac{\partial}{\partial \lambda}\int \psi_n^* \hat{H}\psi_n d\tau$$

$$\frac{\partial E_n}{\partial \lambda} = \int \psi_n^* \frac{\partial \hat{H}}{\partial \lambda}\psi_n d\tau$$

When replacing the parameter λ into nuclear coordinates X_α, one gets the electrostatic Hellmann-Feynman force (or simply Feynman force), where the force exerted on the nucleus α, F_α, is given by:

$$\frac{\partial E_n}{\partial X_\alpha} = F_\alpha = \int \psi_n^* \left(-\nabla_\alpha \hat{V}\right)\psi d\tau$$

The Feynman force has two components: force between protons from different nuclei, $F_{\alpha n}$, and the force of the electron interacting on the nucleus, $F_{\alpha e}$ (Hernández-Trujillo et al. 2007).

$$F_\alpha = F_{\alpha e} + F_{\alpha n}$$

Let us define r as the electron coordinate, and R_A and R_B as the nuclear coordinates of nuclei A and B, respectively. Then, the Feymann force on the nucleus A can be expressed as:

$$F_{\alpha(A)} = F_{\alpha(A)e} + F_{\alpha(A)n} =$$

$$F_{\alpha(A)} = -Z_{eff(A)} \int \frac{\rho(r)(R_A - r)}{|R_A - r|^3} dr + Z_{eff(A)} \sum_{B \neq A} \frac{Z_{eff(B)}(R_A - R_B)}{|R_A - R_B|^3}$$

It is important to add that *at an equilibrium geometry, the sum of all Feynman forces is zero.*

FEYNMAN FORCES AND POTENTIAL ENERGY SURFACE

By knowing the inverse relation between energy and potential force (see chapter six):

$$F = -\frac{dU}{dr} \therefore F_\alpha = -\frac{\partial E_n}{\partial X_\alpha}$$

When potential energy of a molecular system is at a minimum of the potential energy surface, then dU/dR is zero (from definition of derivative of a given function), and so resultant of Feynman forces, F_α, is also zero.

The equilibrium geometry of a given molecular system (with $i = 1$ to $3N$-5(6) degrees of freedom, where N is the amount of atoms) is the one where all dU/dR_i are zero (more details in chapter eight). Then, at equilibrium geometry, the sum of all Feynman forces is zero. Outside the equilibrium geometry, resultant of Feynman forces is not zero.

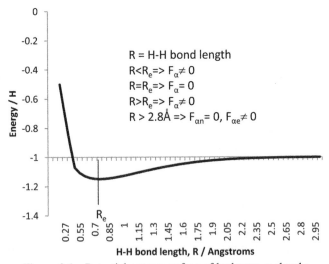

Figure 3.1 Potential energy surface of hydrogen molecule.

The simplest case to analyze is the potential energy surface of hydrogen molecule (see Fig. 3.1), where $i=1$, i.e., it has only one vibrational degree of freedom, and then $X_\alpha = R$ (reaction coordinate relative to the H-H bond). The minimum of this potential energy surface is at 7.4Å, the so-called R_e, where the sum of all Feynman forces is zero. For $R \neq R_e$, $F_\alpha \neq 0$ because dU/dR_i is only zero at the minimum of PES (and also at maximum of PES– see chapter eight). For $R > 2.8$Å, the Feynman component from interacting nuclei, $F_{\alpha n}$, is null because the hydrogen atoms are completely separate.

EHRENFEST FORCE AND VIRIAL THEOREM

The quantum virial theorem can be obtained from the hypervirial theorem for an arbitrary hypervirial operator (linear time-independent operator), F, depending only on the coordinates and momenta of the system where its expectation value, $\langle F \rangle$, is time-invariant (see the previous chapter).

$$\frac{d}{dt}\langle F \rangle = \frac{i}{\hbar}\langle [H,F] \rangle = 0$$

By considering F as the product of coordinates and linear momentum, F is:

$$F = r \cdot p = \frac{\hbar}{i}\left(x\frac{\partial}{\partial x} + y\frac{\partial}{\partial y} + z\frac{\partial}{\partial z} \right) = \frac{\hbar}{i} r \cdot \nabla$$

And H is given by:

$$H - -\frac{\hbar^2}{2m}\nabla^2 + V(r)$$

Then, the commutator $[H,F]$ is (Pilar 1990):

$$[H,F] = \frac{\hbar}{i}\left(-\frac{\hbar^2}{m}\nabla^2 - r \cdot \nabla(V) \right) = \frac{\hbar}{i}(2T - r \cdot \nabla(V))$$

Since F is a hypervirial operator, its expectation value is time-invariant. Then, the last expression below is the quantum mechanical virial theorem (more details in next subsection).

$$\frac{d}{dt}\langle F \rangle = \frac{i}{\hbar}\langle [H,F] \rangle = 0$$

$$\frac{\hbar}{i}(2\langle T \rangle - \langle r \cdot \nabla(V) \rangle) = 0$$

$$2\langle T \rangle = \langle r \cdot \nabla(V) \rangle$$

If $V = kr^n$ where k is a constant, then we have:

$$2\langle T \rangle = \langle r \cdot \nabla(kr^n) \rangle = \langle r \cdot nkr^{n-1} \rangle = n\langle V \rangle$$

Where $\langle T \rangle$ and $\langle V \rangle$ are the expected values of kinetic and potential energy operators.

For a system of interacting electrons, the potential energy term contains the term $\dfrac{1}{r_{ij}}$, where $n = -1$. Then, the virial theorem becomes:

$$2\langle T \rangle = -\langle V \rangle$$

The molecular (or quantum) virial theorem provides the relation of the kinetic energy in N-particle system bonded by potential forces (Slater 1933).

$$2\langle T \rangle = -\sum_{k=1}^{N} \langle F_k \cdot r_k \rangle = \langle v \rangle$$

$$V = \sum_i \left(-r_i \cdot \nabla_i V\right) + \sum_\alpha \left(-X_\alpha \cdot \nabla_\alpha V\right)$$

$$\langle v \rangle = \left\langle V_{en} + V_{ee} + V_{nn} - \sum_\alpha X_\alpha \cdot F_\alpha \right\rangle$$

$$\langle V_e \rangle = \langle r_i \cdot \nabla_i V \rangle = \langle r_i \cdot F_i \rangle$$

Where r_i are the electron coordinates.

At equilibrium geometry and at infinite nuclear separation, where the Hellmann-Feynman forces (or Feynman forces), F_α, are zero, the virial theorem relates the total kinetic energy to the total potential energy (Hernández-Trujillo et al. 2007).

$$\sum_\alpha X_\alpha \cdot F_\alpha = 0 \therefore \langle v \rangle = \langle V \rangle$$

$$2T = -V$$

Thus, at equilibrium geometry, two important considerations exist: (1) there appears the virial relation $2\langle T \rangle = -\langle V \rangle$, *and (2) the only force that operates is the Ehrenfest force, acting on all electrons in an atomic basin.* The Ehrenfest force acting on all electrons from atom A equals the force exerted on its surface, which is similar to the integration of the divergence of the stress tensor, $\nabla \sigma(r)$. The stress tensor, $\sigma(r)$, is the quantum mechanical equivalent for the pressure of a force acting on a surface. Table 3.2 shows the Ehrenfest force from C-C and C-H surfaces. This value is obtained from integration of the surface between two bonding atoms. We observe linearity only for Ehrenfest force from C-H bond according to carbon hybridization.

$$F(A) = \int_A F(r)dr = -\int_A dr \nabla \cdot \sigma(r) = -\oint dS(r;A) \cdot \sigma(r) \cdot n(r)$$

Table 3.2 Total force in electrons (Ehrenfest force) from C-C and C-H surfaces.

Molecule	Ehrenfest force (total force in electrons) from C-C surface/C-H surface/au
Ethane	0.088/−0.051
Ethene	0.062/−0.043
Ethyne	0.074/−0.038

QUANTUM VIRIAL THEOREM

As previously stated, the quantum virial theorem can be obtained from the hypervirial theorem

$$\frac{d}{dt}F = \frac{i}{\hbar}[H,F]$$

Whose corresponding expectation value vanishes in stationary states. By considering hypervirial F operator as a product of position and linear momentum operators of particle n:

$$F = X_n \cdot P_n \therefore H = V(X_i) + \sum_n \frac{P_n^2}{2m}$$

$$[H,F] = [H,X_nP_n] = X_n[H,P_n] + [H,X_n]P_n$$

Proof

$$[H,F] = X_n(HP_n - P_nH) + (HX_n - X_nH)P_n$$
$$\therefore X_nHP_n - X_nP_nH + HX_nP_n - X_nHP_n = [H,X_nP_n]$$

Then

$$[H,F] = X_n[H,P_n] + [H,X_n]P_n \therefore P_n = i\hbar\frac{d}{dX_n}$$

$$[H,F] = i\hbar X_n\frac{dV}{dX_n} - i\hbar\frac{P_n^2}{m}$$

$$\frac{i}{\hbar}[H,F] = \frac{i}{\hbar}\left(i\hbar X_n\frac{dV}{dX_n} - i\hbar\frac{P_n^2}{m}\right) = -\frac{P_n^2}{m} + X_n\frac{dV}{dX_n}$$

Summing over all particles,

$$F = \sum_n X_nP_n \therefore T = \sum_n \frac{P_n^2}{2m}$$

Then

$$\frac{i}{\hbar}[H,F] = 2T - \sum_n X_n\frac{dV}{dX_n}$$

$$\therefore \frac{d\langle A\rangle}{dt} = \frac{i}{\hbar}[H,F] = 0 \therefore \text{ for stationary states}$$

$$2T = \sum_n X_n\frac{dV}{dX_n}$$

The last equation is the quantum virial theorem.

The quantum virial theorem is an important tool for quantum chemistry calculations of molecular systems on the equilibrium geometry, where the ratio $-\frac{2\langle T\rangle}{\langle V\rangle}$ should be unitary (Pilar 1990). Quantum virial theorem is also important in

QTAIM to devise the virial partioning scheme to obtain the atomic energies of subsystems of a molecule (Bader 1994).

ELECTROSTATIC INTERPRETATION OF THE CHEMICAL BOND

In an orbital-free picture, a molecule can be regarded as a polynuclear system immersed in a static electron density. As previously stated, at equilibrium geometry, repulsive (interaction between protons from different nuclei, F_{an}) and attractive (electron interacting on nucleus, F_{ae}) forces acting on the nuclei, the Feynman forces, vanish and the only operative force is the Ehrenfest force on each atom of the molecule. Bader used the electrostatic formula to represent the interaction between one proton and one electron in a hydrogen molecule (see the Acknowledgment section).

$$F = -\frac{e^2}{r^2}$$

We shall use the same electrostatic interpretation to evaluate the strength of chemical bonds and intermolecular interactions, and also to relate this strength to the corresponding molecular energy. For atoms bigger than hydrogen, we have to approximate the sum of all protons as a nuclear charge, Z. More precisely, we shall use effective nuclear charge (or atomic number), Z_{eff} (Clementi and Raimondi 1963). Hereafter, for simplicity, Zq_e will be Z, $Z_{eff}q_e$ will be Z_{eff}, and q_e will be e.

$$F_{elect} = K\frac{(Zq_e)q_e}{r^2} = K\frac{Ze}{r^2}$$

There are two important electrostatic force expressions used in this book: (1) the electrostatic interaction between one electron, e, and the nucleus, Z_{eff}, given by:

$$F_{elect} = K\frac{eZ_{eff}}{r^2} \quad \therefore K = \frac{1}{4\pi\varepsilon_0}$$

Where ε_0 is the vacuum permittivity and r is the distance between electron and nucleus, and (2) the electrostatic force between two atoms, A and B, with opposite partial charges, δ^- and δ^+, given by the expression:

$$F_{elect} = K\frac{\delta_A^-\delta_B^+}{r^2}$$

As we will see in chapter eleven regarding hydrogen bonds, the second equation for electrostatic force is not always an appropriate approach, although it is an intuitive and simple equation.

Alternatively, by considering the overall amount of shared electrons in a chemical bond as one particle (in the same sense we consider all protons in the nucleus as Zq_e), one can use the delocalization index, DI, from QTAIM in the electrostatic force equation. Since DI is the amount of shared electrons between two atoms, the use of DI in the electrostatic equation gives more precise information of the electrostatic interaction in chemical bonds and inter/intramolecular interactions, although it is less straightforward than other previous equations, since it is needed to optimize the geometry of the molecule for further QTAIM calculation. Let us assume $(DI)q_e$ as DI in the electrostatic force equation, for simplicity.

$$F_{elect} = K \frac{DI_{(A-B)} Z_{eff(A)}}{r^2} + K \frac{DI_{(A-B)} Z_{eff(B)}}{r^2}$$

Another alternative is to change the DI into the charge density of the bond critical point between the atoms, ρ_b.

$$F_{elect} = K \frac{\rho_b Z_{eff(A)}}{r^2} + K \frac{\rho_b Z_{eff(B)}}{r^2}$$

Where A and B in both equations are the interacting atoms.

The electrostatic force approach will be used in the following chapters to interpret the electropositivity of the elements, the strength of covalent bonds (chapter six), the strength of intermolecular/intramolecular interactions (chapter eleven), and the stability of delocalized systems (chapter seven).

EXERCISES

1. By assuming 2 electrons in a heteropolar covalent $2c$-$2e$ bonding system, give the electrostatic equation to describe this bond using generic atoms A and B, where $r_{iA} \neq r_B$.
2. Use the expression from the previous exercise to give the order of bond strength for the C-N, C-O, C-F bonds in H_3C-NH_2, H_3C-OH, H_3C-F, respectively. Give the corresponding expressions for each bond.

Data: $Z_{eff(C)} = 3.13; Z_{eff(N)} = 3.83; Z_{eff(O)} = 4.45; Z_{eff(F)} = 5.1$

Acknowledgments

We appreciate Bader's effort for providing the mathematical approach of the forces in molecules and its simplified treatment which can be found in his website (http://www.chemistry.macmaster.ca/esam/Chapter_6/section_1.html)

REFERENCES CITED

Bader, R.F.W. 1994. Atoms in Molecules: A Quantum Theory. Oxford University Press, New York.

Bader, R.F.W. and Fang, D.-C. 2005. Properties of atoms in molecules: caged atoms and the Ehrenfest force. J. Chem. Theory and Comput. 1: 403-414.

Clementi, E. and Raimondi, D.L. 1963. Atomic screening constants from SCF functions. J. Chem. Phys. 38: 2686-2689.

Hernández-Trujillo, J., Cortés-Guzmán, F., Fang, D.-C. and Bader, R.F.W. 2007. Forces in molecules. Faraday Discuss. 135: 79-95.

Levine, I.N. 1999. Quantum chemistry. Prentice Hall. Upper Saddle River.

Pilar , F.L. 1990. Elementary Quantum Chemistry. McGraw Hill Publishing Co., Singapore.

Slater, J.C. 1933. The virial theorem and molecular structure. J. Chem. Phys. 1: 687-691.

Chapter Four

Notions of Thermodynamics, Molecular Energy, and Use of Theoretical Thermodynamic Data

PRINCIPLES OF CLASSICAL THERMODYNAMICS

Thermodynamics is the study of energy transformations in a chemical reaction or in a physical change of a state. In turn, the energy of a system (accumulated in the form of potential and kinetic components) is its capacity to perform work (energy transfer in an ordered manner) or to produce heat (energy transfer in a disordered manner). A thermodynamic process is the transfer of matter (e.g., in a chemical reaction) and/or energy within the system, or between the system and the surroundings. When no transformation occurs in the system and between the system and surroundings, the system reaches a thermodynamic equilibrium, i.e., no change in temperature, pressure, mass of individual phases, and chemical composition with time. Thermodynamic potentials are different forms to evaluate the stored energy of a system. The ones most important for organic chemists are internal energy (U), enthalpy (H), and Gibbs free energy (G), all of them being interrelated.

The first principle of thermodynamics states that if a system is subject to any cyclic transformation (for example, in a piston), the work done on the neighborhood is equal to the heat extracted from the neighborhood (Castellan 1983).

$$\oint dw = \oint dq$$
$$\oint (dq - dw) = 0$$

Internal energy of a substance represents the total energy as kinetic and potential energies of atoms and chemical bonds in the absence of external fields. When there is an infinitesimal change in internal energy, dU, it is a result of work done, w, and/or heat supplied or absorbed, q. The change in internal energy of a closed system is equal to the energy that enters or leaves the system as heat or work.

$$dU = dq - dw$$

$$\int_i^f dU = \int_i^f dq - \int_i^f dw$$

$$\Delta U = q - w$$

In the absence of non-expansion work, that is, in constant volume, the internal energy change equals the heat absorbed or released.

$$\Delta U = q_v$$

Where q_v means heat at constant volume.

For a state change (for example, forcing down the piston from volume V_1 to V_2) at constant pressure, we have the following infinitesimal change in internal energy (Castellan 1983):

$$dU = dq_p - dw_p \therefore dw_p = pdV$$

$$dU = dq_p - pdV$$

$$\int_1^2 dU = \int_1^2 dq_p - \int_{V_1}^{V_2} pdV$$

$$U_2 - U_1 = q_p - p(V_2 - V_1)$$

$$(U_2 + pV_2) - (U_1 + pV_1) = q_p$$

Internal energy is related to enthalpy by means of:

$$H = U + pV$$

$$H = U + nRT \therefore pV = nRT$$

$$H = U + RT \therefore n = 1 \text{ mol}$$

The difference between enthalpy and internal energy is negligible for condensed phases, except at high pressure or very high temperature. At constant pressure and no work performed, the change of heat equals enthalpy.

$$\Delta H = q_p$$

Where q_p means heat at constant pressure.

In 1854, Clausius formulated entropy change as the passage of heat through a fluid in a heat engine from T_1 to T_2.

$$\Delta S = \frac{q}{T_1} - \frac{q}{T_2}$$

Henceforth, the classical concept of entropy (S) at a constant temperature is given by the ratio between heat released or absorbed in a reversible process (or path) and temperature. An example of reversible process is an isothermal expansion of an ideal gas. A reversible process is always close to equilibrium, which requires ideal infinitely slowly changes (in pressure, volume, or temperature, for example).

$$\Delta S = \frac{q_{rev}}{T} \therefore T = \text{const.}$$

The original Clausius definition of reversible cyclic process from Carnot machine was:

$$\oint \frac{dq}{T} = 0$$

The heating of liquid at a constant pressure in a reversible process, i.e., in close equilibrium in each infinitesimal step has the following entropy equation. Provided the initial and final steps are in equilibrium, even in an irreversible heating, the same equation can be applied.

$$\Delta S = \int_{T_1}^{T_2} \frac{C_p}{T} dT \therefore dq_{rev} = C_p dT$$

$$\Delta S = C_p \ln(T_2/T_1) \therefore C_p = \text{const.} \therefore C_p = \left(\frac{\partial H}{\partial T}\right)_p$$

Where C_p is the heat capacity.

Since all real processes are irreversible, when a process occurs in an isolated (or closed) system, its entropy is always increasing. This idea of an isolated system can be expanded to the system and its surroundings as being in a closed system as well, where they do not interact with the rest of the world. Then, for irreversible processes, the entropy of the system and closed surrounding always increases. Clausius formulation of irreversible cyclic process from Carnot machine was:

$$\oint \frac{dq}{T} > 0$$

Then, entropy can always be created while energy is always conserved. Another important consequence of entropy is that *an isolated system in thermodynamic equilibrium reaches the maximum entropy, provided no decreasing entropy reaction occurs.* The increased entropy trend ($\Delta S \geq 0$) is also found in the improvements of Boltzmann's H-theorem (Tolman 1938).

Entropy and entalphy are interrelated by Gibbs free energy, while entropy and internal energy are interrelated by Helmholtz free energy, A.

$$G = H - TS \therefore p = \text{const.}$$
$$A = U - TS \therefore V = \text{const.}$$

These equations come from Clausius inequality of entropy which, in turn, is derived from the yield of a Carnot machine (Castellan 1983).

$$dS - \frac{dq}{T} \geq 0$$
$$TdS - dq \geq 0$$
$$TdS - dU \geq 0 \therefore V = \text{const.}$$
$$TdS - dH \geq 0 \therefore p = \text{const.}$$

In chemistry, since most processes occur at constant pressure rather than constant volume, Gibbs free energy is more useful. For constant pressure and temperature, Gibbs free energy indicates the spontaneity (or not) of chemical processes. Then, Gibbs energy change in a general reaction is given by:

$$\Delta G = G_{prod} - G_{reag}$$
$$\Delta H = H_{prod} - H_{reag}$$
$$\Delta S = S_{prod} - S_{reag}$$
$$\Delta G = \Delta H - T\Delta S$$

Enthalpy, entropy, and Gibbs energy change can have negative values or positive values or nearly zero values. At constant pressure and volume, enthalpy change is related to heat released ($\Delta H < 0$) or absorbed ($\Delta H > 0$). Entropy change is related to increase in degrees of freedom ($\Delta S > 0$) or decrease in degrees of freedom ($\Delta S < 0$). Gibbs energy change is related to spontaneous process ($\Delta G < 0$) or non-spontaneous process ($\Delta G > 0$). A better discussion about degrees of freedom is in the penultimate section.

$$\Delta H < 0 \Rightarrow \text{exothermic (heat released)}$$
$$\Delta H > 0 \Rightarrow \text{endothermic (heat absorbed)}$$
$$\Delta G < 0 \Rightarrow \text{exergonic (spontaneous)}$$
$$\Delta G > 0 \Rightarrow \text{endergonic (non-spontaneous)}$$

With respect to enthalpy and entropy changes, the following situations can take place:

(i) $\Delta H < 0$ and $\Delta S > 0$. In this case, $\Delta G < 0$ and the reaction is spontaneous at any temperature;

(ii) $\Delta H < 0$ and $\Delta S = 0$. In this case, $\Delta G < 0$ and the reaction is spontaneous at any temperature;

(iii) $\Delta H = 0$ and $\Delta S > 0$. In this case, $\Delta G < 0$ and the reaction is spontaneous at any temperature;

(iv) $\Delta H > 0$ and $\Delta S < 0$. In this case, $\Delta G > 0$ and the reaction is non-spontaneous at any temperature;

(v) $\Delta H > 0$ and $\Delta S = 0$. In this case, $\Delta G > 0$ and the reaction is non-spontaneous at any temperature;

(vi) $\Delta H = 0$ and $\Delta S < 0$. In this case, $\Delta G > 0$ and the reaction is non-spontaneous at any temperature;

(vii) $\Delta H < 0$ and $\Delta S < 0$. In this case, $\Delta G < 0$ or $\Delta G > 0$ depending on the reaction temperature;

(viii) $\Delta H > 0$ and $\Delta S > 0$. In this case, $\Delta G < 0$ or $\Delta G > 0$ depending on the reaction temperature.

Then, for a reaction (or other chemical process) to be spontaneous at any temperature, an exothermic reaction and increase in degree of freedom or near zero enthalpy change with increase in degree of freedom or the opposite (cases i to iii) is needed. When enthalpy and entropy change have the same sign (cases vii and viii), the reaction temperature will modulate the spontaneity of the reaction or other chemical processes.

For a decrease in the enthalpy of a reaction (i.e., $\Delta H < 0$) the products have lesser energy than the reagents, that is, the products are stabler than reagents.

$$\text{Reagents} \rightarrow \text{Products} \therefore \text{if } \Delta H < 0$$
$$\text{Then } H_{prod} < H_{reag}$$

For an increase in the degree of freedom of a reaction (i.e., $\Delta S > 0$), the products have a higher entropy than the reagents.

$$\text{Reagents} \rightarrow \text{Products} \therefore \text{if } \Delta S > 0$$
$$\text{Then } S_{prod} > S_{reag}$$

In organic chemistry, sometimes we are more interested in evaluating qualitatively some physical or chemical behavior. Then, with respect to reaction enthalpy change, one must evaluate changes (from reagents to products) in the strength of chemical bonds, resonance effect, intramolecular interaction, and so on. Then, for a better evaluation of enthalpy change, the understanding of the next chapters related to molecular stability is needed.

As to reaction entropy change, the qualitative evaluation is simpler, since there are basically three types of reaction.

(i) $A + B \rightarrow C \therefore \Delta S < 0$
(ii) $A \rightarrow B + C \therefore \Delta S > 0$
(iii) $A + B \rightarrow C + D \therefore \Delta S \approx 0$

The reaction (i) is a type of addition reaction (for example, electrophilic addition in alkenes), where there is a decrease of the degree of freedom. The reaction (ii) is a type of elimination reaction (for example, the dehydration of an alcohol), where the degree of freedom increases. And, the reaction (iii) is a type of substitution reaction (for example, electrophilic aromatic substitution), where there is nearly no net change in the degree of freedom.

EQUILIBRIUM CONSTANT AND ITS RELATION WITH GIBBS FREE ENERGY

The concept of equilibrium requires the knowledge of microscopic and macroscopic properties. We will discuss the quantum statistical approach of the microscopic system later in the chapter. As of now, we will focus on the differences between the idea of the microscopic and macroscopic in a chemical reaction. *Both views (visualization of single molecules reacting and visualization of n moles of molecules reacting) are necessary when dealing with a chemical reaction.* The microscopic standpoint of a reaction is the understanding of the stoichiometric *molecular* relation between reactants and products. For example, in the addition of bromine to ethene, one molecule of bromine is added to one molecule of ethene to produce one molecule of 1,2-dibromoethane. The macroscopic standpoint of a reaction is the understanding of the number of moles, N, of the substance (reagent and product) involved, following a specific stoichiometry, in this reaction. Then,

Microscopic view of a reaction: molecules involved

Macroscopic view of a reaction: n moles (N) of molecules involved

Most of the chemical reactions reach a chemical equilibrium. Few of them pass through an irreversible process, such as combustion and corrosion reactions. In an equilibrium condition, there are a specific number of moles of every product and reagent.

$$A + B \underset{\leftarrow}{\overset{\rightarrow}{\rightleftharpoons}} C + D$$

$$A + B \xrightarrow{\quad direct \quad} C + D$$

$$A + B \xleftarrow{\quad inverse \quad} C + D$$

In equilibrium: $v_{dir} = v_{inv}$

In a simplified perspective, in which the reactions are considered to occur in very dilute solutions (which implies the use of a solvent in a reaction in large amount), one finds an expression for the equilibrium condition associated with the molar concentration of reagents and products. Supposing an unitary stoichoimetric relation in the generalized reaction: $A + B \rightarrow C + D$. Then, there is an equilibrium constant, K, given by:

$$K = \frac{[C][D]}{[A][B]} \therefore \text{For very dilute solutions}$$

A general formula of K for any concentration of reactants and products involve the use of molar activity, a, in the equilibrium constant expression.

$$K = \frac{a_C a_D}{a_A a_B}$$

$$a_A = \gamma_A [A]$$

$$K = \frac{\gamma_C \gamma_D}{\gamma_A \gamma_B} \frac{[C][D]}{[A][B]}$$

Where γ_A is activity coefficient of A.

If a reaction in an equilibrium condition is perturbed by any external influence (for example, removal of one of the products), the reaction goes off the equilibrium condition, but will attempt to return to this condition again, which means it will try to recover its associated value of equilibrium constant.

$$A + B \overset{K_1}{\underset{\leftarrow}{\rightarrow}} C + D$$

$$C + D \overset{1/K_1}{\underset{\leftarrow}{\rightarrow}} A + B$$

$$A \overset{K_2}{\underset{\leftarrow}{\rightarrow}} B + C$$

$$A + C \overset{K_3}{\underset{\leftarrow}{\rightarrow}} E$$

Moreover, each reaction has a specific equilibrium constant at a given temperature. This is because of the relation between equilibrium constant and Gibbs energy change developed in 1911 (Laidler and King 1983). *Since each molecule has its own specific Gibbs energy, any specific reaction has its own specific Gibbs energy change and, as a consequence, its own specific equilibrium constant* (see Appendix 2).

$$\Delta G = -RT \ln K \therefore R = 8.31 \text{JK}^{-1}\text{mol}^{-1}$$

This relation is universal and it is valid for any reaction condition. It is a linear relation between ΔG and $\ln K$. In Isaac's book, there is a table containing pK_a, ΔG^0, ΔH^0, and ΔS^0 of several organic acids and bases. Figure 4.1 shows the plot of pK_a versus ΔG^0, in kJ.mol^{-1}, of 29 organic acids and bases, whose data was collected from Isaacs' book (Isaacs, 1995). We observe that the coefficient of determination ($R^2 = 0.9999$) represents an universal law underlying ΔG^0 and K.

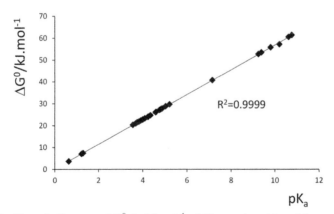

Figure 4.1   Plot of pK_a versus ΔG^0, in kJ.mol^{-1} of 29 organic acids and bases. Data from Isaacs' book (Isaacs, 1995).

By considering the reactions whose entropy change is nearly zero, one can associate product stability with the magnitude of equilibrium constant. When products and reagents have nearly the same energy, $\Delta H = \Delta G = 0$, and the equilibrium constant is nearly one. For products more stable than reagents, $\Delta H < 0$ and $\Delta G < 0$, and the equilibrium constant is higher than one. As to products less stable than reagents, $\Delta H < 0$ and $\Delta G < 0$, and then, $0 < K < 1$.

$$K = e^{-\frac{\Delta G}{RT}}$$

(1) $\Delta G \approx 0 \therefore K = 1$

(2) $\Delta G < 0 \therefore K > 1$

(3) $\Delta G > 0 \therefore 0 < K < 1$

RELATION BETWEEN YIELD AND EQUILIBRIUM CONSTANT

Let us suppose the general reaction in condensed phase:

$$_nA + {_m}B \underset{\leftarrow}{\overset{K}{\rightarrow}} {_o}C + {_p}D$$

$$K = \frac{[C]^o [D]^p}{[A]^n [B]^m}$$

Where it is considered that the reaction solution is very dilute.

Let us suppose that at the end of the reaction (when it reaches the equilibrium), the number of moles at equilibrium, N_e, of each reagent and product is:

$$N_{e(A)} = z \therefore N_{e(B)} = w \therefore N_{e(C)} = x \therefore N_{e(D)} = y$$

Then, the equilibrium constant can be rewritten as:

$$K = \frac{\left(N_{e(C)}\right)^o \left(N_{e(D)}\right)^p}{\left(N_{e(A)}\right)^n \left(N_{e(B)}\right)^m} V^{(n+m-o-p)}$$

$$K = \frac{x^o y^p}{z^n w^m} V^{(n+m-o-p)}$$

Where V is the solution volume. The reaction yield, ε, can be expressed as:

$$\varepsilon = \frac{x}{o} = \frac{y}{p} \therefore x = o\varepsilon \therefore y = p\varepsilon$$

$$0 \le \varepsilon \le 1$$

Then, the equilibrium constant becomes:

$$K = \frac{(o\varepsilon)^o (p\varepsilon)^p}{z^n w^m} V^{(n+m-o-p)}$$

Therefore, *there is a direct relation between K and ε.*

RELATION BETWEEN GIBBS ENERGY CHANGE AND EQUILIBRIUM CONSTANT

Let us take one exergonic and another endergonic reaction. The relations with K are shown as (1) and (2), respectively:

$$(1) \ K = e^{-\Delta G/RT} \ \therefore \ \Delta G < 0 \therefore K = e^{-(-\Delta G)/RT} = e^{\Delta G/RT}$$

$$(2) \ K = e^{-\Delta G/RT} \ \therefore \ \Delta G > 0 \therefore K = e^{-\Delta G/RT}$$

From the first relation, the more exergonic the reaction (i.e., the more negative the Gibbs energy change), the higher K (and, as a consequence, the higher ε as well). From the second relation, the less endergonic the reaction (i.e., the less positive the Gibbs energy change), the higher K and ε. Therefore, *the more spontaneous a reaction (more exergonic or less endergonic), the higher the equilibrium constant and the higher the reaction yield.*

NOTIONS OF THERMODYNAMIC STATISTICS: MOLECULAR ENERGY

While classical thermodynamics can only evaluate the change of a thermodynamic property, thermodynamic statistics gives the absolute value of each molecule in any condition. Computational chemistry uses the thermodynamic statistics to obtain the absolute value of a certain thermodynamic property which can be used to obtain overall change of this property in a certain chemical process. Depending on the quality of theoretical level, it is possible to obtain very close values of a thermodynamic change between experimental and theoretical data (see the next section).

Thermodynamic statistics (founded by Gibbs, Maxwell, and Boltzmann) starts with the Boltzmann distribution equation, in which the molecular partition function used to obtain the absolute values of entropy and internal energy for a specific ensemble is derived.

The Boltzmann distribution measures the occupation probability (or probability distribution) of the particles at a given energy state, ε, the so-called $P(\varepsilon)$. In this book, we show two distinguished forms to obtain the Boltzmann distribution. The first one is very straightforward and the second one is more complicated, but it is important to understand the concept of statistical weight of configuration.

Let us introduce the first methodology. The energy components are added, but probabilities of energy components are multiplied (Engelhardt et al. 2015). Let us consider a system with only two energy components for simplicity.

$$E = \varepsilon_i + \varepsilon_j \ \therefore \ P(E) = P(\varepsilon_i) \times P(\varepsilon_j)$$

By assuming there is a function to describe the relation between energy and probability:

$$P_i(\varepsilon) = A_i f(\varepsilon)$$
$$f(\varepsilon_i + \varepsilon_j) = a f(\varepsilon_i) f(\varepsilon_j)$$

Where A_i is the normalization factor in order to assure that the sum of all probabilities is one. Then, we realize that the Boltzmann function must be exponential in order to satisfy both additive and multiplicative properties of energy and probability, respectively, at the same time.

$$f(\varepsilon) = e^{c\varepsilon}$$

As previously stated, the total probability of all microstates has to be a unity. Then, let us sum over all probabilities:

$$\sum_n P_i(\varepsilon_n) = A_i \sum_n e^{c\varepsilon_n}$$

$$A_i = \frac{1}{\sum_n e^{c\varepsilon_n}}$$

Then, the probability of energy is:

$$P_i(\varepsilon) = \frac{e^{c\varepsilon}}{\sum_n e^{c\varepsilon_n}}$$

By knowing that $c = -1/k_B T$, then:

$$P_i(\varepsilon) = \frac{e^{(-\varepsilon/k_B T)}}{\sum_n e^{(-\varepsilon_n/k_B T)}}$$

From $P(\varepsilon)$, it is possible to obtain the average energy value:

$$\bar{\varepsilon} = \frac{\int_0^\infty \varepsilon . P(\varepsilon) d\varepsilon}{\int_0^\infty P(\varepsilon) d\varepsilon}$$

Let us now use the second methodology, which is based on the original work of Boltzmann in 1877, where he used the notion of the statistical weight of configuration, W.

One particular configuration of a system is a specific distribution of number of particles at each distinguished energy, let us say $\{n_0, n_1, n_2, ...\}$, where n_0 represents $N-x$ particles at fundamental state, n_1 represents $x-y$ particles at excited state, and so on. Then, the statistical weight of configuration, W, gives the information of the distribution of the particles in each energy state (Boltzmann 1877).

$$W = \frac{N!}{n_0! \, n_1! \, n_2! ...} \quad \therefore N = \sum_i n_i$$

For example, let us assume a total number of particles, $N=20$. If all 20 particles are in one single energy state, the configuration is $\{20, 0, 0..., 0\}$ and

$W=1$ is the minimum value. If each particle occupies one distinguished state, the configuration is $\{1,1,1,...1\}$ and $W=20!$ is the maximum value. Atkins mentions the "instantaneous configuration of the system which fluctuates with time because of population change" as a result of temperature, molecular degrees of freedom, collisions, etc. (Atkins 1998). Then, the bulk of molecules is usually arranged in a series of distinguished instantaneous configurations, for example, $\{N,0,0...0\}$ and $\{N-2,2,0...0\}$, each one with its own W.

By using natural logarithm of W and Stirling's approximation ($\ln X! = X \ln X - X$), we have:

$$\ln W = \ln N! - \sum_i \ln n_i!$$

$$\ln W = (N \ln N - N) - \left(\sum_i n_i \ln n_i - \sum_i n_i \right)$$

$$\ln W = N \ln N - \sum_i n_i \ln n_i$$

$$\ln W = \sum_i n_i \ln N - \sum_i n_i \ln n_i$$

$$\ln W = -\sum_i \left(n_i \ln n_i - n_i \ln N \right)$$

$$\ln W = -\sum_i n_i \ln \frac{n_i}{N}$$

For a particular fluctuation of configurations, there is one with largest W. *The configuration with largest W is dominant and usually the properties of the system are determined by the dominant configuration.* In our example, it is $\{N-2,2,0...0\}$. Then, it is very important to search for the largest W. At maximum value of W (which is also a maximum in $\ln W$), it is a critical point where $d\ln W=0$.

$$d \ln W = \sum_i \left(\frac{\partial \ln W}{\partial n_i} \right) dn_i$$

Which has to follow two restrictions:

$$\sum_i \varepsilon_i dn_i = 0 \therefore \sum_i dn_i = 0$$

Where ε_i is the energy of a particular ith state. These two restrictions are related to the fact that the total sum of $n_i \varepsilon_i$ (total energy, E) and total sum of n_i are constant.

$$\sum_i n_i \varepsilon_i = E \therefore \sum_i n_i = N$$

Where E is the total energy of the system and N is the total number of particles. To find the maximum of the differential equation above, we use the Lagrange multiplier method, which inserts new variables (Lagrange multiplier), α and β, in the function.

$$d \ln W = \sum_i \left(\frac{\partial \ln W}{\partial n_i} \right) dn_i + \alpha \sum_i dn_i - \beta \sum_i \varepsilon_i dn_i$$

$$d \ln W = \sum_i \left\{ \left(\frac{\partial \ln W}{\partial n_i} \right) + \alpha - \beta \cdot \varepsilon_i \right\} dn_i$$

$$\left(\frac{\partial \ln W}{\partial n_i} \right) + \alpha - \beta \cdot \varepsilon_i = 0$$

Further details to the derivation of the above equation can be found elsewhere (Atkins 1998). The solution of this Lagrange equation gives the Boltzmann distribution equation:

$$p_i = \frac{n_i}{N} = \frac{\exp\left(\frac{-\varepsilon_i}{k_B T} \right)}{\sum_j \exp\left(\frac{-\varepsilon_j}{k_B T} \right)}$$

$$q = \sum_i \exp(-\beta \cdot \varepsilon_i) = \sum_i \exp\left(\frac{-\varepsilon_i}{k_B T} \right)$$

$$k_B = 1.38 \times 10^{-23} \, \text{JK}^{-1} \text{molecule}^{-1}$$

Where k_B is the Boltzmann constant, T is the absolute temperature, $k_B T$ is the thermal energy, and q is the molecular partition function.

It is important to mention that the numerator part of p_i is similar to the exponential part of the Arrhenius equation.

$$p_i = \frac{\exp\left(\frac{-\varepsilon_i}{k_B T} \right)}{q} \quad \therefore k_B = \frac{R}{N_A}$$

$$R = 8.314 \text{JK}^{-1} \text{mol}^{-1} \quad \therefore N_A = 6.02 \times 10^{23} \, \text{molecules}$$

The Boltzmann distribution equation establishes the most probable populations of a system for each temperature, and the molecular partition function indicates the number of thermally accessible states for a given particle n_i at a given temperature. For example, at $T=0$, only fundamental state is accessible and then $q=1$.

From the Boltzmann distribution equation the total energy of the system, E, can be determined as (Atkins 1998):

$$E = \sum_i n_i.\varepsilon_i \quad \therefore n_i = N \frac{\exp(-\beta \cdot \varepsilon_i)}{q}$$

$$E = \frac{N}{q} \sum_i \varepsilon_i \exp(-\beta \cdot \varepsilon_i)$$

Then, it is possible to establish the relation between molecular partition function and internal energy by applying a simple derivative below (Atkins 1998):

$$\frac{d}{d\beta}e^{-\beta\varepsilon_i} = -\varepsilon_i e^{-\beta\varepsilon_i}$$

$$E = \frac{N}{q}\sum_i \varepsilon_i \exp(-\beta.\varepsilon_i) = -\frac{N}{q}\frac{d}{d\beta}\sum_i \exp(-\beta\cdot\varepsilon_i)$$

$$E = -\frac{N}{q}\frac{dq}{d\beta} \therefore \frac{dx}{x} = \ln x \therefore E = -N\frac{d\ln q}{d\beta}$$

$$U = U(0) - N\left(\frac{\partial\ln q}{\partial\beta}\right)_V$$

From the expression $U=U(0)+E$, it is also possible to obtain the statistical entropy (Atkins 1998) from its derivative:

$$dU = dU(0) + \sum_i n_i d\varepsilon_i + \sum_i \varepsilon_i dn_i$$

Which is simplified to:

$$dU = \sum_i \varepsilon_i dn_i$$

By assuming heat under constant volume.

From the thermodynamic relation between internal energy and entropy:

$$dU = dq_{rev} = TdS$$

One finds the corresponding relation between Boltzmann constant and entropy (Atkins 1998):

$$dS = \frac{dU}{T} = k_B\beta\sum_i \varepsilon_i dn_i$$

From this point on, one obtains the statistical expression for entropy (Boltzmann 1877):

$$S = k_B \ln W$$

And the relation between entropy and molecular partition function:

$$S = k_B\beta[U - U(0)] + Nk_B \ln q$$

By knowing that:

$$\beta = \frac{1}{k_B T}$$

Then, S can be arranged to:

$$S = \frac{[U - U(0)]}{T} + Nk_B \ln q$$

According to the second approach with respect to the relation between entropy and molecular partition function, we have:

$$S = k_B \ln W = k_B \left(N \ln N - \sum_i n_i \ln n_i \right)$$

$$n_i = p_i N$$

$$S = k_B \ln W = k_B \left(N \ln N - \sum_i p_i N \ln p_i N \right)$$

$$S = k_B N \ln N - k_B N \sum_i p_i \ln p_i - k_B N \ln N \sum_i p_i$$

$$S = -k_B N \sum_i p_i \ln p_i$$

$$p_i = \frac{e^{-\beta e_i}}{q}$$

$$S = -k_B N \sum_i \frac{e^{-\beta e_i}}{q} \ln \left(\frac{e^{-\beta e_i}}{q} \right)$$

$$S = k_B N \sum_i \frac{e^{-\beta e_i}}{q} (-\beta e_i - \ln q)$$

$$S = \frac{U}{T} + N k_B \ln q$$

Then, for each type of ensemble, it is possible to obtain one exact expression for absolute value of entropy and internal energy from the molecular (or individual) partition function. *The molecular partition function, q, is associated with non-interacting molecules or individual molecules. In order to consider interacting molecules or the entire molecular system, we have to use the partition function, Q. All molecules have four types of (molecular) partition functions: electronic (q_{elect}), vibrational (q_{vib}), translational (q_{transl}), and rotational (q_{rot}).* For more information, see chapter ten and Appendix 3.

$$q = q_{elect} \cdot q_{vib} \cdot q_{transl} \cdot q_{rot}$$

Ensemble is defined in a geometrical box containing a certain numbers of particles (or molecules) in a periodic boundary condition (which is mirrored throughout the three Cartesian coordinates), which can be understood as a large number of virtual copies of a real system where certain thermodynamic properties are constant. The choice for a specific ensemble depends on the conditions imposed to a certain real system. For example, a great majority of chemical reactions (in an industry or laboratory) occur in a reactor where the total number of particles (or molecules), N, is constant, the volume of the reactor is obviously constant, and the reaction temperature is nearly constant. In order to represent this real system one has to use the canonical ensemble in which N, V, and T are constant. The partition function for the canonical ensemble, $Q(N,V,T)$ is given by Boltzmann distribution:

$$Q(N,V,T) = \sum_i e^{-E_i(N,V)/k_B T}$$

By assuming that the molecules or particles within the ensemble do not react among themselves, it is possible to associate the partition function for the canonical ensemble with the molecular partition function (Cramer 2004).

$$Q(N,V,T) = \frac{1}{N!}\left[\sum_{j(1)} e^{-\varepsilon_{j(1)}/k_B T}\right]\left[\sum_{j(2)} e^{-\varepsilon_{j(2)}/k_B T}\right]...\left[\sum_{j(N)} e^{-\varepsilon_{j(N)}/k_B T}\right]$$

$$Q(N,V,T) = \frac{1}{N!}\left[\sum_{k}^{niveis} e^{-\varepsilon_k/k_B T}\right]^{N} = \frac{[q(V,T)]^{N}}{N!}$$

$$Q = \frac{q^{N}}{N!}$$

From this point on, it is very important to understand that the molecular energy, or the energy of the molecule, has four components. *Every molecule has three types of degrees of freedom: translation movement, rotation movement, and vibration movement. Each of these movements has an associated energy, namely, ε_{transl}, ε_{rot}, and ε_{transl}.* Besides these three energy components associated with the molecular movement, *each molecule has two other energy components associated with the stationary state: electronic energy, ε_{elect} and potential nuclear energy, ε_{nuc}.* For a vast majority of the cases, we are interested in the stationary condition (no movement of nuclei) in order to obtain the stationary wave function and its properties. Then, it is considered adiabatic approximation, where the kinetic nuclear energy vanishes and the nuclear energy comes from the sum of the electrostatic potential between two nuclei from the optimized geometry. The potential nuclear energy can also be regarded as part of the electronic energy. *Then, the total molecular energy is given by sum of the energy components of translational, rotational, and vibrational movement plus the electronic energy.*

$$E = \varepsilon_{transl} + \varepsilon_{rot} + \varepsilon_{vib} + \varepsilon_{elect}$$

The total energy of the system as a function of the canonical partition function, Q, is given by:

$$E = N\bar{\varepsilon} = k_B T^2 \left(\frac{\partial \ln Q}{\partial T}\right)_{N,V}$$

The total entropy as a function of the canonical partition function is:

$$S = k_B T\left(\frac{\partial \ln Q}{\partial T}\right)_{N,V} + k_B \ln Q$$

The corresponding total molecular partition function is decomposed as a product of four molecular partition functions: translational partition function, rotational partition function, vibrational partition f)unction, and electronic partition function.

$$q(V,T) = \left[\sum_{j(1)}^{transl} e^{-\varepsilon_{j(1)}/k_B T}\right]\left[\sum_{j(2)}^{rot} e^{-\varepsilon_{j(2)}/k_B T}\right]\left[\sum_{j(3)}^{vib} e^{-\varepsilon_{j(3)}/k_B T}\right]\left[\sum_{j(4)}^{elect} e^{-\varepsilon_{j(4)}/k_B T}\right]$$

$$q(V,T) = q_{transl}(V,T)q_{rot}(T)q_{vib}(T)q_{elect}(T)$$

The energy of each molecular partition function is the solution of the Schrödinger equation with respect to the particle in a box (energy of translational movement), rigid rotor (rotational movement), harmonic oscillator (vibrational movement), and stationary wave function (electronic energy).

$$\varepsilon_{transl}(n_x, n_y, n_z) = \frac{h^2}{8Ma^2}\left(n_x^2 + n_y^2 + n_z^2\right)$$

$$\varepsilon_{rot} = \frac{J(J+1)h^2}{8\pi^2 I} \therefore J = 0, 1, 2, \ldots$$

$$\varepsilon_{vib} = h\nu(n+1/2) \therefore n = 0, 1, 2, \ldots \therefore \nu = \frac{1}{2\pi}\left(\frac{k}{\mu}\right)^{1/2}$$

From these energy expressions for each type of degree of freedom, it is possible to obtain the corresponding molecular partition function. For instance, the vibrational partition function for a polyatomic system with α vibrational degrees of freedom is:

$$q_{vib}(T) = \sum_n e^{-\beta\varepsilon_n} = \sum_n e^{-\beta[(n+\frac{1}{2})h\nu]} = e^{-\beta h\nu/2}\sum_{n=0}^{\infty} e^{-\beta h\nu n}$$

$$q_{vib}(T) = \prod_{j=1}^{\alpha} \frac{e^{-h\nu_j/2k_B T}}{\left(1-e^{-h\nu_j/k_B T}\right)}$$

Then, the total vibrational energy, E_{vib}, which is the so-called vibrational internal energy, U_{vib}, is:

$$E_{vib} = U_{vib} = Nk_B T^2 \frac{d\ln q_{vib}}{dT}$$

$$U_{vib} = Nk_B \sum_{j=1}^{\alpha}\left(\frac{h\nu_j}{2k_B} + \frac{(h\nu_j/k_B)e^{h\nu_j/k_B T}}{1-e^{-h\nu_j/k_B T}}\right)$$

$$S_{vib} = k_B \sum_{j=1}^{\alpha}\left(\frac{h\nu_j}{2k_B} + \frac{(h\nu_j/k_B)e^{h\nu_j/k_B T}}{T\left(1-e^{-h\nu_j/k_B T}\right)} - \ln\left(\frac{e^{-h\nu_j/2k_B T}}{\left(1-e^{-h\nu_j/k_B T}\right)}\right)\right)$$

The same procedure is done for translational and rotational movements. At the end, one has the translational and rotational entropy and internal energy. The total internal energy is given by the expression:

$$U(T) = U_{transl}(T) + U_{rot}(T) + U_{vib}(T) + U_{vib}(0)$$

$$U_{vib}(0) = E_{elect} + ZPE$$

$$ZPE = \sum_j^{\alpha} \frac{h\nu_j}{2}$$

The last term in $U_{vib}(0)$ is called zero-point energy, ZPE, which is the quantum mechanical solution of the harmonic oscillator made of two particles, for example, hydrogen molecule vibrating in its fundamental state (at $n=0$).

Henceforth, *it is possible to obtain the absolute internal energy (as a consequence, the absolute enthalpy), the absolute entropy and, as a consequence, the absolute Gibbs energy for any molecule, provided that its geometry was previously optimized.*

BOLTZMANN FACTOR AND EQUILIBRIUM CONSTANT

In the last section the Boltzmann distribution equation was shown, which is a probability distribution of particles (or molecules) in a system over all possible states.

$$\frac{n_i}{N} = \frac{\exp(-\varepsilon_i / k_B T)}{\sum_i \exp(\varepsilon_i / k_B T)}$$

Another important relation is the Boltzmann factor, which establishes the ratio of two populations based on the energy difference between the corresponding states of these two populations. The energy difference can be, for example, the difference between two activation barriers from nucleophilic attack onto two different electrophilic faces of the same reagent, or the formation energy difference between two isomers. Simple derivation of Boltzmann factor is given by McDowell (McDowell 1999).

$$\frac{N_i}{N_j} = \exp\left[-\left(E_i - E_j\right)/RT\right]$$

The equilibrium constant can be regarded as the ratio of population between two products (final state) and the reagent (initial state).

$$K = \frac{[C]^o \cdot [D]^p}{[A]^n \cdot [B]^m} \equiv \frac{N_i}{N_j}$$

Then, the equilibrium constant can be expressed as a Boltzmann factor.

$$\frac{N_i}{N_j} \equiv K = \exp\left[-\left(E_i - E_j\right)/RT\right]$$

COMPARISON BETWEEN CLASSICAL AND STATISTICAL ENTROPY

The classical entropy was introduced by Clausius as a transfer of energy (as heat or work) between two bodies which is dependent on the temperature, for example, the black body radiation which was the starting point for quantum mechanics. From classical entropy, it is only possible to obtain the entropy change of a reversible process.

$$\Delta S = \frac{q_{rev}}{T} \therefore T = \text{const.}$$

Boltzmann, in turn, demonstrated that the entropy is a function of the weight of the configuration, W.

$$S = k_B \ln W$$

This expression gives the so-called statistical entropy, which can be rewritten in canonical ensemble as a function of the partition function.

$$S = k_B T \left(\frac{\partial \ln Q}{\partial T} \right)_{N,V} + k_B \ln Q$$

From this point on, it is possible to obtain the absolute entropy for any molecule using the entropy expression from the translational, vibrational, and rotational movements.

$$S = S_{transl} + S_{vib} + S_{rot}$$

Although classical and statistical entropies have distinguished formulas, it is possible to demonstrate that *the calculated entropy change (which uses statistical entropy) is equivalent to the experimental entropy change (which is derived from classical entropy).*

Table 4.1 shows the experimental standard molar entropy at 298 K, obtained from a thermochemical table of classical entropy, and the corresponding theoretical values (from statistical entropy) obtained theoretically by this author of methane, ethane, propane, and butane. One can see that both values of classical and statistical entropy for each molecule are very similar, which shows *(i) that both entropy concepts are similar even though they have different expressions, and (ii) the reliability of computational chemistry to obtain thermodynamic values (besides other observables).*

Table 4.1   Experimental and theoretical standard molar entropy at 298 K, $S°$.

Molecule	Experimental $S°$ (J mol^{-1} K^{-1})	Theoretical $S°$ [a] (J mol^{-1} K^{-1})
Methane	186.2	185.94
Ethane	229.5	226.85
Propane	269.9	272.84
Butane	310.1	304.30

(a) From CCSD/6-31G(d,p) level of theory

The results of Table 4.1 also show that the entropy increases as the size of the alkane chain increases. That is because the degree of freedom of the molecule increases as its size increases. *Then, any entropy change (from a reaction) or entropy difference (between two different molecules) can be evaluated from the contributions of translational, vibrational, and rotational movements to the total degree of freedom of the molecular system. The higher the degree of freedom (from translational and/ or rotational and/or vibrational movements), the higher is the entropy.*

EXPERIMENTAL AND THEORETICAL ENTHALPY AND GIBBS ENERGY OF FORMATION

We have seen a very good agreement between experimental and theoretical standard molar entropy. Let us compare experimental and theoretical enthalpy and Gibbs energy change. The experimental thermodynamic data is called standard enthalpy of formation and standard Gibbs free energy of formation. The corresponding theoretical data does not own the term "standard", since it is obtained from absolute values and not from standardized conditions.

Table 4.2 Experimental Standard enthalpy and Gibbs free energy of formation, $\Delta_f H°$ and $\Delta_f G°$, in kcal mol^{-1}, and their corresponding absolute theoretical values (enthalpy and Gibbs free energy of formation), in Hartree, from MP2/6-311G++(d,p) level of theory.

Thermodynamic Property	$H_2 + {}_{1/2}O_2 \rightarrow H_2O_{(g)}$				
	H_2	O_2	$H_2O_{(g)}$	-	Δ
$\Delta_f H°$/kcal mol^{-1}	0	0	−57.79		−57.79
$\Delta_f G°$/kcal mol^{-1}	0	0	−54.63		−54.63
H/Hartree[a]	−1.1466	−150.0232	−76.2494		−0.0912
G/Hartree	−1.1614	−150.0465	−76.2709		−0.08625
	${}_3H_2 + N_2 \rightarrow {}_2NH_{3(g)}$				
	H_2	N_2	$NH_{3(g)}$	-	Δ
$\Delta_f H°$/kcal mol^{-1}	0	0	−11.02	-	−11.02
$\Delta_f G°$/kcal mol^{-1}	0	0	−3.93	-	−3.93
H/Hartree	−1.1466	−109.2933	−56.3769	-	−0.0207
G/Hartree	−1.1614	−109.3151	−56.3987	-	0.0019
	$3NO_{2(g)} + H_2O_{(l)} \rightarrow 2HNO_{3(aq)} + NO_{(g)}$				
	$NO_{2(g)}$	$H_2O_{(l)}$	$HNO_{3(aq)}$	$NO_{(g)}$	Δ
$\Delta_f H°$/kcal mol^{-1}	7.93	−68.31	−49.56	21.58	−33.03
$\Delta_f G°$/kcal mol^{-1}	12.26	−56.67	−26.59	20.94	−12.35
H/Hartree	−204.6534	−76.2576	−280.3039	−129.6137	−0.0037
G/Hartree	−204.6806	−76.2790	−280.3341	−129.637	0.0156
	$H_2O_{(l)} + CO_{2(aq)} \rightarrow H_2CO_{3(aq)}$				
	$H_2O_{(l)}$	$CO_{2(aq)}$	$H_2CO_{3(aq)}$	-	Δ
$\Delta_f H°$/kcal mol^{-1}	−68.31	−94.05	−167.22	-	−4.85
$\Delta_f G°$/kcal mol^{-1}	−56.67	−94.26	−148.92	-	2.02
H/Hartree	−76.2576	−188.1914	−264.4329	-	0.0161
G/Hartree	−76.2790	−188.2157	−264.4635	-	0.0312

(a) For practical reasons, it was chosen to use Hartree unit instead of kcal mol^{-1} unit. In order to convert the Hartree into kcal mol^{-1} the corresponding value of the former must be multiplied by 627.51.

As previously stated, except for the standard molar entropy, classical thermodynamics can only provide values from the change of a thermodynamic property in a certain process. Then, in order to obtain the experimental enthalpy and Gibbs energy of formation of the molecules, the standard state conditions in classical thermodynamics (pure liquids or gases at 1 atm and at 298 K in 1M solution and the energy of formation of an element in its normal state having zero value) was established. By assuming zero value for the energy of the elements in their most natural occurrence, it is possible to obtain the standard enthalpy and Gibbs energy of formation of the molecules formed directly from these elements (first two reactions in Table 4.2), and the enthalpy and Gibbs energy of formation of the molecules formed from molecules which were directly formed from the elements in their natural state (last two reactions in Table 4.2). Therefore, it is possible to obtain a table with the enthalpy of formation of almost any molecule.

Since theoretical approach provides absolute energy values of the molecules, one cannot compare the standard enthalpy of formation of any molecule (e.g., H_2O in Table 4.2) to the absolute theoretical enthalpy of this molecule. Nonetheless, it is possible to compare the standard enthalpy of formation of any molecule to the corresponding theoretical enthalpy of formation of this molecule (the Δ values in Table 4.2).

From the results of Table 4.2, one can see that for reactions that occur exclusively in the gas phase (the first two reactions), the agreement between experimental and theoretical enthalpy and Gibbs free energy of formation (Δ values) is usually very good. For $H_2O_{(g)}$ formation, theoretical enthalpy and Gibbs free energy of formation are -57.23 and -54.12 kcal mol^{-1} (where 1 Hartree equals 627.51 kcal mol^{-1}), whose errors from experimental are 1% for both enthalpy and Gibbs free energy of formation. For $NH_{3(g)}$ formation, theoretical enthalpy and Gibbs free energy of formation are -12.99 and 1.19 kcal mol^{-1} whose errors from experimental are 15% and 70% for enthalpy and Gibbs free energy of formation, respectively. Only theoretical Gibbs free energy of formation of $NH_{3(g)}$ has a very high discrepancy.

For the last two reactions in Table 4.2, the discrepancy between theoretical and experimental values is large. The main reason is the explicit interaction through hydrogen bonds among water molecules and the solvated molecule (e.g., $HNO_{3(aq)}$ and $H_2CO_{3(aq)}$), which were not considered in our theoretical calculations. For these solvated molecules, we have only used the implicit model (Tomasi et al. 2005), more specifically the Polarizable Continuum Model using Integral Equation Formalism, IEFPCM (Klamt et al. 2015), which proved to be ineffective for thermodynamic property calculations involving protic solvents such as water, i.e., solvents having H-X bond, where X is an electronegative atom. To improve the agreement between theoretical and experimental results for reactions involving solvated molecules in water (or other protic solvent), it is needed to include water molecules interacting through hydrogen bonds with these solvated molecules. This is a very good exercise for those interested in theoretical calculations.

EXERCISES

1. By knowing the pK_a's of the following acids HCOOH, CH$_3$COOH, CHCl$_2$COOH (3.752, 4.756, and 1.30, respectively), obtain Gibbs free energy change of their dissociation reactions in water at 298.15 K. Answers: 21.4 ; 26.9 ; 7.4 kJ mol^{-1}.

2. By knowing that the electrostatic force of π-bond is weaker than that of a

σ-bond (see chapter six), predict the sign of the enthalpy change for the reaction of ethene (with one π-bond and five σ-bond) with hydrogen molecule (with one σ-bond) giving ethane (with seven σ-bonds). Predict the sign of the entropy change of this reaction as well.

REFERENCES CITED

Atkins, P.W. 1998. Physical Chemistry. Oxford University Press, Melbourne.

Boltzmann, L. 1877. "Über die Beziehung zwischen dem zweiten Hauptsatz der mechanischen Wärmetheorie und der Wahrscheinlichkeitsrechnung respektive den Sätzen über das Wärmegleichgewicht." English translation: "On the relationship between the second main theorem of mechanical heat theory and the probability calculation with respect to the results about the heat equilibrium". Sitzungsberichte der Kaiserlichen Akademie der Wissenschaften in Wien, Mathematisch-Naturwissenschaftliche Classe LXXVI Abt. II, 76: 373–435.

Castellan, G.W. 1983. Physical Chemistry. Addison Wesley Publishing Company, Inc.

Cramer, C.J. 2004. Essentials of Computational Chemistry—Theories and Models. John Wiley and Sons Ltd., The Atrium.

Engelhardt, L., del Puerto, M.L. and Chonacky, N. 2015. Simple and synergistic ways to understand Boltzmann distribution function. A. J. Phys. 83: 787-793.

Isaacs, N.S. 1995. Physical Organic Chemistry; Longman, Harlow.

Klamt, A., Moya, C. and Palomar, J. 2015. A comprehensive comparison of the IEFPCM and SS(V)PE continuum solvation methods with the COSMO approach. J. Chem. Theory Comput. 11: 4220–4225.

Laidler K.J. and King, M.C. 1983. The development of transition state theory. J. Phys. Chem. 87: 2657-2664.

McDowell, S.A.C. 1999. A simple derivation of the Boltzmann distribution. J. Chem. Ed. 76: 1393-1394.

Tolman, R.C. 1938. The Principles of Statistical Mechanics. Clarendon Press. Oxford.

Tomasi, J., Mennucci, B. and Cammi, R. 2005. Quantum mechanical continuum solvation models. Chem. Rev. 105: 2999–3094.

Chapter Five

Quantum Mechanics and Periodic Table

BRIEF HISTORY OF PERIODIC TABLE

In 1829, Johann W. Döbereiner was the first to group the atomic elements according to their atomic masses and densities. He observed that several elements could be grouped in triads, and each triad (containing three elements) had similar chemical and physical properties. In 1864, John Newlands suggested a different grouping of chemical elements, according to the increasing order of their atomic masses, where he noticed that within these series there were repeated similar properties in elements separated by eight elements (law of octaves). This was the first model which recognized the periodicity of the elements. In 1862, Alexandre-Émile de Chancourtois made the first graphical attempt of grouping the elements in a cylindrical model-so-called telluric helix due to the tellurium atom in the middle of the graph. The second attempt to group elements according to periodic properties was credited to Julius Lothar Meyer in 1862. He grouped the elements according to their valence instead of atomic mass and arranged the elements in horizontal or vertical form. In 1868, Dimitri Mendeleev wrote his textbook, where he classified the elements according to their atomic mass and valence and he could predict new elements. In 1871, Mendeleev presented a new form of his periodic table where similar elements were arranged in the same column, resembling the representative group of modern periodic table for the first rows of this table. Both Mendeleev and Meyer are regarded as the authors of the periodic table, although the highest credit is given to Mendeleev, owing to the prediction capacity of his model.

QUANTUM CHEMISTRY AND ELECTRON CONFIGURATION

In 1923, Bohr constructed a table giving the number of electrons of a specific element in a specific "orbit" (as recognized at that time), which represents another form of grouping the atomic elements (Bohr 1923). Bohr also developed the aufbau principle (or building-up principle), which states the low-lying orbitals are filled before the higher energy ones. Then, Bohr was the first to group the atomic elements according to their electronic configuration based on two quantum numbers n (principal quantum number) and k (subordinate quantum number as Bohr stated). Curiously, in the same year, H.G. Deming organized the extended Mendeleev table with 18 columns, which became very famous at that time.

The next advance in the construction of a table of elements from electronic configuration came from Stoner in 1924, where he used Landé's classification of X-ray levels (which includes a third quantum number) to design his proposed table (Lande 1923, Stoner 1924). Improvement of Stoner's table of elements came from Pauli's works, where he found the fourth quantum number and stated the exclusion principle (see chapter one). Pauli's results on spin and exclusion principle were used to explain the lengths of the periods (2, 8, 18, 32 ...) in the periodic tables of Bohr and Stoner, and represented important contributions to the electronic configuration of atoms and the construction of the modern periodic table. However, the well-defined electronic configuration of an atom is incorrect due to the lack of applicability of adiabatic principle for atoms larger than hydrogen (Ehrenfest 1917, Scerri 1991).

In addition, by using the equation below:

$$\frac{d}{dt} A = \frac{i}{\hbar}[H, A]$$

If operator A commutes with Hamiltonian, then A is time-independent, as it is regarded as hypervirial operator, and its expectation value is time-invariant (known as the hypervirial theorem). However, if the operator A does not commute with H, then A is time-dependent and its expectation value is not time-invariant.

For many-electron system, it can be shown that angular momentum, L^2, and angular momentum projection, L_z, of *individual electrons* do not commute with Hamiltonian. Then, these operators for individual electrons are time-dependent (and hypervirial theorem does not apply to them), where relativity effects should be taken into account. This statement defies Bohr's assumption of stationary state for individual electrons (Bohr 1913, Pilar 1990, Scerri 1991). *As a consequence, giving electrons individual quantum numbers but not taking into account relativity effects (i.e., by using old quantum mechanics or the Schrödinger equation instead of the Dirac equation) is not correct.* From a theoretical perspective, it seems very hard to obtain the total electron configuration, since for its precise calculation post-Hartree-Fock method (configuration interaction or coupled-cluster) and large basis set are needed. Then, the picture of single atomic orbitals in ground state for describing electronic configuration from post-HF methods and large basis set is unreal (Scerri 2004). Although the total electron configuration cannot be obtained precisely from quantum chemistry, experiments from photoelectron spectroscopy (giving the ionization potential) partly supports the Madelung rule (Gillespie et al. 1996).

Further enhancement of Bohr's aufbau principle came from the Madelung rule, where orbitals are filled according to the $n+l$ rule: orbital with lower $n+l$ value is filled first and when two orbitals have same $n+l$ value, the one with lower n is filled first. The Madelung rule is empirical and based on atomic spectra analysis.

General valence configurations:

- Groups 1–2: $n\,s^{1\text{-}2}$
- Groups 13–18: $n\,s^2\,p^{1\text{-}6}$
- Transition metals: $(n-1)\,d^{1\text{-}10}/\,n\,s^{1\text{-}2}$
- Inner transition (f‑block): $(n-2)\,f^{1\text{-}14}/(n-1)\,d^{0\text{-}1}\,n\,s^2$

1	2	3	4	5	6	7	8	9	10	11	12	13	14	15	16	17	18
																H $1s^1$	He $1s^2$
Li $2s^1$	Be $2s^2$											B $2s^2p^1$	C $2s^2p^2$	N $2s^2p^3$	O $2s^2p^4$	F $2s^2p^5$	Ne $2s^2p^6$
Na $3s^1$	Mg $3s^2$											Al $3s^2p^1$	Si $3s^2p^2$	P $3s^2p^3$	S $3s^2p^4$	Cl $3s^2p^5$	Ar $3s^2p^6$
K $4s^1$	Ca $4s^2$	Sc $3d^1 4s^2$	Ti $3d^2 4s^2$	V $3d^3 4s^2$	Cr $3d^5 4s^1$	Mn $3d^5 4s^2$	Fe $3d^6 4s^2$	Co $3d^7 4s^2$	Ni $3d^8 4s^2$	Cu $3d^{10} 4s^1$	Zn $3d^{10} 4s^2$	Ga $4s^2p^1$	Ge $4s^2p^2$	As $4s^2p^3$	Se $4s^2p^4$	Br $4s^2p^5$	Kr $4s^2 4p^6$
Rb $5s^1$	Sr $5s^2$	Y $4d^1 5s^2$	Zr $4d^2 5s^2$	Nb $4d^4 5s^1$	Mo $4d^5 5s^1$	Tc $4d^6 5s^1$	Ru $4d^7 5s^1$	Rh $4d^8 5s^1$	Pd $4d^{10} 5s^0$	Ag $4d^{10} 5s^1$	Cd $4d^{10} 5s^2$	In $5s^2p^1$	Sn $5s^2p^2$	Sb $5s^2p^3$	Te $5s^2p^4$	I $5s^2p^5$	Xe $5s^2p^6$
Cs $6s^1$	Ba $6s^2$	*La – Lu $5d^1 6s^2$	Hf $5d^2 6s^2$	Ta $5d^3 6s^2$	W $5d^4 6s^2$	Re $5d^5 6s^2$	Os $5d^6 6s^2$	Ir $5d^7 6s^2$	Pt $5d^9 6s^1$	Au $5d^{10} 6s^1$	Hg $5d^{10} 6s^2$	Tl $6s^2p^1$	Pb $6s^2p^2$	Bi $6s^2p^3$	Po $6s^2p^4$	At $6s^2p^5$	Rn $6s^2p^6$
Fr $7s^1$	Ra $7s^2$	#Ac – Lr $6d^1 7s^2$	Rf	Db	Sg	Bh	Hs	Mt	Ds	Rg	Cn	Nh	Fl	Mc	Lv	Ts	Og

*Lanthanides

*La	Ce	Pr	Nd	Pm	Sm	Eu	Gd	Tb	Dy	Ho	Er	Tm	Yb	Lu
$4f^0 5d^1 6s^2$	$4f^1 5d^1 6s^2$	$4f^3 5d^0 6s^2$	$4f^4 5d^0 6s^2$	$4f^5 5d^0 6s^2$	$4f^6 5d^0 6s^2$	$4f^7 5d^0 6s^2$	$4f^7 5d^1 6s^2$	$4f^9 5d^0 6s^2$	$4f^{10} 5d^0 6s^2$	$4f^{11} 5d^0 6s^2$	$4f^{12} 5d^0 6s^2$	$4f^{13} 5d^0 6s^2$	$4f^{14} 5d^0 6s^2$	$4f^{14} 5d^1 6s^2$

#Actinides

#Ac	Th	Pa	U	Np	Pu	Am	Cm	Bk	Cf	Es	Fm	Md	No	Lr
$5f^0 6d^1 7s^2$	$5f^0 6d^2 7s^2$	$5f^2 6d^1 7s^2$	$5f^3 6d^1 7s^2$	$5f^4 6d^1 7s^2$	$5f^6 6d^0 7s^2$	$5f^7 6d^0 7s^2$	$5f^7 6d^1 7s^2$	$5f^9 6d^0 7s^2$	$5f^{10} 6d^0 7s^2$	$5f^{11} 6d^0 7s^2$	$5f^{12} 6d^0 7s^2$	$5f^{13} 6d^0 7s^2$	$5f^{14} 6d^0 7s^2$	$5d^1 6s^2$

Figure 5.1 Valence electron configuration of the elements.

The Madelung rule has several exceptions, all of them involving transition metals (Scerri 2013). The exceptions to Madelung's rule are indicated in Fig. 5.1, and are related to Nb (group 5), Cr, Mo (group 6), Tc (group 7), Ru (group 8), Rh (group 9), Cu, Ag, Au (all group 11), Pd, Pt (group 10), plus Pa, U, Np, Gd, Cm from inner transition metals. Scerri also includes No, La, Ce, Ac, and other transition metals as exceptions to the Madelung rule. Then, the Madelung rule works well for representative elements (mainly used in organic chemistry). Theoretical justification of the Madelung rule was given by Wong using the Fermi-Thomas statistical model as an approximation of the quantum atom (Wong 1979).

To sum up, *the electron configurations of chemical elements come from experimental line spectra of monoatomic elements (noble gases) or pure homonuclear substances from atomic emission (or absorption) spectroscopy. The Madelung rule is a theoretical model that predicts experimental results with a reasonable precision, although there are several exceptions.*

ATOMIC RADIUS, NUCLEAR EFFECTIVE CHARGE, AND ELECTRONEGATIVITY

The second most important periodic property after electronic configuration is the atomic radius, which measures the size of the atom. It can be defined from Van der Waals radius, covalent radius, or ionic radius, and the size of neutral atoms varies from 0.3 to 3.0 Å. In the same period of the periodic table, the atomic radius decreases from the left to the right since the number of electrons increase in the same electron shell along with protons in the nucleus. Therefore, the electrostatic force increases, because the negative and positive charge increases without considerable change in the distance between them. As a consequence, atomic radius slightly decreases. On the other hand, within the same group, the atomic radius increases considerably from top to bottom since the electron shell increases, yielding a considerable increase in atomic radius (see Fig. 5.2).

$$F_{elect} = K \frac{e_{ext} Z_{eff}}{r^2}$$

Where e_{ext} is the electron outside the atom which is being attracted by the nucleus of the reference atom with nuclear effective charge Z_{eff}, and r is the distance between the nucleus and the external electron.

The atomic radius directly influences another very important periodic property: the electronegativity (or its inverse, the electropositivity). The electronegativity is the tendency of a neutral atom or ion to attract electrons or charge density which depends inversely on its atomic radius. Smaller atoms tend to attract electrons more easily and bigger atoms tend to be less electronegative. The attraction between a reference atom and external electron or charge density depends solely on the electrostatic force involving protons in the reference atom and the charge of the external electrons. Supposing the latter being constant, then electronegativity of each element can be compared.

1	2	3	4	5	6	7	8	9	10	11	12	13	14	15	16	17
Li 1.23	Be 0.89											B 0.80	C 0.77	N 0.74	O 0.74	F 0.72
Na 1.57	Mg 1.36											Al 1.25	Si 1.17	P 1.10	S 1.04	Cl 0.99
K 2.03	Ca 1.74	Sc 1.44	Ti 1.32	V 1.22	Cr 1.17	Mn 1.17	Fe 1.17	Co 1.16	Ni 1.15	Cu 1.17	Zn 1.25	Ga 1.25	Ge 1.22	As 1.21	Se 1.14	Br 1.14
Rb 2.16	Sr 1.91	Y 1.62	Zr 1.45	Nb 1.34	Mo 1.29	Tc ----	Ru 1.24	Rh 1.25	Pd 1.28	Ag 1.34	Cd 1.41	In 1.50	Sn 1.40	Sb 1.41	Te 1.37	I 1.33
Cs 2.35	Ba 1.98	Lu 1.69	Hf 1.44	Ta 1.34	W 1.30	Re 1.28	Os 1.26	Ir 1.26	Pt 1.29	Au 1.34	Hg 1.44	Tl 1.55	Pb 1.46	Bi 1.52		

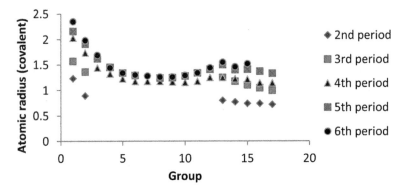

Figure 5.2   Covalent atomic radius in the periodic table and according to each period.

The protons in the nucleus of the reference atom are somewhat interacting with their own electrons. Then, their influence on external electrons must be discounted from the proton-internal electrons interaction, the so-called screening constant, σ. There are different ways to calculate the screening constant. The most appropriate and accepted is the one from Clementi and Raimondi's work (Clementi and Raimondi 1963), where σ is obtained quantum mechanically. The nuclear effective charge, Z_{eff} is:

$$Z_{eff} = Z - \sigma$$

Where Z is the atomic number or the total nucleus charge.

There is a Z_{eff} and σ for each subshell of each atom. As we approach the nucleus, Z_{eff} tends to Z. As we go further away from the nucleus, Z_{eff} decreases because σ increases. Table 5.1 shows Z_{eff} according to σ obtained quantum mechanically of the last subshell of the first representative elements. We can see that Z_{eff} increases from left to right in each column. The same trend occurs with the electronegativity. Then, for each row of the periodic table, one can see that electronegativity and Z_{eff} are linearly related (see Fig. 5.3), with the exception of noble gases. In this case, both negative and positive charges of the electrostatic formula are the main factors to account for the trend of electronegativity within each period (row).

Table 5.1 Z_{eff} of the last subshell.

Atom	Li	Be	B	C	N	O	F	Ne
Subshell	2s	2s	2p	2p	2p	2p	2p	2p
Z_{eff}	1.28	0.95	2.42	3.13	3.83	4.45	5.1	5.75

Atom	Na	Mg	Al	Si	P	S	Cl	Ar
Subshell	3s	3s	3p	3p	3p	3p	3p	3p
Z_{eff}	2.51	3.31	4.06	4.28	4.89	5.48	6.11	6.76

1	2	3	4	5	6	7	8	9	10	11	12	13	14	15	16	17
H 2,1																H 2,1
Li 1,0	Be 1,5											B 2,0	C 2,5	N 3,1	O 3,5	F 4,1
Na 1,0	Mg 1,2											Al 1,5	Si 1,7	P 2,1	S 2,4	Cl 2,8
K 0,9	Ca 1,0	Sc 1,2	Ti 1,3	V 1,5	Cr 1,6	Mn 1,6	Fe 1,6	Co 1,7	Ni 1,8	Cu 1,8	Zn 1,7	Ga 1,8	Ge 2,0	As 2,2	Se 2,5	Br 2,7
Rb 0,9	Sr 1,0	Y 1,1	Zr 1,2	Nb 1,2	Mo 1,3	Tc 1,4	Ru 1,4	Rh 1,4	Pd 1,4	Ag 1,4	Cd 1,5	In 1,5	Sn 1,7	Sb 1,8	Te 2,0	I 2,2
Cs 0,9	Ba 1,0	Lu 1,1	Hf 1,2	Ta 1,3	W 1,4	Re 1,5	Os 1,5	Ir 1,6	Pt 1,4	Au 1,4	Hg 1,4	Tl 1,4	Pb 1,5	Bi 1,7	Po 1,8	At 1,9
Fr 0,9	Ra 1,0															

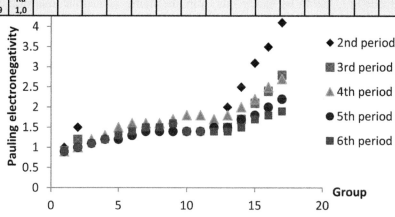

Figure 5.3 Pauling electronegativity in the periodic table and according to each period.

On the other hand, Z_{eff} increases in each group as we go downwards. The opposite trend occurs with the electronegativity. Then, for each group, the most important factor is the atomic radius, which decreases the electrostatic interaction between the nucleus of the reference atom and the external electron in a second power decrease.

To sum up, Z and Z_{eff} are directly related. As the Z increases, Z_{eff} increases as well and vice-versa. Important to add that Z_{eff} is always smaller than Z for the same atom.

Electronegativity is a very important property that must always be recalled by the students in order to explain a series of physical and chemical properties. For example, it is used to explain the difference of bond strength between C=C and C=O bonds. One sees that the latter is stronger because oxygen atom is more electronegative and decreases the bond length, which decreases the r_{Cl}. Moreover, the atomic radius of oxygen is smaller than that from carbon and Z_{eff} of oxygen is higher than that from carbon. All these factors contribute for a higher electrostatic force in C=O bond. The electronegativity is also important to explain" the difference of the strength of the hydrogen bond between alcohol molecules and amine molecules, or the difference of polarizability between oxygen atom and sulfur atom, or the higher stability of a specific resonance structure, etc. *There are several phenomena in chemistry that can be reasoned by means of electronegativity. Keep this in mind.*

ATOMIC RADIUS AND BOND LENGTH

From the atomic radius of the bonding atoms, one can have a good qualitative (or even good or reasonable quantitative) idea of the corresponding bond length, L, by simply adding both atomic radii which form the chemical bond (summation of atomic radii). For example, take for instance silicon, carbon, and hydrogen atoms. Silicon atom is the largest atom and hydrogen the smallest. When forming C-H, Si-H, and Si-C bonds, we can predict (from the atomic radius) the following trend of bond length: $L_{Si-C} > L_{Si-H} > L_{C-H}$ (Fig. 5.4).

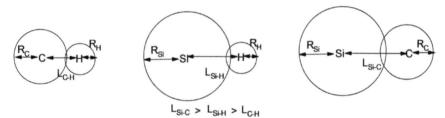

$L_{Si-C} > L_{Si-H} > L_{C-H}$

Figure 5.4   Pictorial representation C-H, Si-H, and Si-C bonds from their corresponding atomic volume.

For these selected examples, the summation of bonding atomic radii and experimental bond length gave very close values, except for Si-C bond (Table 5.2). However, the trend of bond length is the same for both cases (summation of atomic radii and experimental bond length).

Table 5.2   Summation of atomic radii and experimental bond, in Angströms, from selected bond types.

Bond type	Summation of atomic radii/Å	Experimental bond length/Å
C-H	1.08	1.09
Si-H	1.48	1.48
Si-C	1.94	1.85

(a) Hydrogen covalent atomic radius: 0.31 Å
(b) Carbon (sp^3) covalent atomic radius: 0.77 Å
(c) Silicon covalent atomic radius: 1.17 Å

EXERCISE

1. Explain the bond strength trend: HF > HCl > HBr > HI, based on the electrostatic force model.

REFERENCES CITED

Bohr, N. 1913. On the constitution of atoms and molecules. Phil. Mag. 26: 476-502

Bohr, N. 1923. The structure of the atom. Nature 112: 29-44.

Clementi, E. and Raimondi, D.L. 1963. Atomic screening constants from SCF functions. J. Chem. Phys. 38: 2686-2689.

Ehrenfest, P. 1917. XLVIII. Adiabatic invariant and the theory of quanta. Phil. Mag. 33: 500-513.

Gillespie, R.J., Spencer, J.N. and Moog, R.S. 1996. Demystifying introductory chemistry, Part 1: Electron configurations from experiment. J. Chem. Ed. 73: 617-622.

Landé, A. 1923. Termstruktur un zeemaneffekt der multipletts. Z. Phys. 15: 112-123.

Pilar, F.L. 1990. Elementary Quantum Chemistry. McGraw Hill Publishing Co., Singapore.

Scerri, E.R. 1991. The electronic configuration model, quantum mechanics and reduction. The British J. Phil. Sci. 42: 309-325.

Scerri, E.R. 2004. Just how ab initio is ab initio quantum chemistry? Found. Chem. 6: 93-116

Scerri, E.R. 2013. The trouble with the aufbau principle. Ed. Chem. 50: 24-26.

Stoner, E.C. 1924. LXXIII. The distribution of electrons among atomic levels. Phil. Mag. 48: 719-736.

Wong, D.P. 1979. Theoretical justification of Madelung's rule. J. Chem. Ed. 6: 714-718.

Chapter Six

Quantum Mechanical Resonance, Chemical Bond, and Hybridization

BOHR'S ATOM AND LEWIS'S IDEAS ON VALENCE AND CHEMICAL COMBINATIONS

In 1913, two very important papers were published for chemistry and physics: "On the constitution of atoms and molecules" (Bohr 1913) and "Valence and tautomerism" (Lewis 1913). The former describes the electron in a stationary state in fixed circular or elliptical orbits. For example, for hydrogen, the electron around the nucleus with angular momentum of entire multiples of $(h/2\pi)$ and frequencies:

$$v = \frac{2\pi^2 e^4 m}{h^2}\left(\frac{1}{\tau_2^2} - \frac{1}{\tau_1^2}\right)$$

Which has succeeded to explain the hydrogen spectrum and given birth to the first model of the quantum atom.

On the other hand, Lewis introduced the idea of valence as "number of positions, or regions, or points (bond-termini) on the atom at which attachment to corresponding points on other atoms occurs" (Lewis 1913). Lewis defines the polar and non-polar combinations, KCl and CH_4, respectively, where he ascribed to methane a fixed arrangement of the atoms in the molecule, and to potassium chloride a higher degree of freedom. Lewis states an important rule for the modern resonance (as discussed later): "In the strictly non-polar type of compounds the phenomenon of tautomerism (i.e., modern resonance) between two forms (two resonance structures), without the intervention of a polar form, is probably unknown". He also constructed his model of structural organic molecules and chemical bond. He cited Bray and Branch's work with arrows to indicate polar covalent bond or ionic bonds (from the electropositive atom towards the electronegative atom), and upwards and downwards arrows to symbolize apolar covalent bond:

$$K \rightarrow Cl$$

$$H \rightarrow O \uparrow\downarrow O \leftarrow H$$

Lewis suggests changing arrows into straight lines to represent (polar and non-polar covalent bonds):

H-O-O-H

To our knowledge, Lewis was the first to classify homopolar covalent, ionic/heteropolar covalent, and metallic bonds. He stated: "To the polar and non-polar types of chemical compound we may add a third, the metallic. In the first type, the electrons occupy fixed positions within the atom. In the second type, the electrons move freely from atom to atom within the molecule. In the third or metallic type, the electron is free to move even outside the molecule". These concepts were not based on quantum mechanics still under development at that time, but on the experimental electric and electronic properties (mobility, dielectric constant, ionization, reactivity, etc.) of the studied compounds.

LEWIS'S CUBIC MODEL, OCTET RULE, ELECTRON PAIR, AND STRUCTURAL CHEMISTRY

In Lewis's seminal paper in 1916, he establishes the basis of structural organic chemistry. After reviewing non-polar and polar molecules from his 1913 work, he introduces the *cubical atom model*, where each vertex is filled with one valence electron represented by a circle around the vertex. Then, lithium has one circle in one vertex (one electron in valence shell), boron has three (three electron in valence shell), oxygen has six circles in six vertexes (six electrons in valence shell), and neon has all vertexes filled with circles (eight electrons in valence shell). It is worth mentioning that a cube has eight vertexes which can represent the number of valence shell of representative elements (Lewis 1916). He used this model to represent an ionic compound where the cation has a cube without valence electrons and the anion has eight electron in the cube, which leads to the *octet rule* (first elaborated by Mendeleev). He also used the cubic atom model to represent covalent molecules. For example, the iodine molecule formed from two iodine atoms (each one with seven valence electrons represented by seven circles in the vertexes of the cube), where one covalent bond is represented by the junction of two vertexes of the two iodine cubic atoms, and now each iodine cube has eight electrons, again following the octet rule (Fig. 6.1(A)). *The cubic atomic model showed the importance of the octet rule.* But, alternatively to the cubic model, he developed another model to represent molecules which became everlasting by replacing the cubes with the atomic symbols and circles in the vertexes by dots to represent the valence electrons (Fig. 6.1(B)). *This representation led to another important conclusion: each covalent bond is made of an electron pair.* Lewis could also use his model to represent a chemical bond, e.g., ammonia reacting with a proton to form an ammonium ion (Lewis 1916). Lewis's model was also able to describe double and triple covalent bonds (Fig. 6.1(C)), for example, in ethene and ethyne, although in most cases it failed to predict the correct geometry.

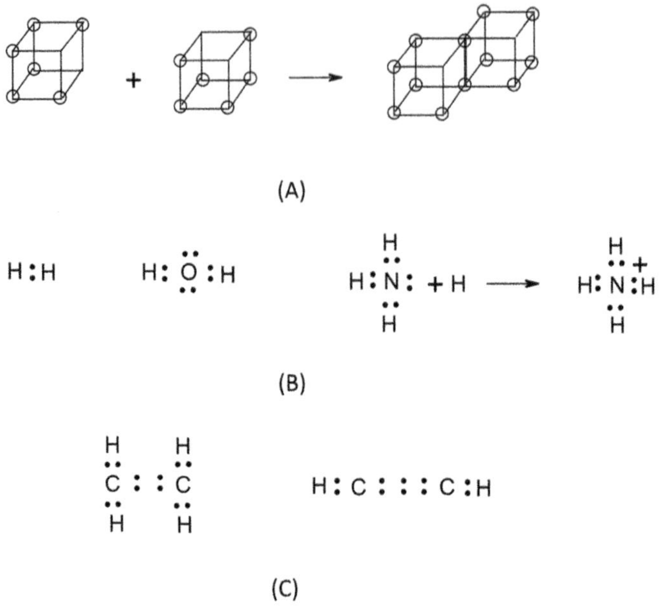

(A)

H:H   H: O :H   H:N: +H ⟶ H: N:H

(B)

(C)

Figure 6.1 (A) Lewis cubic model; Lewis model for representation of molecules with (B) single bonds, and (C) double and triple bonds.

QUANTUM MECHANICAL RESONANCE AND CHEMICAL BOND

Resonance in chemistry (or quantum chemistry) began when Heisenberg incorporated Bose-Einstein statistics and Pauli's exclusion principle in his Quantum Matrix (which appeared in Schrödinger Wave Mechanics) using the many-body problem for this purpose. Then, he used two particle coupled oscillators in the Hamilton basis leading to the "principle of resonance frequency" (Heisenberg 1926). As Heisenberg usually highlighted the similarities among classical physics, old quantum theory and his Matrix Mechanics, he stated: "In classical mechanics one finds that two periodic oscillating systems can come into resonance if the frequencies of the separate systems are independent of the energy and are approximately equal.(...) In QM (Matrix Mechanics), in accord with general laboratory observations, two atomic systems resonate when the absorption frequency of one system coincides with the emission frequency of the other" (Heisenberg 1926). Then, from the interaction between these two oscillators (which execute the same motion in different phases), an additional energy W^1 of the perturbed systems is produced. After the development of matrix diagonalization and solution of linear equations, he found the solution for the interaction between n and m oscillators in terms of W^1_{nm} and W^1_{mn}. Afterwards, he incorporated the resonance phenomenon in Schrödinger quantum mechanics using normalized wave functions φ^a_n and φ^b_n in his Matrix Mechanics. Then, the equivalent solutions were:

$$W_{nm}^1 = \frac{1}{\sqrt{2}\left(\varphi_n^a \varphi_m^b + \varphi_m^a \varphi_n^b\right)}$$

$$W_{mn}^1 = \frac{1}{\sqrt{2}\left(\varphi_n^a \varphi_m^b - \varphi_m^a \varphi_n^b\right)}$$

These equations were the basis for the formulation of the chemical bond from Heitler and London. They used Heisenberg's reasoning, changing two interacting oscillators into two electrons in a bonding region between two hydrogen nuclei, yielding the two guess wave functions for H_2: $\psi = \phi_A(1)\phi_B(2)$ and $\psi = \phi_A(2)\phi_B(1)$, where numbers are electrons and letters are the hydrogen nuclei (Heitler and London 1927). When considering the indistinguishability of identical particles and doing normalization and orthogonalization processes of these wave functions, two final wave functions, α and β, were obtained:

$$\alpha = \frac{1}{\sqrt{2+2S}}\left[\phi_A(1)\phi_B(2) + \phi_A(2)\phi_B(1)\right]$$

$$\beta = \frac{1}{\sqrt{2-2S}}\left[\phi_A(1)\phi_B(2) - \phi_A(2)\phi_B(1)\right]$$

Where S is the overlap integral, α is the H_2 the symmetric repulsive wave function, and β the H_2 antisymmetric attractive wave function. These equations are equivalent to the solution for the interaction between n and m oscillators (W_{nm}^1 and W_{mn}^1) in terms of wave mechanics. *Then, resonance is the basis of the formation of both H_2 wave functions.*

They also found the energies of both wave functions, E_α and E_β as a function of the interatomic distance (or reaction coordinate), R, where the energy components take into account the resonance phenomenon, giving the plot in Fig. 6.2. At the equilibrium geometry (R_e) of antisymmetric hydrogen, there lies the minimum energy of singlet system. At the dissociation limit, the energy of both wave functions converges to the sum of the atomic energies of two isolated hydrogen atoms. *The energy difference between the minimum at equilibrium geometry and the sum of energy of the isolated atoms at the dissociation limit is the bond dissociation energy, BDE, which is associated, in great part, with the resonance.*

Pauling pointed out that the bond energy for the H_2 antisymmetrical wave function from Heitler and London's work was 67% from the experimental value of 102.6 kcal mol^{-1}, but the R_e was very close to the experimental value of 0.74 Å (Pauling 1960). By applying a slightly different mathematical approach where the value of Z was changed in order to find the minimum energy value, Wang's H_2 wave function gave a more precise value (80% with respect to the experimental value) of the hydrogen chemical bond (Wang 1928). *Therefore, resonance cannot be considered the unique factor associated to the chemical bond, but certainly is the main reason for the stable molecule formation.* According to Pauling, the other 15% of the hydrogen molecule chemical bond is derived from what he called deformation (i.e., expansion of the wave function). James and Coolidge have introduced a more complicated variation function (one method to

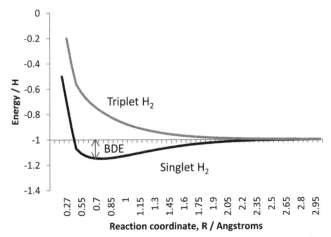

Figure 6.2   Plot of energy versus atomic distance (reaction coordinate)
of singlet and triplet H_2.

solve many-body systems), including the elliptical coordinates r_{A1}, r_{A2}, r_{B1}, r_{B2} (distance between hydrogen nucleus, A or B, and electron, 1 or 2), r_{AB} (internuclear distance), and the new (at that time in 1933) r_{12} (the interelectronic distance), where the variation function with a higher number of terms (or more expanded) led to results very close to experimental values (Pauling and Wilson 1935, James and Coolidge 1933). In modern quantum chemistry, similar (or better yet) results can be obtained from a complete basis set (CBS) and full-configuration interaction (Full-CI). Then, *resonance effect on non-expanded wave function accounted for 85% of the hydrogen chemical bond (including 5% of ionic resonance structures) and the remaining 15% arose from the expansion of the hydrogen wave function.*

Both authors also studied the diatomic He system and found that no chemical bond could be done for this case (Heitler and London 1927, Esposito and Naddeo 2014).

In terms of chemical structure, the α wave function of (stable) singlet hydrogen molecule can be understood as:

$$\alpha \equiv H \uparrow\downarrow H \leftrightarrow H \downarrow\uparrow H$$

Where we removed the subscripts A and B for the hydrogen atoms (being replaced by left hydrogen and right hydrogen), and changed the numbers 1 and 2 for electrons into spin "up" and spin "down". Both are just a matter of convention. The important thing is to notice the mathematical result of α wave function in terms of chemical structure, that is, *the stable molecule is formed by the interchange of two resonance (or canonical) structures where the electrons interchange their spins and the real molecule is neither of them, but the result of this composition represented by the double arrow (↔), which is the graphical representation of the resonance. And, as a consequence of the resonance between two resonance structures, the real molecule (α), the so-called resonance hybrid, has a lower energy in comparison with the isolated atoms and this is the chemical bond energy, which also can be called the resonance energy.*

London went beyond, by saying that: "The forms of operation of the homopolar valence forces can be modeled on the symmetry properties of the Schrödinger eigenfunctions of the atoms of the periodic system and can be interpreted as quantum mechanical resonance effects" (London 1928). In this work he calculated the optimized bond length of several diatomic molecules with very good precision and also calculated the contour plots of the electron density of singlet hydrogen molecule and diatomic helium, showing the distinguished nature of a bonded system and non-bonded system using the equation for electron density:

$$\int \left| \psi\left(r_1, r_2\right) \right|^2 d\tau_3$$

In Fig. 6.3, we show our calculated electron density (from CCSD/6-311G level of theory) of optimized geometries of singlet hydrogen molecule (A), diatomic helium (B), and triplet hydrogen molecule (C). Figures 6.3(A) and (B) are very similar to that obtained by London (London 1928). Then, London was the first to give a clear picture (an image) of a chemical bond (Fig. 6.2(A)) and also of van der Waals interaction (Fig. 6.3(B)). Here we have included the triplet hydrogen (Fig. 6.3(C)), which clearly shows the effect of Pauli's exclusion principle when two electron have the same spin. In Fig. 6.3, the optimized internuclear distance in Angströms is also depicted.

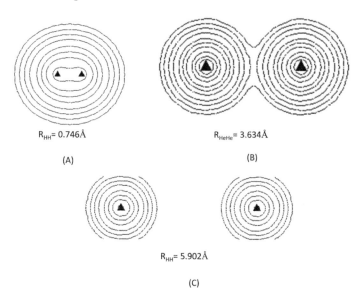

$R_{HH}= 0.746\text{Å}$

(A)

$R_{HeHe}= 3.634\text{Å}$

(B)

$R_{HH}= 5.902\text{Å}$

(C)

Figure 6.3   Calculated isodensity curves of the electron density of (A) singlet hydrogen molecule, (B) diatomic helium, (C) triplet hydrogen molecule.

QTAIM CONCEPT OF CHEMICAL BOND

The quantum theory of atoms in molecules, QTAIM, analyzes the gradient of the charge density (besides mainly the Laplacian of the charge density and the potential energy density) of a wave function from an expected optimized geometry obtained from an expected high level of theory in order to provide reliable observables that can be comparable to experimental ones. For more details, the reader should go (back) to chapter two. In brief, QTAIM is an excellent tool to provide information of a chemical bond, its type (covalent, ionic, metallic, and even van der Waals interactions), and its strength in terms of charge density. From the density matrix generated from an optimized wave function, there is an algorithm to obtain critical points (the same we learn from Calculus courses) from gradient of the charge density. Where the gradient is zero, there is a critical point. The most important is the bond critical point, BCP, where its topological information can tell us the type and strength of the chemical bond. Please see Table 2.2 in chapter two. Here we provide the molecular graph (set of critical points of optimized geometry united by bond paths where there is maximum density in comparison with its transverse neighborhood) of singlet hydrogen (A), diatomic helium (B), and triplet hydrogen (C) in Fig. 6.4 along with their corresponding value of charge density of the BCP, ρ_b. In singlet hydrogen, there is a covalent bond and it has the highest value of ρ_b. In diatomic helium, there is a weak van der Waals interaction and the lowest value of ρ_b. And in the triplet hydrogen there is no chemical bond or van der Waals interaction, and as a consequence, there is no ρ_b.

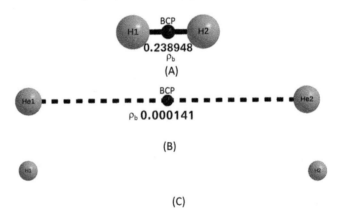

Figure 6.4   Molecular graph of (A) singlet hydrogen molecule, (B) diatomic helium, (C) triplet hydrogen molecule.

ELECTROSTATIC FORCE AND COVALENT BOND

Before Bader (as shown in chapter three), Pauling also used the Hellmann-Feynman theorem to interpret the chemical bond in terms of the electrostatic force, although in a more basic manner. He stated: "This theorem states that the force acting on each nucleus in a molecule is exactly that calculated by the principles of classical electrostatic theory from the charges and positions of the other nuclei and of the electrons; the electrons are taken to have the spatial distribution given by the square of the electronic wave function. At the equilibrium configuration of a molecule the resultant force acting on each nucleus vanishes; hence for this configuration the repulsion of a nucleus by the other nuclei is just balanced by its attraction by the electron" (Pauling 1960).

It is important to highlight that all chemical bonds (covalent bond, ionic bond, coordination bond, metallic bond) and all inter/intramolecular interactions have the same electrostatic nature because the unique operating force for uniting atoms and molecules is the attraction between electrons and protons from the fundamental electromagnetic force. Another physical factor derived from quantum mechanics observed by Fock is the exchange operator in the Schrödinger equation (Fock 1930). *However, in a qualitative fashion, it suffices to use the simple electrostatic equation to compare different types of bonds and even different types of intermolecular interactions.*

In organic chemistry, we usually deal with single, double, and triple bonds, mainly from carbon, nitrogen, and oxygen atoms. We will use the electrostatic equation (see chapter three) to reason the following increasing bond strength: single CC bond < double CC bond < triple CC bond. For these electrostatic equations we consider that: (1) by using Lewis model, there are two bonding electrons, four bonding electrons, and six bonding electrons in single, double, and triple CC bonds, respectively; (2) the effective nuclear charge of the carbon atom, $Z_{eff(C)}$, does not alter significantly from hydrocarbons with single, double, and triple CC bonds; and (3) the distance between carbon nucleus and electron, r_{Cl}, decreases as the bond length decreases from single to triple bond, that is, $r_{Cs1} > r_{Cd1} > r_{Ct1}$, where C_s, C_d, and C_t represent carbon atom from single bond, double bond, and triple bond, respectively. In single CC bond, two electrons interacting with two carbon atoms give the four-fold the electrostatic force of one σ electron interacting with an effective nuclear charge of the carbon atom. In double CC bond, two σ electrons interacting with two carbon atoms give the four-fold σ electrostatic force of one electron interacting with an effective nuclear charge of the carbon atom and two π electrons interacting with two carbon atoms give four-fold π electrostatic force, and so on. Then, we have the equations below, where 1_σ and 1_π represent electron σ and π, respectively.

$$F_{single\ CC} = 4\left(K\frac{eZ_{eff(C)}}{r_{C_s 1\sigma}^2} \right)$$

$$F_{double\ CC} = 4\left(K\frac{eZ_{eff(C)}}{r_{C_d 1\sigma}^2} \right) + 4\left(K\frac{eZ_{eff(C)}}{r_{C_d 1\pi}^2} \right)$$

$$F_{triple\ CC} = 4\left(K\frac{eZ_{eff(C)}}{r_{C_t 1\sigma}^2} \right) + 8\left(K\frac{eZ_{eff(C)}}{r_{C_t 1\pi}^2} \right)$$

$$r_{C1\sigma}^2 < r_{C1\pi}^2$$

Then
$$F_\sigma > F_\pi$$

$$r_{C_t 1\sigma}^2 < r_{C_d 1\sigma}^2 < r_{C_s 1\sigma}^2$$

Moreover
$$F_{triple\ CC} > F_{double\ CC} > F_{single\ CC}$$

INVERSE RELATION BETWEEN POTENTIAL ENERGY AND FORCE AND ITS IMPORTANCE FOR CHEMISTRY

Taking for granted the law of conservation of energy at a closed system:

$$\Delta K + \Delta U = 0$$

Where ΔU is the potential energy change and ΔK is the kinetic energy change. This law states that no energy is generated or destroyed, only transformed.

By knowing the relation between work and kinetic energy:

$$W = \Delta K$$

And replacing it in the previous equation, one gets the following relation:

$$\Delta U = -W = -\int_{r_0}^{r} F dr$$

Then, the force equation in terms of energy is given by the formula:

$$F = -\frac{dU}{dr}$$

As a consequence, there is an inverse relation between force and potential energy. One simple example is the gravitational force and its corresponding potential energy: regarding one body with m mass attracted to the gravitational force of Earth with M mass, the higher the body, the smaller its gravitational force and the higher its potential energy. This inverse relation between potential energy and force is very important for chemistry as well, since the total energy of a molecule is its potential energy regarding Bohr-Oppenheimer approximation and the stationary state. *The total energy of a molecule can be predicted indirectly (mainly for comparison reasons in a qualitative fashion) by the use of electrostatic force of the corresponding chemical bonds and/or inter/intramolecular interactions in the analyzed molecule.*

Then, for example, the electrostatic force can be used to predict or to rationalize the following energy trend: σ bond (σ) < π bond (π) < lone pair electrons (n), which can be experimentally observed from ultra-violet spectroscopy or photoelectron spectroscopy (Turro 1991). Let's take formaldehyde for example. The energy trend observed in photoelectron spectroscopy are: $\sigma(CO) < \pi(CO) < n(O)$. The corresponding electrostatic equations are:

$$F_{\sigma(CO)} = 2\left(K \frac{eZ_{eff(C)}}{r_{C1\sigma}^2} \right) + 2\left(K \frac{eZ_{eff(O)}}{r_{O1\sigma}^2} \right)$$

$$F_{\pi(CO)} = 2\left(K \frac{eZ_{eff(C)}}{r_{C1\pi}^2} \right) + 2\left(K \frac{eZ_{eff(O)}}{r_{O1\pi}^2} \right)$$

$$F_{n(O)} = 2\left(K \frac{eZ_{eff(O)}}{r_{O1n}^2} \right)$$

$$r_{C1\sigma} < r_{C1\pi} \therefore r_{O1\sigma} < r_{O1\pi}$$

Where 1σ and 1π refer to σ and π electrons, respectively,. The only significant difference between $F_{\sigma(CO)}$ and $F_{\pi(CO)}$ is that $r_{C1\sigma} < r_{C1\pi}$ and $r_{O1\sigma} < r_{O1\pi}$ because the σ electrons are closer to the nuclei than the π electrons. Figure 6.5 shows the CO π orbital (A) and the CO σ orbital (B) of formaldehyde using ADNDP-MO method (see chapter two), whose results are similar to GVB for localized orbitals. One can infer that CO π electrons are, in average, more distant from the nuclei than CO σ electrons. As for the n electrons (lone pair electrons) from oxygen atom, they are attracted by only one nucleus, while the pair of σ and π bond electrons are attracted by two nuclei. Then $F_{\sigma(CO)} > F_{\pi(CO)} > F_{n(O)}$. Since there is the inverse relation between potential energy and force, the energy trend of $\sigma(CO)$, $\pi(CO)$, and $n(O)$ can be predicted from the electrostatic equation and this trend is the same as that from experimental results.

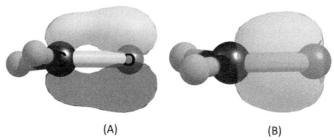

(A)   (B)

Figure 6.5   (A) CO π orbital and (B) CO σ orbital of formaldehyde using ADNDP-MO method.

As we have already highlighted, *the difference between the molecular energy at the minimum of the potential energy surface, PES, and the sum of the isolated atoms after dissociation gives the bond energy which is also called bond dissociation energy, BDE.* Table 6.1 gives experimental BDE in kcal mol^{-1} in terms of enthalpy of some bonds obtained at 0 K after homolysis (homolytic dissociation) of the corresponding chemical bond. *Experimentally, it is not possible to measure the bond strength directly. Then, BDE is one alternative way to measure the bond strength in an indirect manner, as it has a direct relation with bond strength.*

One can see that double bond is not a two-fold of one single bond and triple bond is not a three-fold of one single bond, owing to the fact that π bond is weaker than σ bond (as previously discussed). One can also see that BDE of C-halogen bond decreases as we go down the halogen group in the periodic table. The bond length increases in the order: C-F<C-Cl<C-Br, since the atomic radius of the halogen atoms have the same order: F<Cl<Br. Once again, we can use electrostatic force to explain the trend in their BDEs: C-F>C-Cl>C-Br. As the atomic radius increases in the halogen atoms, the r_{A1} also increases to the second power, where A is the halogen atom and 1 is one of the electrons of the C-halogen bond, and then the electrostatic force decreases, which gives the following bond strength: C-F<C-Cl<C-Br. Important to emphasize that both Z_{eff} and atomic radius increase in the same order (F<Cl<Br) in the electrostatic force equation, but this increase in Z_{eff} is smaller than that from atomic radius because the latter increases to the second power.

Table 6.1 Bond dissociation energy, in kcal mol^{-1}, of selected bonds.

Bond	BDE (kcal mol^{-1})
C-F	117
C-Cl	79
C-Br	69
C-C	83-85
C=C	146-151
C≡C	200

MULTIPLICITY

According to IUPAC Gold Book, multiplicity (M) is the number of possible orientations (according to the equation below) of the spin angular momentum corresponding to a given total spin quantum number (S). It indicates whether the substance is a radical or not, and gives the amount of unpaired electron-spins. The total spin quantum number is a sum of all one electron spin functions $\alpha(\xi)$ and $\beta(\xi)$ of an atom or molecule. The core electron and lone pair electrons are always spin-paired and their partial S is zero. Only unpaired electrons count for S.

$$M = 2S + 1$$
$$Singlet : 2(0) + 1 = 1$$
$$Doublet : 2(1/2) + 1 = 2$$
$$Triplet : 2(1) + 1 = 3$$

In a singlet molecule/atom, all electrons are spin-paired, which is more stable than other multiplicities (doublet, triplet, so on) in most cases. Doublet atom or molecule is a radical with one unpaired electron. Triplet atom or molecule is a diradical with two unpaired electrons.

HYBRIDIZATION

The concept of hybridization was simultaneously created by Pauling and Slater (Slater 1931, Pauling 1931). In Pauling's own words: "The type of bond formed by an atom is dependent on the ratio of bond energy to energy of penetration of the core (*s-p* separation). When this ratio is small, the bond eigenfunctions are *p* eigenfunctions, giving rise to bonds at right angles to one another; but when it is large, new eigenfunctions especially adapted to bond formation can be constructed" (Pauling 1931). When Pauling says "When this ratio is small, the bond eigenfunctions are *p* eigenfunctions, giving rise to bonds at right angles to one another", he is mentioning the formation of π-bonds from "pure atomic *p* orbitals". where no hybridization process occurs, and when he says "but when it is large, new eigenfunctions especially adapted to bond formation can be constructed", he is talking about "mixed atomic orbitals" to form σ bonds.

Recall that electron configurations of chemical elements come from atomic emission (or absorption) spectroscopy of monoatomic elements or pure homonuclear substances, where it is possible to obtain the energy level diagram of a given element. See, for example, the lithium energy level diagram in comparison with the hydrogen level (Fig. 6.6).

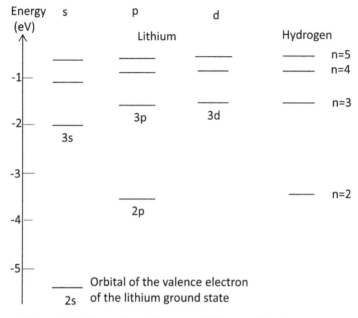

Figure 6.6 Energy level diagram of lithium and hydrogen atoms.

The energy levels of each orbital (following aufbau principle of Madelung rule) give the lithium electron configuration of the ground state and even of its excited states. The same procedure is done for each element. Then, we obtain the fundamental configuration of the chemical elements. As for the carbon atom, its fundamental electron configuration is: $1s^2\, 2s^2\, 2p^2$, which corresponds to the triplet ^3P state. However, from the carbon atomic orbitals of its fundamental configuration, it is not possible to construct any of its compounds, since the geometric parameters (bond angle and bond lengths) of these molecules do not fit those from hypothetical equivalent molecule constructed from carbon fundamental atomic orbitals. At this point, Pauling gave an important commentary (Pauling 1960): "There are four orbitals in the valence shell of the carbon atom. (…) These are, however, not orbitals used directly in bond formation by the atom. (They are specially suited to the description of the free carbon atom; if quantum theory had been developed by the chemist rather than the spectrocopist, it is probable that the tetrahedral orbitals described below would play a fundamental role in the theory (…) We expect this hybridization to take place in order that the bond energy may be a maximum". When Pauling refers to the tetrahedral orbital, he is mentioning the sp^3 hybrid orbitals.

Pauling developed the equations for the hybridized orbitals (Pauling 1931). For example, the four hybridized orbitals (ψ_1 to ψ_4) which describes the sp^3 hybridization are:

$$\psi_1 = \frac{1}{2}s + \frac{\sqrt{3}}{2}p_x$$

$$\psi_2 = \frac{1}{2}s - \frac{1}{2\sqrt{3}}p_x + \frac{\sqrt{2}}{\sqrt{3}}p_z$$

$$\psi_3 = \frac{1}{2}s - \frac{1}{2\sqrt{3}}p_x - \frac{1}{\sqrt{6}}p_z + \frac{1}{\sqrt{2}}p_y$$

$$\psi_4 = \frac{1}{2}s - \frac{1}{2\sqrt{3}}p_x - \frac{1}{\sqrt{6}}p_z - \frac{1}{\sqrt{2}}p_y$$

These expressions give rise to the shape of sp^3 orbitals, each occupying one corner of a tetrahedron which corresponds to the geometry of alkanes for example. *All of them have the same shape (only distinguished orientations) and same energy (they are degenerate).*

Then, hybridization is a quantum mechanical approach (rationalized by Pauling and Slater) based on classical VB theory to construct new atomic orbitals (hybrid orbitals) from linear combination of the fundamental atomic orbitals so that these new orbitals (ψ_1 to ψ_4) could be used to form covalent bonds with other atoms yielding valence bonding orbitals. It is important to note that the original hybridization method (from Classical VB) uses only the minimum basis set (fundamental atomic orbitals). However, Modern Valence Bond methods can provide similar bonding orbitals than those from original hybridization method (Penotti et al. 1988).

Shaik and collaborators developed a modern version of the hybridization method (with new expressions for the hybridized orbitals) where they proved that there is an overlap between hybrid atomic orbitals, and this overlap lowers the energy of hybrid orbitals in comparison with those from original hybrid orbitals. The energy cost to go from carbon fundamental electron configuration to sp^3 hybrid orbitals, that is, to go from ^3P state to ^5S state, is nearly 96.4 kcal mol^{-1}. The overlap between hybrid atomic orbitals decreases this energy cost (Shaik et al. 2017). However, for practical reasons, the original hybridization method is a reasonable tool to account for chemical bonds and geometric parameters.

The hybridization process occurs in two steps: as mentioned above, the excitation of the electron from the ground state of a given atom, for example, one electron from the triplet fundamental carbon atom, ^3P, to the quintuplet hybridized carbon atom, ^5S state. Next, the formation of the new degenerate orbitals with the same shape. The excitation itself does not change the shape of the orbitals and does not lead to the correct geometry. Take for example, only the first step of the hybridization of carbon atom and its subsequent formation of C-H bonds in methane without the hypothetical occurrence of the second step of hybridization (Fig. 6.7(A)). One can see that bond angles (at least involving p orbitals) are right ($90°$), which does not correspond to the correct geometry of methane. Then, a second step is needed to assure the correct geometry of the formed molecule (Fig. 6.7(B)). Both steps are shown in sequence in Fig. 6.7(C), which yields the four sp^3 orbitals. In Fig. 6.7(D), we show the subsequent formation of the bonding orbitals between the sp^3 carbon atom and the hydrogen atom to form methane with its correct tetrahedral geometry. *At this point (after the formation of the molecule from its isolated atoms), new VB orbitals are generated, and the bonding orbitals are called* $\sigma(sp^3$-s) *bonding orbitals.*

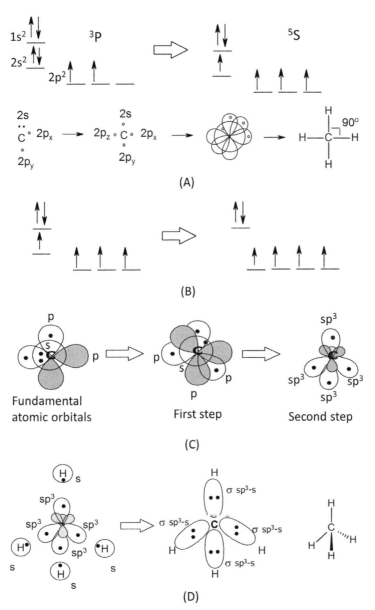

Figure 6.7 (A) First step of hybridization of carbon atom to yield the hypothetical methane without orbital degeneration; (B) schematic energy representation of second step of carbon hybridization; (C) orbital representation of sp^3 carbon atom hybridization showing both steps; (D) orbital representation of formation of methane from completely hybridized carbon atoms.

An important rule in hybridization and chemical bond relation is: when there are one or two π bond(s) in the molecule, the one or two fundamental p orbitals are kept untouched during the hybridization process, so that during the bonding formation these p orbitals (one or two from each atom) will generate π bond. *In another words, π bond is formed from pure p fundamental atomic orbitals even if other orbitals pass through the hybridization process.*

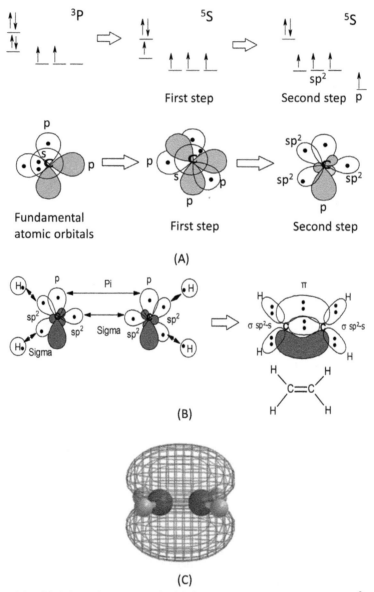

Figure 6.8   (A) Schematic energy and orbital representation of both steps of sp^2 carbon hybridization process; (B) orbital representation of ethane formation; (C) VB π-bond orbital of ethane.

Take for example, the ethene molecule. The carbon atoms that will form a double bond in ethene will pass through a slightly different process than that to form alkane molecules. After the first step of the hybridization process (the excitation from ^3P to ^5S), during the second step, one of the p orbital remains unaltered (Fig. 6.8(A)). After the formation of two sp^2 carbon atoms, the three sp^2 orbitals from each carbon atom form three σ bonds: two $\sigma(sp^2\text{-}s)$ bonding orbitals and one $\sigma(sp^2\text{-}sp^2)$ bonding orbital. The remaining p orbital of each carbon atom forms the π bond (Fig. 6.8(B)). GVB ethene π orbital is shown in Fig. 6.8(C).

In the case of ethyne, two p orbitals remain unaltered after the hybridization process of each carbon atom in which they become sp hybridized with two sp orbitals and two p orbitals.

Oxygen and nitrogen atoms also hybridize before forming bonds and molecules. However, they do not pass through excitation step (first step of the hybridization process in carbon atom). The fundamental electron configurations of oxygen and nitrogen are: $1s^2\,2s^2\,2p^4$ and $1s^2\,2s^2\,2p^3$, respectively, having two or three unpaired electrons, respectively. Since oxygen and nitrogen are divalent or trivalent, respectively (i.e., they form two or three chemical bonds, respectively), they use their original unpaired electrons for their bonds with no need to excite their electrons to increase their valence, as it happens with carbon atom.

Two very important bits of information that chemistry students should always bear in mind: (1) the association between hybridization and bond angle plus geometry; (2) the association between hybridization and electronegativity (see Table 6.2). As for sp^3 hybridization, take alkanes for example. Each sp^3 carbon atom is a single tetrahedral joined to other neighbor tetrahedral(s). See, for example, ethane in Fig. 6.9. As for sp^2 hybridization, take for example alkenes. In alkenes, one and just one double bond exists, where there are two vicinal sp^2 carbon atoms and each of them is planar with respect to their three bonding atoms, giving the trigonal geometry. *The bond angle is the maximum possible angle in a plane to avoid valence shell electron pair repulsion, VSEPR.* In the case of sp^2 carbon atoms, the bond angles in a plane with three bonds from a central atom are 120°. As for the sp hybridization, take alkyne for example. In alkynes, one and just one triple bond exists where there are two vicinal sp carbon atoms, and each of them has a linear geometry with respect to its two bonding atoms. The maximum possible angle to avoid VSEPR in a plane for two bonds in a central atom is 180°.

Regarding electronegativity and hybridization, since s orbital is closer to the nucleus with respect the p orbital and since hybridized orbitals are a linear combination of fundamental atomic orbitals, in a rough approximation (not taking for granted Pauling's hybridized wave functions, but only the number of participating orbitals), sp hybridization has the higher s character. And what is the relation to electronegativity? Since s orbital is closer to the nucleus, it is more attracted to it, yielding a smaller r value (distance between s electron and nucleus) and due to the square inverse relation between distance (r) and electrostatic force, it enables a higher electrostatic interaction between the nucleus and external electron(s) from another atom. Then, the higher the s character in the hybridized atom, the higher is its electronegativity.

Table 6.2   Relation of hybridization type with geometry, bond angle, and s character.

Hybridization	Geometry	Bond angle/°	s character %
sp^3	Tetrahedral	109.5	25
sp^2	Trigonal planar	120	33
sp	Linear	180	50

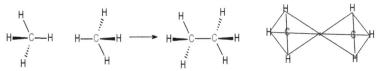

Figure 6.9   Hypothetical formation of ethane from two methane molecules and the tetrahedric view of ethane molecule.

EXERCISES

1. Do the hybridization process (using energy diagram level and orbitals) of sp hybridization and subsequent chemical bond formation involving sp carbon atoms in ethyne. (hint: use three-dimensional space to represent each bond, σ and two π bonds, in each axis).

2. Give the fundamental and hybridized electron configuration for sp^3, sp^2, and sp oxygen and nitrogen atoms, and afterwards do subsequent bonding formation of sp^3, sp^2, and sp oxygen atoms in water, formaldehyde, and carbon monoxide, respectively. (Tip for the formation of carbon monoxide: first put the valence electrons around each atom; in the case of carbon atom, put all four valence electrons unpaired; make ellipsis linking unpaired electrons from both atoms and check whether they have a complete octet or not; in case they do not have a complete octet proceed with one electron transfer from one atom to the other; make a new bond between the unpaired electrons; give the hybridization according to number of π bonds formed; give the atomic orbitals before forming the molecule, and the valence bonding orbitals after forming the molecule). See the note below.

3. Give the atomic orbitals (before forming the molecule) and the valence bonding orbitals (after forming the molecule) of NO. Follow similar hints given in exercise 2. See the note below.

4. Nitrogen dioxide, NO_2, reacts with water to form nitric acid, which is a reagent for aromatic nitration. When it reacts with sulfuric acid it forms the nitronium ion, NO_2^+, the electrophile for aromatic nitration (chapter twenty). Give the orbitals before and after the formation of the bonds between oxygen and nitrogen in NO_2 and NO_2^+, by knowing that the nitrogen in NO_2 has sp^2 hybridization and one oxygen atom does not follow the octet rule in NO_2.

5. Do the hybridization process (using energy diagram level and orbitals) of sp hybridization of carbon atom and sp^2 hybridization of oxygen atom in carbon dioxide molecule (O=C=O). Show the valence bond orbitals of this molecule.

Note

Octet rule: For each atom, one counts its non-bonding and bonding valence electrons, plus the shared electrons from its bonded atom(s), but does not count the core electrons. In some cases, one (or more) atom(s) do not follow the octet rule in a molecule, for example, one oxygen atom in NO_2.

Formal charge: For each atom one counts its non-bonding and bonding valence electrons, plus the core electrons, but does not count the shared electrons from its bonded atom(s).

REFERENCES CITED

Bohr, N. 1913. On the constitution of atoms and molecules. Phil. Mag. 26: 476-502.

Heisenberg, W. 1926. Mehrkörperproblem un resonanz in der quantenmechanik. Z. f. Phys. 38: 411-426 (translated by F.A. Kracklauer in 2006).

Heitler, W. and London, F. 1927. Wechselwirkung neutraler atome und homoopolare bindung nach der quantenmechanik. Z. Phys. 44: 455-472.

Lewis, G.N. 1913. Valence and tautomerism. J. Am. Chem. Soc. 35: 1448-1455.

Lewis, G.N. 1916. The atom and the molecule. J. Am. Chem. Soc. 38: 762-785.

London, F. 1928. Zur quantentheori der homöopolaren valenzzhahlen. Z. Phys. 46: 455-477.

Esposito, S. and Naddeo, A. 2014. The genesis of the quantum theory of the chemical bond. Adv. Hist. Stud. 3: 229-257.

Fock, V.A. 1930. Näherungsmethode zur lösung des quantenmechanischen mehrkörperproblems. Z. Phys. 61: 126-148.

James, H.M. and Coolidge, A.S. 1933. The ground state of the hydrogen molecule. J. Chem. Phys. 1: 825-835.

Pauling, L. 1931. The nature of the chemical bond. Application of results obtained from the quantum mechanics and from a theory of paramagnetic susceptibility to the structure of molecules. J. Am. Chem. Soc. 53: 1367-1400.

Pauling, L. 1960. The Nature of the Chemical Bond and the Structure of Molecules and Crystals: An Introduction to Modern Structural Chemistry. Cornell University Press. Ithaca, New York.

Pauling, L. and Wilson, E.B. 1935. Introduction Quantum Mechanics with Application to Chemistry. McGraw-Hill Book Company Inc., New York.

Penotti, F., Gerrat, J., Cooper, D.L. and Raimondi, M. 1988. The ab initio spin-coupled description of methane: hybridization without preconceptions. J. Mol. Struct. (Theochem) 169: 421-436.

Shaik, S., Danovich, D. and Hiberty, P.C. 2017. To hybridize or not to hybridize? This is the dilemma. Comp. Theor. Chem. 1116: 242-249.

Slater, J.C. 1931. Directed valence in polyatomic molecules. Phys. Rev. 37: 481-489.

Turro, N.J. 1991. Modern Molecular Photochemistry. University Science Books, Mill Valey.

Wang, S.C. 1928. The problem of the normal hydrogen molecule in the new quantum mechanics. Phys. Rev. 31: 579-586.

Chapter Seven

Electron Delocalization, Resonance Types, and Resonance Theory

ELECTRON LOCALIZATION AND ELECTRON DELOCALIZATION

Localized electrons are confined within one atom or one bond with no possibility to move away to other atoms or bonds. They are core electrons (for example, from $1s^2$ orbital in C, N, O, F atoms), lone pair electrons in heteroatom (oxygen and nitrogen, for example) which is bonded to carbon atom by single bond (for example, in alcohols, amines, ethers), and electrons from single bond or double/triple bond having no vicinal sp^2 or sp carbon atom, (for example, double bond in alkene and triple bond in alkyne). Core electrons and electrons from single bonds are always localized. Then, localized electrons (or localized orbitals) belong to one or maximum two atoms. Saturated molecules (whose non hydrogen atoms are sp^3 hybridized), such as alkanes, ethers, amines, and alcohols have exclusively localized electrons. Unsaturated molecules (in which there is, at least, one π bond), such as alkenes, alkynes, aldehydes, ketones have localized electrons as well. These molecules are simply known as localized molecules or localized systems with $2c$-ne, where $n = 2$, 4, or 6, e is electron, and c is atomic center.

Delocalized electrons, on the other hand, are valence electrons interacting to (and moving around), at least, three atoms. Aromatic compounds, conjugated polyenes, alkadienes, and carboxylate are examples of organic compounds with delocalized bonding electrons (or delocalized bonds). Carbocations with multi-center bonds (carbonium ions) also have delocalized electrons. Any delocalized system can be regarded as having a multicenter bonding, for example, $4c$-$4e_\pi$ (in butadiene), $6c$-$6e_\pi$ (in hexatriene), and $8c$-$8e_\pi$ (in octatetraene). Delocalized systems are very important in coloring chemistry and organic electronic materials.

Alkenes have just one double bond, while alkadienes and conjugated polyenes have two or more conjugated double bonds. Conjugated systems have alternating double bond and single bond in a sp^2 (and/or sp, if there is triple bond) hydrocarbon chain. Alkenes have the general formula C_nH_{2n}, while alkadienes and conjugated polyenes systems have the general formula C_nH_{n+2}.

By using the G4 method, or Gaussian-4 theory (Curtiss et al. 2007), we obtained the combustion enthalpy change, ΔH_{comb}, of ethene, butadiene, hexatriene, and octatetraene. Since the stoichiometric coefficients of the studied reactions (in Table 7.1) are different, the combustion enthalpy change needs to be parametrized for comparison, yielding the parametrized combustion enthalpy change, $\Delta H_{comb(p)}$. The reference reaction is the combustion of ethene. The combustion enthalpy changes of other reactions have to be multiplied by $2/x$, where x is the CO_2 stoichiometric coefficient of the corresponding combustion reaction. As $\Delta H_{comb(p)}$ becomes less exothermic, the more stable is the molecular system (de Freitas and Firme 2013). One can see in Table 7.1 the following stability trend: ethene < butadiene < hexatriene < octatetraene. *Then, alkadienes and conjugated polyenes are more stable than ethene. As the molecular system becomes more conjugated, i.e., when going from butadiene to octatetrane, the more stable is the system.*

Table 7.1 Combustion enthalpy change, ΔH_{comb}, and parametrized combustion enthalpy change, $\Delta H_{comb(p)}$, of ethene, butadiene, hexatriene, and octatetraene.

Reaction	ΔH_{comb} / kcal mol^{-1}	$\Delta H_{comb(p)}$ / kcal mol^{-1}
Ethene + 3 $O_2 \rightarrow$ 2 CO_2 + 2 H_2O	−317.96	−317.96
Butadiene + 11/2 $O_2 \rightarrow$ 4 CO_2 + 3 H_2O	−579.98	−289.99
Hexatriene + 8 $O_2 \rightarrow$ 6 CO_2 + 4 H_2O	−841.18	−280.11
Octatetraene + 21/2 $O_2 \rightarrow$ 8 CO_2 + 5H_2O	−1102.06	−275.52

This stability trend (ethene < butadiene < hexatriene < octatetraene) can be understood from the bond strength of the whole π bonding system (since the whole π bonding system is the main stability factor) using the electrostatic force equation, which is based on the interaction between two charged particles separated by r distance vector. In the case of ethene, butadiene, and conjugated polyenes, the most important stability factor is the π bonding system, and then the electrostatic force analysis is based only on π bonding system. Since all carbon atoms are sp^2 hybridized, we can assume that $r_{C1\pi}$ (the distance vector uniting electron 1 from π-bond and carbon atom) is nearly the same for all studied molecules, where distances higher than a double bond are excluded. Another reasonable approximation refers to the effective nuclear charge of carbon atom which is practically the same for all analyzed systems. Moreover, the π bonding system in ethene, butadiene, hexatriene, and octatetraene involves $2c$-$2e$ (two centers and two electrons), $4c$-$4e$, $6c$-$6e$, and $8c$-$8e$, respectively. Then, in ethene, butadiene, hexatriene, and octatetraene, there are 4 (2×2), 16 (4×4), 36, and 64 times the electrostatic force between one π electron (e_π) and one carbon effective nuclear charge (Z_{eff}) separated at $r_{C1\pi}$ distance vector. As a consequence, the bond strength of the π bonding system is: ethene < butadiene < hexatriene < octatetraene, which explains the stability order from combustion reaction.

$$F_{\pi\text{-bond(ethene)}} = 4\left(K \frac{e_\pi Z_{\text{eff}(C)}}{r_{C1\pi}^2} \right)$$

$$F_{\pi\text{-bond(butadiene)}} = 16\left(K \frac{e_\pi Z_{\text{eff}(C)}}{r_{C1\pi}^2} \right)$$

$$F_{\pi\text{-bond(hexatriene)}} = 36\left(K \frac{e_\pi Z_{\text{eff}(C)}}{r_{C1\pi}^2} \right)$$

$$F_{\pi\text{-bond(octatetraene)}} = 64\left(K \frac{e_\pi Z_{\text{eff}(C)}}{r_{C1\pi}^2} \right)$$

Regarding the experiment of ultra-violet/visible (or photoelectron) spectroscopy, the following energy trend is known: σ bond < π bond < n electron, because σ bond is stronger than π bond and n electron has the smallest electrostatic interaction (as seen in the previous chapter). It is already known that the energy excitation of conjugated polyenes is lower than that from localized π bond systems. Figure 7.1 depicts the σ bond excitation to σ^* antibond (electron jump from its bonding orbital to its antibonding orbital), π bond excitation to π^* antibond, and the n electrons (lone pair of electrons) excitation to π^* antibond (a most probable transition when there is hetereoatom in a double bond). The HOMO energies (in Table 7.2 and footnotes) follow the trend: ethane < ethene < butadiene < hexatriene < octatetraene, and the LUMO-HOMO difference, Δ(L-H), has an opposite trend in accordance with the corresponding excitation energies. But one should bear in mind that in butadiene and conjugated polyenes, only one of their π bonds are being represented, not the whole π bonding system.

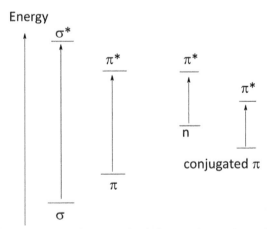

Figure 7.1   Schematic energy level diagram for σ-σ^*, π-π^*, n-π^*, and conjugated π-π^* electron excitation.

There is a straightforward relation between HOMO and bond type. The HOMO reveals the energy of σ bond in ethane and localized π bond in ethene, but no single association of π bond with HOMO can be done for butadiene and conjugated polyenes, since only one of the π bonding electrons of the delocalized system is related to HOMO (then it is a partial representation of the whole π bonding system for them). Recall that in a molecular orbital, there are, at most, two electrons, and butadiene, hexatriene, and octatetraene have four, six, and eight π bonding electrons, respectively. *Then, in order to relate the stability trend, electrostatic force, and orbital or bonding energy for alkadienes and conjugated polyenes, we cannot use their single HOMO, because stability is related to the whole π bonding system, and not to partial π bond.* Otherwise, since their HOMO have higher energy than HOMO of ethene (footnotes in Table 7.2), we would wrongly assume that the π bonding system in alkadienes and conjugated polyenes are weaker (owing to the inverse relation between force and potential energy) than that in ethene. This is wrong! The stability trend in Table 7.1 and the electrostatic force equation of the whole π bond show that the π bonding system in alkadienes and conjugated polyenes are stronger than that in ethene. To avoid the misleading association between single HOMO for butadiene and conjugated polyenes and their stability trend, we opted to use π bonding averaged HOMO, A_{HOMO} (see footnotes of Table 7.2). Then, the π bonding averaged HOMO of butadiene and conjugated polyenes ranges from -0.276 to -0.279 Hartree, and they are lower in energy than HOMO of ethene (Table 7.2). According to the inverse relation between electrostatic force and potential energy, π bond strength (as whole π bonding system) in butadiene and conjugated polyenes is stronger than that in ethene, which is in accordance with the stability order (Table 7.1). Hence, averaged HOMO from the whole π bonding system, its inverse relation of the electrostatic force of the whole π bonding system, and stability are inter-related.

Table 7.2   HOMO, LUMO, and HOMO-LUMO difference of ethane, ethene, butadiene, hexatriene, and octatetraene.

Molecule	HOMO/H	LUMO/H	Δ(L-H)/H
Ethane	−0.339	0.104	0.443
Ethene	−0.268	0.017	0.285
Butadiene	−0.276[a]	−0.024	0.252
Hexatriene	−0.280[b]	−0.045	0.235
Octatetraene	−0.280[c]	−0.058	0.222

(a) Average HOMO (A_{HOMO}) of butadiene (average of HOMO=−0.231 and HOMO-1=−0.321 Hartree); (b) A_{HOMO} of hexatriene (average of HOMO=−0.211, HOMO-1=−0.288 , HOMO-2=−0.340 Hartree); and (c) A_{HOMO} of octatetraene (average of HOMO=−0.198, HOMO-1=−0.264, HOMO-2=−0.314 and HOMO-3=−0.343 Hartree).

This approach to A_{HOMO} is actually only valid to analyze the relation among stability, the energy of the whole π bonding system, and its inverse relation to the electrostatic force. As we showed before, by using the electrostatic force equation of the whole π bonding system, we could account for the stability trend as the conjugation increased (Table 7.1). Then, to associate π bond strength with orbital energy of a conjugated system, it is necessary to consider the energy of the whole π bonding system, which is achieved by the approach of π bonding averaged HOMO. Fortunately, by using A_{HOMO} instead of HOMO for butadiene and conjugated polyenes, we also found the same expected Δ(L-H) trend, which decreases as the conjugation increases. Certainly, in terms of electronic excitation, the most probable transition arises from the HOMO (and not from A_{HOMO}) to LUMO. But, as mentioned above, the A_{HOMO} approach suffices to account for the relation involving stability (in great part associated with the whole π bonding system), the energy of the whole π bonding system, and the electrostatic force of the whole π bonding system.

Hereafter, we introduce an important law in chemistry: *delocalized systems are always more stable than reference localized systems.*

ORIGIN AND EVOLUTION OF THE RESONANCE CONCEPT

As already mentioned in previous chapters, the resonance idea was first introduced by Heisenberg to study many-body (actually, two-body) systems in the perspective of his Matrix Mechanics and also in Schrödinger's Wave Mechanics. At that time this work received remarkable recognition, as Birtwistle wrote: "In June 1926 Heisenberg wrote an outstanding paper on resonance in atoms with two electrons, which contained the key to the solution of the spectrum of neutral helium" (Birtwistle 1928). Heitler and London changed the general two-body system into two electrons in H_2 molecule, and found that resonance (the exchange of the two electrons in the two nuclei) was responsible for the chemical bond in H_2, which was later improved in Wang's work by replacing Z into Z_{eff}. Soon afterwards, Pauling used the Heitler-London concept of resonance to devise the properties of electron-pair bond where one finds: "the main resonance terms for a single electron-pair bond are those involving only one eigenfunction from each atom" (Pauling 1931). However, one year later, Pauling gave a more comprehensive definition of resonance. He started to use the concept of resonance to explain that the correct structure of some compounds (HF or SiF_4, for example) cannot be rationalized by either ionic extreme H^+F^- (in HF) or electron-pair bond extreme H:F (in HF). By deriving corresponding ionic and covalent Morse curves (potential energy versus reaction coordinate or interatomic distance) for several molecules, he found that molecules wherein both ionic and covalent Morse curves intersected could not be represented either by pure ionic character or pure covalent character (Pauling 1932). Then, the real molecule (resonance hybrid) lies between them, and Pauling began to represent them quantum-mechanically as (possibly influenced by his previous study on hybridization):

$$\Psi = a\psi_{ionic} + b\psi_{electron\text{-}pair}$$

Pauling also began to use the resonance term between two covalent structures, for example in CO molecule, and later in benzene molecule. In the former, Pauling still used the idea of mixing ionic and covalent structures (Pauling 1932), but for the latter he referred only to two covalent structures (the Kekulé structures). This represented another expansion of the term resonance: firstly to represent in great part a single electron pair bond; secondly to represent mixing of Morse curves from ionic and covalent structures of a real molecule; and finally the mixing of neutral covalent structures (changing the approach of Morse curves into interatomic distance function) to represent the real molecule (Pauling et al. 1935). In his book (Pauling 1960), Pauling amplified the quantum hybrid function for more than two possible structures (since he expanded the resonance concept from ionic and covalent mixing to mixing of different covalent structures):

$$\Psi = a\psi_I + b\psi_{II} + c\psi_{III} + ...$$

Pauling also stated that the stabilization of the system by resonance energy is the most important feature of quantum mechanical resonance (Pauling 1960). This statement might possibly be the connection among the three applications of the resonance. Bent (Bent 1953) called them resonance of type 1 (the original idea of resonance developed by Heitler and London to the chemical bond), resonance of type 2 (resonance between ionic and covalent structures), and resonance type 3 (resonance involving only charged and/or neutral covalent structures).

In IUPAC's Gold Book, the definition incorporates these types of resonance: "In the context of chemistry, the term refers to the representation of the electronic structure of a molecular entity in terms of contributing structures. Resonance among contributing structures means that the wavefunction is represented by "mixing" the wavefunctions of the contributing structures. The concept is the basis of the quantum mechanical valence bond methods. The resulting stabilization is linked to the quantum mechanical concept of "resonance energy". The term resonance is also used to refer to the delocalization phenomenon itself". The first four sentences refer to resonance type 1 and 2, and the last sentence refer to resonance type 3.

MESOMERISM

As Ingold himself defined: "Mesomerism is an extension of valency theory (that is the classical valence bond theory) and, like all valency theory, is founded in the quantum theory (...). The fundamental wave-property in the theory of valency is resonance – the resonance of connected standing waves; their mutual perturbation replaces these waves by new standing waves" (Ingold 1938). Ingold exemplified the resonance theory by using the two resonance structures: $R_2N\text{-}CH\text{=}N^+R_3$ and $R_3{}^+N\text{=}C(H)\text{-}NR_2$ with their own standing waves which resonate to form the real molecule. Ingold stated that the term mesomerism was given to account for the special importance of resonance in organic chemistry. He also said: "When mesomerism was first recognized as a general phenomenon in organic chemistry, it was appreciated as an electron displacement than as an energy disappearance" (Ingold 1938). From this point on, we can say that mesomerim is equivalent to resonance type 3 (resonance involving only charged and/or neutral covalent structures), although IUPAC recognizes mesomerism as essentially synonymous with resonance (which could include all types of resonance). Nonetheless, IUPAC adds that mesomerism is "particularly associated with the picture of π-electrons as less localized in an actual molecule than in a Lewis formula" (IUPAC Gold Book). Then, mesomeric state is the same as resonance hybrid and energy of mesomerism is synonymous to resonance energy (Ingold 1938).

THE RESONANCE THEORY (RESONANCE TYPE 3)

Delocalized systems can be represented by resonance structures where their linear combination yields the resonance hybrid, i.e., the real delocalized molecule. The resonance structure is an unreal localized share of the whole, real molecule. As previously stated, the real delocalized molecule, the resonance hybrid, Ψ, is represented by a linear combination of resonance structures, φ_i, which is assigned to coefficient c_i, which determines the weight of its corresponding resonance structure to make up the real molecule. Then, the sum of all c_i's is a unity. The larger the c_i value, the greater the importance (and "virtual" stability) of the corresponding φ_i to the real molecule. *It is important to note that neutral resonance structures are more stable (highest c_i) than charged resonance structures, and that the more charged the resonance structure the more unstable it is (smallest c_i).*

$$\Psi = c_1\varphi_1 + c_2\varphi_2 + ... + c_i\varphi_i$$
$$c_1 + c_2 + ... + c_i = 1$$

Another important rule on resonance theory: *the more resonance structures (with non-zero c_i's) the more stable the resonance hybrid, and the closer the values of c_i's the more stable the resonance hybrid.*

One can say that in ethene its resonance hybrid is equivalent to its unique resonance structure. In butadiene, there are two resonance structures (one neutral and one charged resonance structure). In hexatriene and octatetraene, there are three and four resonance structures, respectively (Fig. 7.2). It is important to note that electron delocalization between two vicinal atoms (in a double bond) were not regarded for simplicity reasons. We arbitrarily assume the electron displacement "to begin" from the double bond at the right position of the first resonance structure (φ_1) to yield the subsequent resonance structure (φ_2). Taking butadiene as an example in Fig. 7.2, one can see that the π electrons move from one bond (the rightmost double bond) to the vicinal single bond, leaving one carbon (the one which loses the double bond) positively charged. Simultaneously, a double bond is formed in what was a vicinal single bond, forcing the displacement of the subsequent π electrons to the leftmost carbon atom which becomes negatively charged. This same type of electron displacement pattern is used in hexatriene and octatriene. The mixing of these resonance structures yields the real molecule (resonance hybrid shown in Fig.7.2 for butadiene and hexatriene). In Fig. 7.3, the electron movement appears, represented by arrows in a bond line formula (see chapter nine) and the corresponding π and p orbitals in each resonance structure, which give a better picture of the π electron movement in the hydrocarbon skeleton.

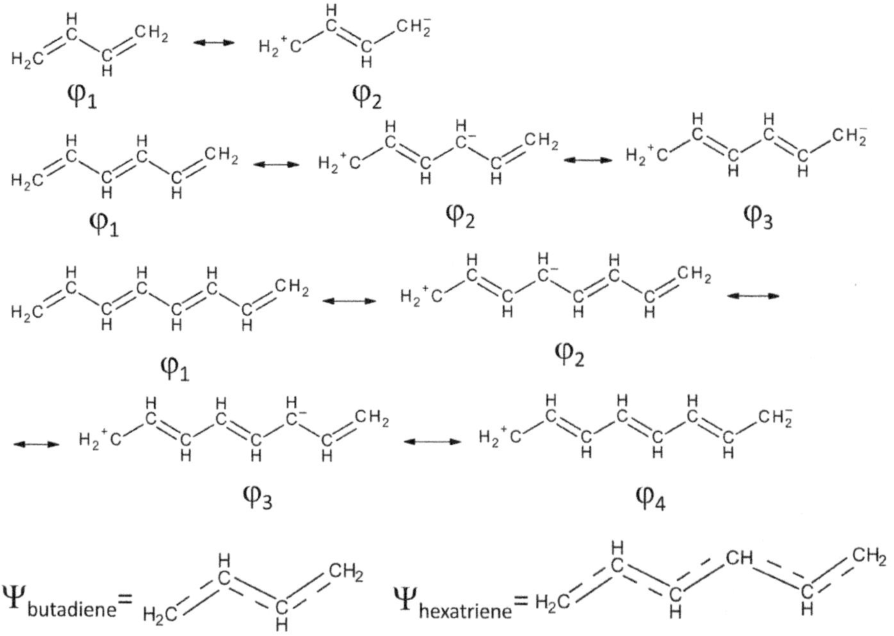

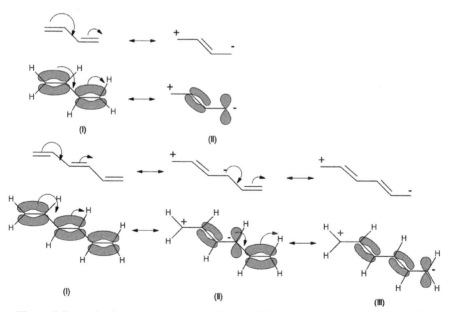

Figure 7.2 Resonance structures of butadiene, hexatriene, and octatetraene, and the resonance hybrid of butadiene and hexatriene.

Figure 7.3 π-electron movement in butadiene and hexatriene represented in bond line formula, and the corresponding π and p orbitals in each resonance structure.

From these electron displacement patterns, butadiene, hexatriene, and octatetraene have two, three, and four resonance structures, respectively. As we have already seen, the stability order is: butadiene < hexatriene < octatetraene. As mentioned above, the higher the number of non-zero (or nearly non-zero c_i's) resonance structures, the more stable the real molecule.

$$\Psi_{butadiene} = c_1\varphi_1 + c_2\varphi_2$$
$$\Psi_{hexatriene} = c_1\varphi_1 + c_2\varphi_2 + c_3\varphi_3$$
$$\Psi_{octatetraene} = c_1\varphi_1 + c_2\varphi_2 + c_3\varphi_3 + c_4\varphi_4$$

EXERCISES

1. Give the two resonance structures of HBr.
2. Give the resonance structures and resonance hybrid of allyl cation and 1-penta-2,4-dienyl cation. In addition, give the equation of hybrid wave function along with values of corresponding coefficient of each resonance structure for both cations.
3. Give the resonance structures of cyclopentadienyl anion and cycloheptatrienyl cation. As for the former, give the equation of the hybrid wave function along with each coefficient of the resonance structures.

REFERENCES CITED

Bent, R.L. 1953. Aspects of isomerism and mesomerism. J. Chem. Educ. 30: 220-228.

Birtwistle, G. 1928. The New Quantum Mechanics. Cambridge University Press, New York.

Curtiss, L.A., Redfen, P.C. and Raghavachari, K. 2007. Gaussian-4 theory. J. Chem. Phys. 126: 084108.

De Freitas, G.R.S. and Firme, C.L. 2013. New insights into the stability of alkenes and alkynes, fluoro-substituted or not: a DFT, G4, QTAIM and GVB study. J. Mol. Model. 19: 5267-5276.

Ingold, C.K. 1938. Resonance and mesomerism. Nature 141: 314-318.

Pauling, L. 1931. The nature of the chemical bond. Application of results obtained from the quantum mechanics and from a theory of paramagnetic susceptibility to the structure of molecules. J. Am. Chem. Soc. 53: 1367-1400.

Pauling, L. 1932. The nature of the chemical bond. III. The transition from one extreme bond type to another. J. Am. Chem. Soc. 54: 998-1003.

Pauling, L., Brockway, L.O. and Beach, J.Y. 1935. The dependence of interatomic distance on single bond-double bond resonance. 57: 2705

Pauling, L. 1960. The Nature of the Chemical Bond and the Structure of Molecules and Crystals: An Introduction to Modern Structural Chemistry. Cornell University Press. Ithaca, New York.

Chapter Eight

Quantum Chemistry of Potential Energy Surface (Geometric Parameters, Energy Derivatives, Optimized Geometries, and Transition States)

GEOMETRIC PARAMETERS

A diatomic molecule has only one geometric parameter: the bond length, which is the distance between the bonding atoms. A triatomic molecule has two geometric parameters: bond length and bond angle. A bond length always involves two atoms, and a bond angle involves three atoms (a central atom bonded to two atoms). A tetraatomic molecule has three geometric parameters: bond length, bond angle, and the dihedral angle, which is the angle between two intersecting planes, and it involves four neighbor atoms. Any molecule with four or more atoms has these three geometric parameters (Figs 8.1 to 8.3).

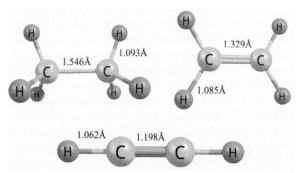

Figure 8.1   Bond lengths, in Angstrom, of the optimized ethane, ethane, and ethyne molecules.

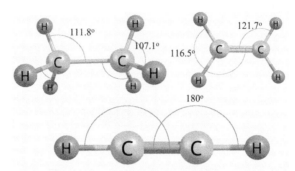

Figure 8.2   Bond angles, in degrees, of the optimized ethane, ethane, and ethyne molecules.

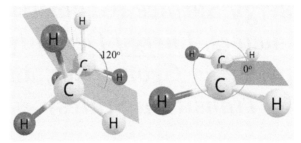

Figure 8.3   Dihedral bond angles, in degrees, of the optimized ethane and ethene molecules.

   The geometric parameters describe the nuclear configuration of a molecule. *The nuclear configuration of a molecule obtained from the experimental single crystal X-ray diffraction is always associated with its optimized structure and it can be compared to its corresponding optimized theoretical structure, usually in good agreement...* Take for example the molecule represented by Fig. 8.4(A). Figures 8.4(B) and 8.4(C) show the crystallographic and theoretical optimized structures

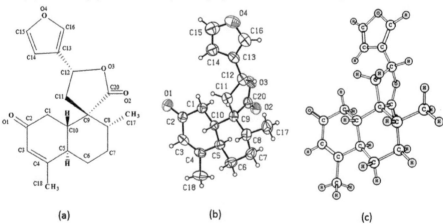

(a)   (b)   (c)

Figure 8.4   (A) Bond-line representation; (B) crystallographic representation; (C) theoretically optimized representation of *t*-DCTN (Soares et al. 2014). See the Acknowledgment section.

of *trans*-Dehydrocrotonin, *t*-DCTN, a bioactive clerodane, respectively. The root mean square deviation, RMSD, which measures the effectiveness of the theoretical data in relation to the experimental one was 0.013 and 0.9333 for bond lengths and bond angles, respectively, which indicates a very good agreement between experimental and theoretical data (Soares et al. 2014). The optimized geometry was generated using the density functional theory, DFT, method which was not discussed in chapter two.

The nuclear configuration of a molecule can be described numerically by means of Cartesian coordinates of each atom of the molecule or by means of the Z-matrix containing the geometric parameters of the molecule. Figure 8.5 shows the Z-matrix and Cartesian coordinates of the optimized water molecule.

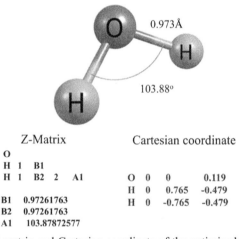

	Z-Matrix				Cartesian coordinate		
O							
H	1	B1					
H	1	B2	2	A1	O	0	0	0.119
					H	0	0.765	-0.479
B1	0.97261763				H	0	-0.765	-0.479
B2	0.97261763							
A1	103.87872577							

Figure 8.5   Z-matrix and Cartesian coordinate of the optimized water molecule.

DEGREE OF FREEDOM AND PROCEDURE TO RUN A QUANTUM CHEMISTRY CALCULATION

The degree of freedom, DOF, can be divided in two types: total degree of freedom (involving translational, rotational, and vibrational movements), DOF_T, and vibrational degree of freedom, DOF_V, which involves only the vibrational modes. The latter is usually called only degree of freedom, but rotational and translational movements are also degrees of freedom!

Any molecule (small or big, H_2 or an enzyme) has three translational movements (one in each x, y, or z axis). Linear molecule (e.g., H_2) has two rotational movements (involving the whole molecule turning around in one of the two axis). A non-linear molecule (e.g., H_2O) has three rotational movements.

The total degree of freedom is $3N$, where N is the total number of atoms in the molecule. The vibrational degree of freedom leaves aside the total number of translational and rotational movements (6 for a non-linear molecule or 5 for a linear molecule) of a molecule.

The vibrational degree of freedom informs the number of vibrational modes in that molecule. There are several types of vibrational modes: symmetrical and asymmetrical stretching, ss and as, respectively (where the bond lengths change during the vibration), bending (scissoring, s) and rocking, r (where the bond angles change during the vibration) and twisting, t, and wagging, w (where the dihedral angle changes during the vibration). These vibrations occur at any temperature. Then, *molecules vibrate all the time at any temperature and the number of vibrations is given by the degree of freedom.* Moreover, each vibration has a specific vibration frequency, usually given in cm^{-1}.

$$\text{DOF}_{(t)} = 3N$$

$$\text{DOF}_{(v)} = 3N - 6 \text{ for non-linear molecules}$$

$$\text{DOF}_{(v)} = 3N - 5 \text{ for linear molecules}$$

Where N is the number of atoms in the molecule. The numbers 6 and 5 refer to the number of translations and rotations of the whole molecule.

Let us take a water molecule (a non-linear molecule) as an example. It has three atoms, and its total degree of freedom is $3(3)=9$. It has three translational movements with frequencies 0.04, 0.01, and 0.04 cm^{-1}. It has three rotational movements with frequencies 14.12, 47.6, and 52.45 cm^{-1}. It has three vibrational movements (vibrational degree of freedom is three) or three modes of vibrations: symmetric and asymmetric stretching (3585 cm^{-1} and 3506 cm^{-1}, respectively) and scissoring at 1885 cm^{-1} (Fig. 8.7). For each type of geometrical parameter there is an associated general type vibrational mode. They can be informally classified as bonding vibration (symmetrical and asymmetrical stretching), angular vibration (scissoring and rocking) and dihedral vibration (twisting and wagging).

All translational, rotational, and vibrational frequencies can also be obtained theoretically by means of the Hessian of the molecular energy, **H**, after diagonalization of this matrix (see the expression in the next subsection) yielding the force constant from which the corresponding frequency is obtained. At first, it is necessary to optimize the geometry of the molecule. After its optimization, it is possible to calculate the frequencies of the total degree of freedom of the molecule in most programs (Gamess, Gaussian, Orca, etc.), and it is also possible to obtain all absolute thermodynamic properties (entropy, internal energy, enthalpy, and Gibbs free energy) of this molecule along with its infrared and Raman spectra. At first, one needs to draw the molecule in an appropriate quantum chemistry program, for example, Avogadro – a free software for visualization, molecule edition, and input creation for a posterior *ab initio* calculation (Hanwell et al. 2012). After molecule edition, Avogadro generates the corresponding Z-Matrix. Afterwards, it is necessary to choose the *ab initio* software. Then in Avogadro itself it is possible to add the information to generate the input file for posterior *ab initio* calculation. It is necessary to inform the type of method (B3LYP, ωB97XD, MP2, CCSD, etc.), the type of basis set (STO-3G, 6-31G(d,p), etc.), the type of SCF calculation (DIIS, EDIIS, CDIIS, etc.), and the type of optimization algorithm (Berny, eigenvalue-following, etc.). At the end of the calculation, one can observe the optimized geometry and its internal coordinates. After that it is possible to calculate the frequencies along with thermodynamic properties, infrared, and Raman spectra.

Alternatively, there are free websites that provide a straightforward platform for molecule edition, input calculations, and to run the calculation from their own servers, and only an internet access is necessary for doing simple *ab initio* calculations. One of these websites is the ChemCompute (Perri and Weber 2014).

OPTIMIZATION AND FREQUENCY CALCULATIONS

From the optimized geometry of a given molecule, it is possible to obtain all theoretical properties, which are in good agreement with the experimental data in a vast majority of the cases. In Fig. 8.6, one can see the very high coefficient of correlations, R^2, between theoretical and experimental from ^1H and ^{13}C chemical shifts (a nuclear magnetic resonance, NMR, property) of *t*-DCTN in excellent agreement between theoretical and experimental data (Soares et al. 2014). The thereotical magnetic property was obtained from a previous optimized geometry of *t*-DCTN.

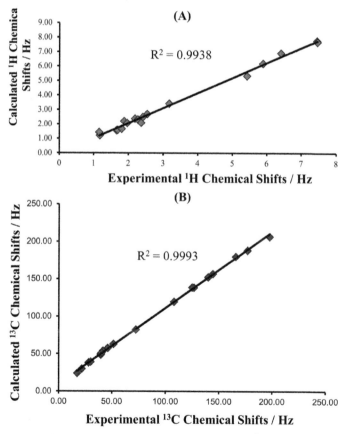

Figure 8.6 Plots of (A) experimental versus calculated ^1H chemical shifts and (B) experimental versus calculated ^{13}C chemical shifts of *t*-DCTN (Soares et al. 2014). See the Acknowledgment section.

Then, *any optical, electrical, magnetic, thermodynamic, electronic, and kinetic property can be obtained theoretically from the optimized geometry of the molecule or set of molecules in order to compare to experimental data, giving very good agreement between them in most cases.* Moreover, due to this very good agreement between experimental and theoretical data, one can obtain the latter (and that is a reliable information!) when it is not possible to obtain the former.

The starting point is to use a specific graphic software to draw the initial geometric structure of the molecule. This same program will transform the image into the corresponding Z-matrix or cartesian coordinates, which will be read in a specific computational chemistry software. The user chooses the type of MO or DFT method and the basis set (see chapter two), plus the type of calculation (optimization, vibrational frequency, excited state, properties, etc). We emphasize that prior to any kinetic, thermodynamic, or other property, it is a must to first perform the optimization calculation. Then, henceforth, if optimization calculation was chosen by the user, the computational chemistry software will pick the initial geometry from Z-matrix or cartesian coordinates, and will generate the optimized geometry after a specific cycle of optimization steps, provided the calculation terminated normally. In order to check whether it is a minimum or maximum in the potential energy surface, a posterior frequency calculation is needed. *If the molecule is a reactant, product, or intermediate, the optimized geometry corresponds to a minimum in the potential energy surface. If the molecule is the transition state of a given reaction, the optimized geometry corresponds to a maximum in the potential energy surface.*

From a given initial nuclear configuration R, it is intended to reach the minimum (or in some cases, the maximum) of the potential energy surface of the corresponding molecule. Then, from a given algorithm (preferably using quadratic harmonic approach), the computational chemistry software changes the initial coordinates of the molecule from R into R_1, afterwards from R_1 into R_2, and so on until reaching R_e (the equilibrium geometry). In each step of the optimization procedure, the computational chemistry software calculates

$$q = -\frac{f(R)}{H(R)}$$

Where $f(R)$ is the first gradient, the force applied to a given geometric coordinate, and $H(R)$ is the Hessian matrix of the molecular energy (the same used in QTAIM calculation, which is limited to 3×3 square matrix).

$$f_i = \frac{\partial E(R)}{\partial R_i}$$

$$H_{ij} = \frac{\partial E(R)}{\partial R_i \partial R_j}$$

Where R_i and R_j refer to all initial nuclear coordinates of the initial nuclear configuration R. The size of the Hessian matrix in a geometry calculation is $3N \times 3N$, where N is the number of atoms of the molecule.

$$\mathbf{H} = \begin{pmatrix} \dfrac{\partial^2 E}{\partial R_1{}^2} & \dfrac{\partial^2 E}{\partial R_1 \partial R_2} & \dfrac{\partial^2 E}{\partial R_1 \partial R_{3N}} \\[2.5ex] \dfrac{\partial^2 E}{\partial R_2 \partial R_1} & \dfrac{\partial^2 E}{\partial R_2{}^2} & \dfrac{\partial^2 E}{\partial R_2 \partial R_{3N}} \\[2.5ex] \dfrac{\partial^2 E}{\partial R_{3N} \partial R_1} & \dfrac{\partial^2 E}{\partial R_{3N} \partial R_2} & \dfrac{\partial^2 E}{\partial R_{3N}{}^2} \end{pmatrix}$$

Where R_1, R_2,...R_{3N} are the geometric coordinates of the molecule, which should not be confused with R_1, R_2,...R_e (the nuclear configuration in each step of the optimization procedure).

Knowing R and q, it is possible to obtain R_e as shown in the equation below. Actually, before the end of the optimization calculation, R_e is R_1, R_2, ...until R_e at the last step of the optimization. Then, q is used to determine the direction of each optimization step towards the minimum or maximum in case of transition state calculation (Szabo and Ostlund 1996).

$$\mathbf{q} = (R_e - \mathbf{R})$$

The optimization calculation stops when the conditions of the thresholds are satisfied, i.e., when the first derivative of molecular energy reaches a value close to zero. This is the same criterion to find the critical points in QTAIM. In order to check whether the optimized geometry is at minimum or maximum in the potential energy surface, it is necessary to calculate the Hessian matrix of the optimized geometry (where the matrix elements are second derivatives of the molecular energy). In this calculation, the Hessian matrix is diagonalized and each diagonal element is the corresponding force constant, K_i, from which the corresponding frequency of vibration, v_i is obtained.

$$K_i = \left(\frac{\partial^2 E}{\partial R^2} \right)_i \therefore \mu_i = \frac{m_1 m_2}{m_1 + m_2}$$

$$v_i = \frac{1}{2\pi} \sqrt{\frac{K_i}{\mu_i}}$$

From Hartree-Fock approximation (see chapter two), and considering a simple case of beryllium atom (with two electron pairs occupying orbitals 1 and 2), its energy E is described by $E = 2h_{11} + 2h_{22} + J_{11} + J_{22} + 4J_{12} - 2K_{12}$, Where $2h_{11}$ and $2h_{22}$ are the kinetic energy and electron-proton potential energy of the two electrons in orbitals 1 and 2, respectively; J_{11} and J_{22} are the electron-electron potential energy of two electrons occupying orbitals 1 and 2, respectively; $4J_{12}$ is the electron-electron potential energy between one electron in orbital 1 and one electron in orbital 2; and $2K_{12}$ is the exchange energy between electrons in orbitals 1 and 2. For a general case of a singlet molecule, its molecular energy is given by (Szabo and Ostlund, 1996):

$$E = \sum_{\mu\nu} P_{\nu\mu} H_{\mu\nu} + \frac{1}{2} \sum_{\mu\nu.\lambda.\sigma} P_{\nu\mu} P_{\lambda.\sigma} \left[(\mu.\nu|\sigma.\lambda) - \tfrac{1}{2}(\mu.\lambda|\sigma.\nu) \right] + V_{NN}$$

$$V_{NN} = \sum_A \sum_{A>B} \frac{Z_A Z_B}{R_{AB}} \therefore \quad P_{\mu.\nu} = 2 \sum_a^{N/2} C_{\mu.a} C_{\nu.a}$$

$$a = \psi_i = \sum_{\mu=1}^{K} C_{\mu i} \phi_\mu$$

Where μ, ν, σ, and λ (short for ϕ_μ, ϕ_ν, ϕ_σ, and ϕ_λ) are spin orbitals coming from a specific basis set, and C is the expansion coefficient of the spin orbitals.

By differentiating the molecular energy equation in terms of nuclear coordinates R, we have (for closed shell systems):

$$\frac{\partial E}{\partial R_A} = \sum_{\mu\nu} P_{\nu\mu} \frac{\partial H_{\mu\nu}}{\partial R_A} +$$

$$+ \frac{1}{2} \sum_{\mu\nu\cdot\lambda\cdot\sigma} P_{\nu\mu} P_{\lambda\cdot\sigma} \frac{\partial \left[(\mu \cdot \nu|\sigma \cdot \lambda) - \tfrac{1}{2}(\mu \cdot \lambda|\sigma \cdot \nu) \right]}{\partial R_A} - \sum_{\mu\nu} Q_{\nu\mu} \frac{\partial S_{\mu\cdot\nu}}{\partial R_A} + \frac{\partial V_{NN}}{\partial R_A}$$

$$Q_{\nu\cdot\mu} = 2 \sum_a^{N/2} \varepsilon_a C_{\mu\cdot a} C_{\nu\cdot a}$$

When the $\dfrac{\partial E}{\partial R_A}$ for all coordinates are zero, that is, when all Feynman forces vanish in the molecule, the optimized geometry was obtained (minimum or maximum at potential energy surface). The same procedure using first derivatives in coordinates of real space (i.e., gradient) of charge density is used to find critical points in QTAIM.

POTENTIAL ENERGY SURFACE AND TRANSITION STATE

When a reaction occurs, the involved molecules (or involved molecule for unimolecular reaction) follow one or more reaction paths, where the variation in their nuclear configuration is followed by the variation of the molecular energy along the reaction path. *The variation of the molecular energy along the reaction path (when there are modifications of the nuclear configuration of the involved molecules in the reaction) is plotted in a graph called potential energy surface.*

Let us suppose one substitution reaction changing the halogen atom bonded to a carbon atom, for example, the reaction described below:

$$NaBr + CH_3Cl \rightarrow NaCl + CH_3Br$$

This reaction follows a reaction path that passes through a transition state where chemical bonds are being broken and/or being formed. In that case, in the transition state, one chemical bond is being broken and one chemical bond is being formed.

For simplicity, we will use transition state to refer to the transition structure or activated complex.

$$\left[{}^{\delta-}\text{Br--CH}_3\text{--Cl}^{\delta-} \right]\text{Na}^+$$

Where dashed lines represent bonds being broken or being formed. In the transition state above, the Br-C bond is being formed and C-Cl bond is being broken. Then, the reaction pathway is:

$$\text{NaBr} + \text{CH}_3\text{Cl} \rightarrow \left[{}^{\delta-}\text{Br--CH}_3\text{--Cl}^{\delta-} \right]\text{Na}^+ \rightarrow \text{NaCl} + \text{CH}_3\text{Br}$$

The reactants and products of this reaction are minima in the potential energy surface (where the first derivatives of the molecular energy are all zero and second derivatives, from diagonalized Hessian matrix, of the molecular energy are all positive).

$$\left(\frac{\partial E}{\partial R_i} \right)_i = 0 \therefore \text{ For all } i = 1, 2, \ldots 3N \therefore \left(\frac{\partial^2 E}{\partial R_i^2} \right)_i > 0 \therefore \text{ For all } i = 1, 2, \ldots 3N$$

This is the same procedure to find critical points, maximum, and minimum of any function in any coordinate. For a minimum critical point an f function with respect to x axis, we have:

$$\frac{df}{dx} = 0 \therefore \frac{d^2 f}{dx^2} > 0$$

In order to compare to the transition state, we calculate the optimization and frequency of the reactant complex ($\text{NaI} + \text{CH}_3\text{Cl}$), where the number of vibrational modes is:

$$\text{DOF}_{(v)} = 3(7) - 6 = 15$$

There are seven atoms in the reactant complex and the number of vibrational modes is fifteen (same as the transition state). In the reactant complex, all fifteen frequencies are positive. See in Fig. 8.7 the optimized structure of the reactant complex, and the corresponding vibrational frequencies in Table 8.1. We have seen in chapter two that when second derivative is positive and first derivative is zero, the critical point in that function is a minimum.

Figure 8.7 Optimized [NaBr CH$_3$Cl] reactant complex.

In the transition state there are also fifteen vibrational modes, but one and just one has imaginary (or also called negative) frequency. This negative frequency is associated with the reaction path, where its vibration shows the Br-C bond being formed and C-Cl bond being broken simultaneously. See in Fig. 8.8 the optimized structure of the transition state, and the corresponding vibrational frequencies in Table 8.1. *These are the two important conditions for the transition state: (1) it has to have one and only one imaginary (or negative) frequency; and (2) this frequency has to be associated with the reaction path (which requires the knowledge of the reaction mechanism).*

$$\left(\frac{\partial E}{\partial R_i}\right)_i = 0 \therefore \text{For all } i = 1,2,...3N$$

$$\therefore \left(\frac{\partial^2 E}{\partial R_1^2}\right)_1 < 0 \therefore \left(\frac{\partial^2 E}{\partial R_i^2}\right)_i > 0 \therefore \text{For all } i = 2,3,...3N$$

Table 8.1   Frequency values, in cm^{-1}, of the reactant complex NaBr+CH$_3$Cl, and its corresponding transition state.

Entry	Frequency values of reactant complex[a]/cm^{-1} (vibrational mode)	Frequency values of transition state[a]/cm^{-1} (vibrational mode)
1	50.37 (t)	−525.66 (ss)
2	67.54 (s)	96.13 (s)
3	106.50 (w)	144.54 (w)
4	133.70 (s)	148.84 (s)
5	146.24 (s)	176.52 (s)
6	286.36 (as)	245.70 (as)
7	696.16 (as)	409.74 (w)
8	1052.46 (r)	803.84 (r)
9	1065.57 (t)	897.59 (t)
10	1408.62 (w)	970.73 (w)
11	1478.85 (s)	1418.34 (s)
12	1482.07 (s)	1488.85 (s)
13	3069.54 (ss)	3174.36 (ss)
14	3191.16 (as)	3344.92 (as)
15	3223.31 (as)	3468.86 (as)

(a) Calculated at ωB97XD/6-311G++(d,p).

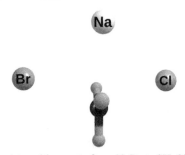

Figure 8.8   Optimized transition state from NaBr + CH$_3$Cl substitution reaction.

The potential energy surface can be constructed in terms of the electronic energy (which also includes the potential energy between the atoms), internal energy, enthalpy, or Gibbs free energy (see chapter four). This type of molecular energy must be specified in the ordinate of the potential energy surface (by considering bi-dimensional cartesian coordinates). *For the sake of simplicity, the potential energy surface of a reaction, in almost all cases, is only associated with the nuclear motion related to the reaction path, disregarding all other vibrational modes, rotational, and translational motions.*

The potential energy surface of a reaction that follows the transition state theory (see chapter ten) has at least two minima (where the reactants lie at one extreme and products at the other extreme) and one maximum (where the transition state lies). This is the case of the reaction described above. In these three points the $\frac{\partial E}{\partial R_i}$ of all coordinates are zero in reactants, products, and transition state. The diagonal elements of Hessian of molecular energy $\frac{\partial^2 E}{\partial R_{ij}^2}$ are all positive for reactants and products, and only one is negative (associated with the reaction coordinate) in the transition state.

Figure 8.9 shows the potential energy surface for the substitution reaction (NaBr + $CH_3Cl \rightarrow$ NaCl + CH_3Br) with three critical points (two minima and one maximum). *The energy difference between the transition state and reactant complex is the reaction energy barrier.*

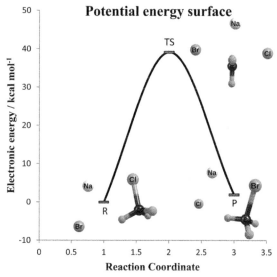

Figure 8.9 Potential energy surface (from electronic energy) of NaBr + $CH_3Cl \rightarrow$ NaCl + CH_3Br reaction.

To sum up, by optimizing the geometry of reactants and products and calculating their frequencies, one obtains their absolute internal energies, enthalpy values, entropy values, and Gibbs free energies. Then, it is possible to obtain the heat of formation of the reaction, the reaction entropy change, and the formation Gibbs free energy, which give us the information of spontaneity, equilibrium constant, and 'thermodynamic yield" of the reaction. Afterwards, one calculates the transition state of the reaction by following a similar procedure for the calculation of reactants and products except for the fact that the input information for TS calculation is somewhat different than that for the minimum of PES calculation. Hence, it is possible to obtain the energy barrier as free energy of activation, enthalpy of activation, internal energy of activation, or electronic energy of activation. At last, it is possible to calculate the rate constant, the velocity of the reaction (by knowing the concentration of the reactants in the rate determining step), and the 'kinetic yield" for a specific time interval.

Further information about the two-dimensional potential energy surface and three-dimensional potential energy surface can be found elsewhere (Carroll 2010).

INTRINSIC REACTION COORDINATE

The intrinsic reaction coordinate, IRC, corresponds to vibrationless-rotationless motion path of the reaction. The IRC calculation uses a previously calculated transition state structure as the starting point in order to search for products or intermediate in forward direction and for reactants or intermediate in reverse direction. Then, from IRC calculation, one can find two local minimum structures, provided the appropriate transition state is given (Maeda et al. 2004). The IRC is based on mass-weighted steepest decent method given by:

$$\frac{d\mathbf{q}}{ds} = \pm \frac{\mathbf{g}}{|\mathbf{g}|}$$

Where $\mathbf{q}$ is the mass-weighted cartesian coordinates and $\mathbf{g}$ is the mass-weighted gradient vector.

The IRC calculation can also be used to confirm whether the previously obtained transition state structure belongs to the reaction coordinate or not.

The method used in our calculations is based on Hessian based predictor-corrector (HPC), which uses an LQA integrator based on a second order Taylor series of PES for the predictor steps, and a fitted distance weighted interpolant surface for the corrector steps (Hratchian and Schlegel 2004). The LQA integrator is given below:

$$E(\mathbf{x}) = E_0 + \mathbf{g}_0^t \Delta\mathbf{x} + \tfrac{1}{2} \Delta\mathbf{x}^t \mathbf{H}_0 \Delta\mathbf{x}$$

Where $\Delta\mathbf{x}$, $\mathbf{g}_0$, and $\mathbf{H}_0$ are displacement vector, gradient vector, and Hessian at x_0.

Several examples and applications of IRC surfaces will be given in subsequent chapters of this book.

EXERCISES

1. Use the Boltzmann factor to obtain the ratio of populations of propene in minimum and maximum of PES (Fig. 8.10), knowing that $R = 8.31 \, \text{JK}^{-1}\text{mol}^{-1}$ and $T = 298.15 \text{K}$.

   Answer: $N_i/N_j = 10$ (which means 10 molecules of propene in minimum for 1 molecule of propene in maximum of PES).

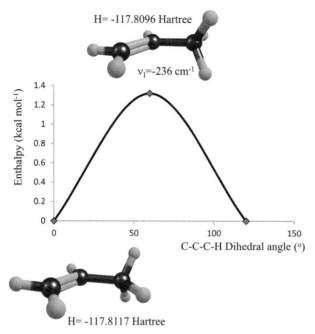

H= -117.8096 Hartree

ν_I=-236 cm^{-1}

H= -117.8117 Hartree

Figure 8.10 Plot of enthalpy versus C-C-C-H dihedral angle of propene along with its optimized geometries of minimum and maximum, and their corresponding enthalpy values calculated at ωB97XD/6-311G(d,p) level of theory.

2. According to the surface described by the function below which has three critical points: $(1,0)$, $(-1,0)$, and $(0,0)$, find which critical point is maximum and minimum at this surface using its 2×2 Hessian matrix.

$$f(x, y) = x^4 + 4x^2 y^2 - 2x^2 + 2y^2$$

$$H = \begin{vmatrix} \dfrac{\partial^2 f(x, y)}{\partial x^2} & \dfrac{\partial^2 f(x, y)}{\partial x \partial y} \\ \dfrac{\partial^2 f(x, y)}{\partial y \partial x} & \dfrac{\partial^2 f(x, y)}{\partial y^2} \end{vmatrix}$$

Acknowledgment

Figures 8.4 and 8.6 were published in our own article, whose copyrights belong to Sociedade Brasileira de Química, and we appreciate the publisher's permission for printing ths figures in this book.

REFERENCES CITED

Carroll, F.A. 2010. Perspectives on Structure and Mechanism in Organic Chemistry. John Wiley & Sons Inc. Hoboken.

Hanwell, M.D., Curtis, D.E., Lonie, D.C., Vandermeersch, T., Zurek, E. and Hutchison, G.R. 2012. Avogadro: an advanced semantic chemical editor, visualization, and analysis platform. J. Cheminformatics 4: 17.

Hratchian, H.P. and Schlegel, B. 2004. Accurate reaction paths using Hessian based predictor-corrector integrator. J. Chem. Phys. 120: 9918-9924.

Maeda, S., Harabuchi, Y., Ono, Y. Taketsugu, T. and Morokuma, K. 2004. Intrinsic reaction coordinate: calculation, bifurcation, and automated search. Int. J. Quantum Chem. 115: 1-12.

Perri, M.J. and Weber, S.H. 2014. Web-based job submission interface for the Gamess computational chemistry program. J. Chem. Educ. 91: 2206-2208.

Soares, B.A., Firme, C.L., Maciel, M.A.M., Kaiser, C.R., Schilling, E., and Bortoluzzi, A.J. 2014. Experimental and NMR theoretical methodology applied to geometric analysis of the bioactive clerodane trans-dehydrocrotonin. J. Braz. Chem. Soc. 25: 629-638.

Szabo, A. and Ostlund, N.S. 1996. Modern Quantum Chemistry - Introduction to Advanced Electronic Structure Theory. Dover Publications, Inc., New York.

Chapter Nine

Representations of Organic Molecules, Atomic Charge, and Formal Charge

FIRST REPRESENTATIONS OF ORGANIC MOLECULES

As already discussed in chapter six, Lewis developed the cubical atom model (Lewis 1913, 1916), where each vertex is filled with one valence electron represented by a circle around the vertex and he developed another model to represent molecules which became everlasting by replacing the cubes with the atomic symbols and circles in the vertexes by dots to represent the valence electrons (Fig. 9.1(A)). Before Lewis, Kekulé developed the structure of benzene made of alternating single and double bonds represented as one or two paralleled traces (Kekulé 1866), henceforth, known as the Kekulé structure (Fig. 9.1(B)). The difference between the Lewis model and the Kekulé structure is that two dots are substituted by a single trace, four dots by double trace, and six dots by triple trace to represent chemical bonds. Even before Kekulé's proposal of the benzene structure, Loschmidt (Bader and Parker 2001, Loschmidt 1861) developed another molecular model based on rings (Fig. 9.1(C)). Also in 1865, Hoffmann developed the stick-and-ball model (Hoffmann 1865) to represent the methane molecule in planar arrangement (Fig. 9.1(D)), which was later corrected by Van't Hoff, showing the tetrahedral structure of saturated carbon atom and its implication in stereoisomery (Van't Hoff 1874). The stick-and ball model is widely used to represent optimized molecules theoretically. Actually, the first organic molecular representation was developed by Couper to represent glucose, oxalic acid, and other organic molecules. (Couper 1858, Perkins 2005).

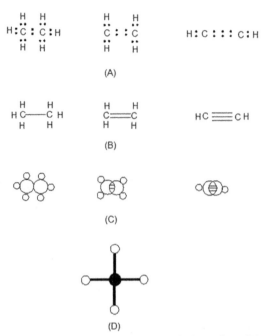

Figure 9.1   Representation of ethane, ethane, and ethyne from (A) Lewis model,
(B) Kekulé model, (C) Loschmidt model, and stick-and-ball
representation of methane (according to Hoffmann).

REPRESENTATION OF ORGANIC MOLECULES

From the first attempts to represent organic molecules graphically, only the Kekulé structure is still employed nowadays. Stick-and-ball model is very useful as well, but it is not a handy method to draw the structures by hand. For representing chemical equations, condensed formulas are more appropriate, although other forms can be used, such as bond-line formula or dashed-wedged line notation (Fig. 9.2).

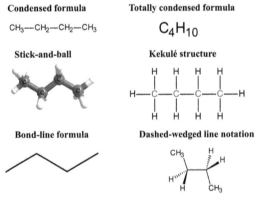

Figure 9.2   The most used molecular representations of butane.

BOND-LINE FORMULA

The bond-line formula is a handy way to represent organic molecules, mainly with long-chained skeleton. It is also the usual formula to represent molecules in a mechanism. Unlike the Kekulé structure, bond-line formula represents the geometry of the saturated or unsaturated carbon chain correctly, except to represent the branching of a saturated carbon correctly, where dashed-wedged formula is more suited. For saturated molecules with three or more carbon atoms, the bond-line formula uses the zig-zag line to represent the carbon chain which corresponds to the most stable conformer of that carbon chain.

The features of a bond-line formula are: (1) hydrogen atoms are not represented explicitly except when bonded to heteroatom (oxygen, nitrogen, etc.) or when it is the hydrogen bonded to carbonyl group in aldehydes; (2) carbon atoms are represented by dots; (3) the saturated carbon chain is represented in zig-zag conformation (the most stable conformation of a carbon chain); (4) it respects the geometry of sp^2 and sp carbon atoms, i.e., trigonal planar and linear geometries; (5) heteroatoms are explicitly represented by their corresponding symbols along with hydrogen atoms where it is the case (Fig. 9.3).

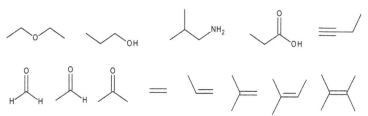

Figure 9.3   Bond-line formula of several molecules.

DASHED-WEDGED LINE NOTATION

The dashed-wedged line notation differs from bond-line formula for the tetrahedral representation of branching or even hydrogen atoms of sp^3 carbon atom. The dashed line represents the bond behind the plane of the paper and the wedged line represents the bond above the plane of the paper. There are situations in which we must use dashed-wedged line notation: (1) for representing molecules with optical stereoisomerism (Fig. 9.4(A)); (2) for representing rotamers in a lateral view where it is also possible to use sawhorse projection (Figure 9.4(B) shows how to draw these structures from the molecule's backbone); and (3) for representing saturated one-carbon molecules (Fig. 9.4(C)). Except for these three cases, it is optional to use bond-line or dashed-wedged line notation. For simplicity, the former is preferred.

The dashed-wedged line notation can be a variation of the bond line formula when representing branched molecules (even when there is no asymmetric carbon) by using same zig-zag line to represent the carbon chain and dashed-wedged bonds to represent the branch. See Fig. 9.4(A), which is used for stereoisomers but could be used for branched molecules with no stereogenic center. On the other hand, dashed-wedged line notation can be used to represent the rotamers of any molecule with two carbon atoms or more (ethane, propane, butane, or larger), where Fig. 9.4(B) shows how to draw this type of dashed-wedged line notation, which is different from the former case.

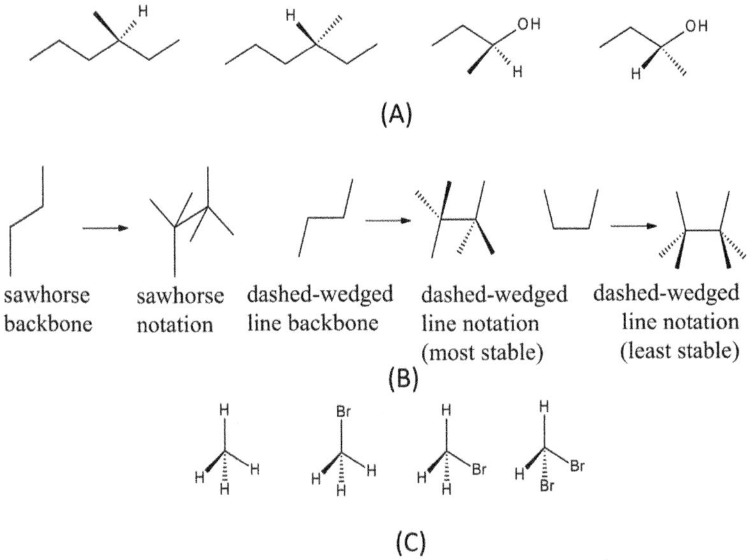

(A)

| sawhorse backbone | sawhorse notation | dashed-wedged line backbone | dashed-wedged line notation (most stable) | dashed-wedged line notation (least stable) |

(B)

(C)

Figure 9.4 (A) Dashed-wedged line notation of two pairs of enantiomers (optical stereoisomers); (B) construction of sawhorse and dashed-wedged line notation from four-carbon backbone; (C) dashed-wedged line notation of methane, bromomethane, and dibromomethane.

The features of dashed-wedged line notation are very similar to those from bond-line formula, except for the representation of branched sp^3 carbon and for the optional explicit representation of hydrogen atoms bonded to sp^3 carbon atom.

NEWMAN PROJECTION

As mentioned above, the lateral view of rotamers (or conformers) can be represented by dashed-wedged line notation. Alternatively, the lateral view of rotamers can be represented by sawhorse projection as well. On the other hand, the front view of rotamers can be represented in a handy scheme, called Newman projection after its inventor (Newman 1955).

In Newman projection, there is a circle with a point in the middle which represents the fore carbon and behind this carbon, the rear carbon is located. Both carbons made up the bond which generates conformers by its rotation. Each of these carbons (front and rear carbons) is bonded to other three atoms. The bonds of front carbons are represented by lines (similar to Kekulé structure) and when there is a bond with a hydrogen, it must by explicitly represented. The bonds of the rear carbon are half represented where the line begins at the line of the circle.

These six bonds (three from the front and three from the rear carbon) are planarized in Newman projection. Then the virtual bond angle between them is 60° which is also the dihedral angle interval between two conformers. Figure 9.5 shows the construction of Newman projection from the lateral view of the two rotamers of ethane, staggered (Fig. 9.5(A)) and eclipsed (Fig. 9.5(B)), both of them in their optimized geometries.

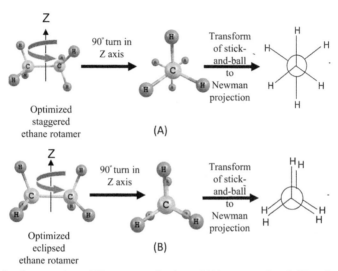

Figure 9.5   Construction of Newman projection of (A) staggered and (B) eclipsed ethane from their corresponding optimized stick-and-ball representations.

FISCHER PROJECTION

Another planarized representation of a molecule containing sp^3 carbons is called Fischer projection due to its inventor Emil Fischer. This German chemist made outstanding contributions to organic synthesis, which led him to win a Nobel prize in chemistry (Forster 1920). While synthesizing several carbohydrates and analyzing their configuration, Fischer developed in 1981 a handy two-dimensional representation of monosaccharides, thanks to the knowledge of the tetrahedral nature of saturated carbon atom from Van't Hoff's work (Nagendrappa 2011). Fischer projection is an important tool to represent stereoisomers and/or to build stereoisomers, mainly with more than one asymmetric carbon.

In Fischer projection, each carbon from the carbon chain is an intersection point of a cross, where its horizontal lines represent the bonds directed towards the viewer (from the central atom) and the vertical lines represent the bonds directed behind the viewer. Figure 9.6(A) shows the four diastereoisomers (each one having two asymmetric carbon atoms whose configurations are represented by *S* or *R*- see more details in chapter thirteen) of threonine, a proteinogenic aminoacid, represented by Fischer projection. Figure 9.6(B) shows how to construct a simple Fischer projection.

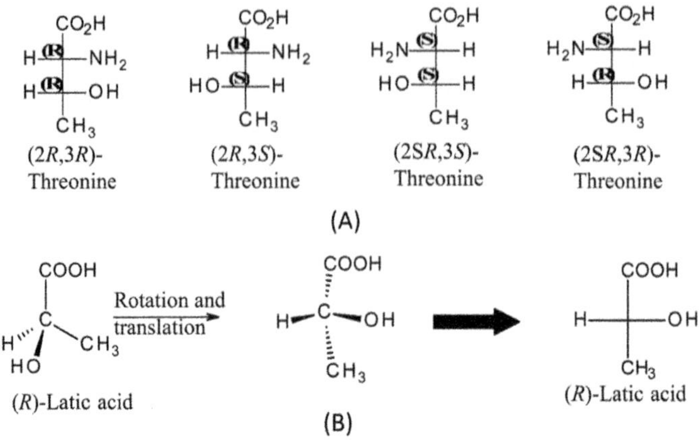

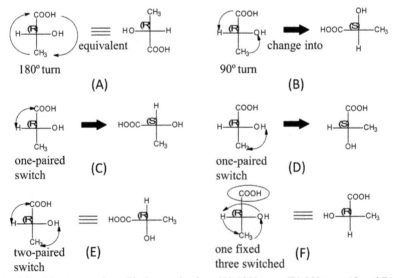

Figure 9.6 (A) Fischer projection of the four diastereoisomers of threonine and (B) construction of Fischer projection from dashed-wedged line notation of (R)-latic acid.

In Fischer projection it is possible to move the molecule (through 180° or 90° rotation) or switch the substituents (one-paired switch, two-paired switch, or one fixed three switched). The 180° turn keeps the same configuration of the asymmetric carbon (yielding the same molecule), while a 90° turn leads to the corresponding stereoisomer by changing the configuration of the asymmetric carbon (Fig. 9.7(A) and Fig. 9.7(B), respectively). Likewise, a one-paired switch of substituents changes the configuration of the asymmetric carbon and a two-paired switch of substituents does not change its configuration (Fig. 9.7(C) and Fig. 9.7(D), respectively). At last, it is also possible to keep any substituent fixed and move clockwise or anti-clockwise. This movement does not change the configuration of the carbon, leaving the molecule unaltered (Fig. 9.7(E) and (F)).

Figure 9.7 Movements from Fischer projection: (A) 180° turn; (B) 90° turn; (C and D) one-paired switch; (E) two-paired switch; and (F) one fixed three switched from (R)-latic acid.

FORMAL CHARGE

The formal charge, FC, is a very easy, useful method to obtain the charge of an atom in a molecule.

$$FC_i = Z_i - Ne_i$$

Where FC_i is the formal charge of atom i, Z_i is the atomic number (number of protons) of atom I, and Ne_i is the number of electrons of atom I, in which each single, double, or triple covalent bond counts only one, two, or three electron(s) for each bonding atom, respectively. See the note on next page.

   According to its formula, formal charge is always an integer number, which is an important limitation of FC since almost always the actual charge is a fractional number, i.e., a partial atomic charge. Nonetheless, it has a practical utility for evaluating the charge flux of a reaction and to evaluate the nucleophilic or electrophilic site in a molecule, that is, the atom where the electron charge is concentrated or dispersed, respectively, in a molecule.

   Since no quantum mechanical calculation is done to obtain the formal charge, it is only a rough approximation of the atomic charge and it is very imprecise for some molecules, e.g., carbocations, carboxylates, CO and $NaBH_4$ or $LiAlH_4$ (see Fig. 9.8). In the case of carbocations, by knowing that electronegativity of hydrogen and carbon atoms is 2.1 and 2.5, respectively, hydrogen atom is more electropositive than carbon atom, then each hydrogen has a fractional positive charge. On the other hand, aluminium is more eletropositive than hydrogen (where electronegativity of aluminium is 1.4), then in AlH_4^- each hydrogen atom has a fractional negative charge. Regarding the carboxylate, both oxygen atoms have a fractional negative charge. Despite these exceptions, formal charge is useful to observe the electrophile site of carbocations (i.e., the carbon atom), the nucleophile site of the carboxylate (i.e., the oxygen atom) and the electrophile site of nitronium ion and nitrosonium ion, NO_2^+ and NO^+, respectively (i.e., the nitrogen atom).

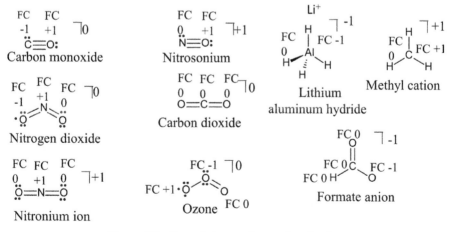

Figure 9.8   Formal charge of several molecules.

Note

Octet rule: One counts for each atom its non-bonding and bonding valence electrons, plus the shared electrons from its bonded atom(s), but does not count the core electrons.

Formal charge: One counts for each atom its non-bonding and bonding valence electrons, plus the core electrons, but does not count the shared electrons from its bonded atom(s).

PARTIAL ATOMIC CHARGE

Partial atomic charge, q, whose atomic unit is e (1.602×10^{-19} Coulomb), occurs in all heteronuclear molecules and it is related to the asymmetric distribution of the charge density along the molecular system having polar covalent bonds. However, there is not an universal best procedure for the calculation of partial atomic charge. Then, several methods have been devised to calculate the partial atomic charge theoretically. One of the most appropriate methods for calculating partial atomic charges for molecules is QTAIM (or AIM) atomic charge (Cramer 2004). Moreover, the partial atomic charge is also dependent upon the method and/or the basis set to obtain the wave function. It is advisable to use the relatively precise *ab initio* method, e.g., couple cluster methods with single and double excitations, CCSD, to calculate the wave function for a subsequent calculation of partial atomic charge. We have observed severe errors using MP2 or DFT methods for some molecular systems, even using the most appropriate method to calculate partial atomic charge. *Then, the user should be aware of using a precise level of theory and a precise partial atomic charge method in order to obtain reliable values of partial atomic charge.*

The QTAIM atomic charge has the following equation (see chapter two):

$$q(\Omega) = Z_\Omega - \int_\Omega d\tau \rho(r)$$

In our calculations we used the optimized wave function from CCSD/6-31G++(d,p) level of theory in order to obtain the matrix density for further QTAIM calculation of the integration of each atomic basin, Ω, in order to obtain its corresponding QTAIM partial atomic charge. Then, Fig. 9.9 shows the theoretical partial atomic charges of the same molecules used in Fig. 9.8 for comparison.

In carbon monoxide, the negative charge ($-1.38e$) is located at the oxygen atom as expected since it is more electronegative than the carbon atom. Then, formal charge gives an opposite and wrong information about the actual atomic charges in carbon monoxide. Likewise, formal charge does not indicate the electrophilic site (carbon atom) in carbon dioxide, as one can see the QTAIM partial atomic charge of carbon atom ($+2.679e$). On the other hand, formal charge correctly indicates the electrophilic site of NO_2, NO_2^+, and NO^+ which is located in nitrogen atom with QTAIM charge $+0.798e$, $+1.004e$, and $+1.443e$, respectively. Again, in ozone, formal charge gives an opposite sign for mid and external oxygen atoms in comparison with the corresponding QTAIM partial atomic charge.

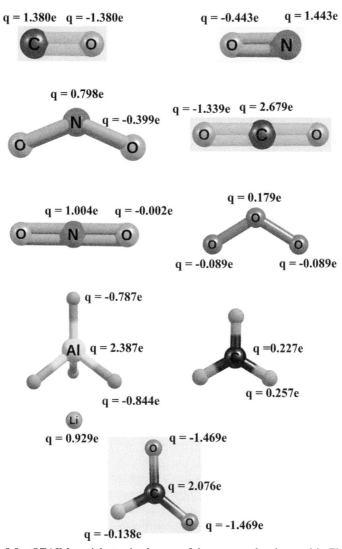

Figure 9.9   QTAIM partial atomic charges of the same molecules used in Fig. 9.8.

In lithium aluminum hydride, as expected, the negative charge is located in hydrogen atoms and not in aluminum atom as informed by formal charge. In methyl cation, the positive charge is distributed over all hydrogen and carbon atoms, being higher in hydrogen because it is more electropositive than carbon. In formal charge approach, the positive charge is solely located in the carbon atom. In the case of formate anion, the Kekulé structure used in Fig. 9.8 is actually one of its resonance structures (which is usually used more to represent organic molecules) rather than the resonance hybrid which would be more similar to its optimized structure. This fact explains why the negative charge is equally distributed in both oxygen atoms in the formate anion and concentrated in one oxygen atom in the formal charge result.

EXERCISES

1. From the molecules represented by stick-and-ball model in Fig. 9.10, pass them to bond-line formula.

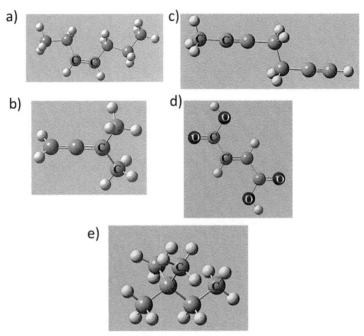

Figure 9.10 Stick-and-ball model of some molecules for Exercise 1.

2. From the numbered carbon atoms in Fig. 9.11, represent the corresponding molecules in dashed-wedged line notation of the most stable rotamer (staggered), where each substituent bonded to these atoms should be represented in a condensed form when they are larger than hydrogen. Take for example the dashed-wedged line notation of butane from its staggered rotamer from carbon atoms 2 and 3 in Fig. 9.2. Finally, from the latter pass them to the corresponding Newman projection.

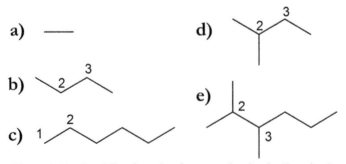

Figure 9.11 Bond-line formula of some molecules for Exercise 2.

3. Represent in Fischer projection the molecules in Newman projection below (Fig. 9.12):

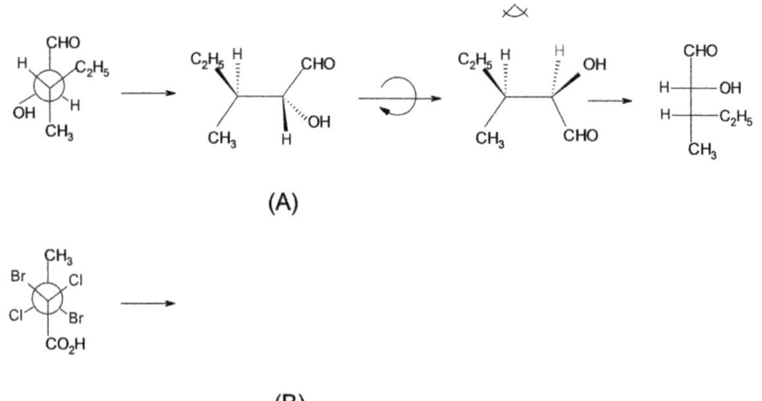

(A)

(B)

Figure 9.12   Newman projection of molecules for exercise 3:
(A) resolved exercise; (B) exercise to be done.

=========== **REFERENCES CITED** ===========

Bader, A. and Parker, L. 2001. Joseph Loschmidt, physicist and chemist. Phys. Today 54: 45-50.

Couper, A.S. 1858. On a new chemical theory. Philosophical Mag. 16: 104-116.

Cramer, C.J. 2004. Essentials of Computational Chemistry: Theories and Models. John Wiley & Sons. Ltd. The Atrium.

Forster, M.O. 1920. Emil fischer memorial lecture. J. Chem. Soc. Trans. 117: 1157-1201.

Hoffmann, A. 1865. On the combining power of atoms. Proc. Royal Inst. 4: 401-430.

Kekulé, A. 1866. Untersuchungen über aromatische verbindungen ueber die constitution der aromatischen verbindungen. i. ueber die constitution der aromatischen verbindungen. Eur. J. Org. Chem. 137: 129-196.

Lewis, G.N. 1913. Valence and tautomerism. J. Am. Chem. Soc. 35: 1448-1455.

Lewis, G.N. 1916. The atom and the molecule. J. Am. Chem. Soc. 38: 762-785.

Loschmidt, J. 1861. Chemische Studien I. Carl Gerold's Sohn, Vienna.

Nagendrappa, G. 2011. Hermann emil fischer: life and achievements. Resonance 16: 606-618.

Newman, M.S. 1955. A notation for the study of certain stereochemical problems. J. Chem. Educ. 32: 344-347.

Perkins, J.A. 2005. A history of molecular representation. Part one: 1800 to the 1960s. J. Biocommun. 31.

Van't Hoff, J.H. 1874. A suggestion looking to the extension into space of the structural formulas at present used in chemistry, and a note upon the relation between the optical activity and the chemical constituents of organic compounds. Archives Neerlandaises des Sciences Exactes et Naturelles 9: 445-454.

Chapter Ten

Kinetics and Mechanism: Notions and the Quantum Statistical Influence

GENERAL INFORMATION ABOUT CHEMICAL REACTIONS

Except for radical reactions, the vast majority of chemical reactions (including Bronsted-Lowry acid-base reaction) occur between an electrophile and nucleophile. An electrophile is a molecule with dispersed electron charge in a specific region called electrophilic site. A nucleophile is a molecule with concentrated electron charge in a specific region called nucleophilic site. When an electrophile encounters a nucleophile and they have sufficient kinetic energy to surpass the energy barrier (chapter eight), they react by forming a new chemical bond between the nucleophilic and electrophilic sites. Depending on the electrophile, whileforming a new bond, the other chemical bond is broken simultaneously.

Electrophiles are generally carbocations, hydronium ion, nitronium ion, halogenoalkanes, carbonyl compounds, where the electrophilic site is the trivalent carbon (in carbocations), carbon bonded to halogen (in halogenoalkanes), and carbonyl carbon (in carbonyl compounds). Nucleophiles are generally carbanions (in organometallic substances), hydrides, alcohols, amines, alkenes, and alkynes where the nucleophilic site is located in one atom (carbon atom with negative charge in carbanion, oxygen, or nitrogen atom in alcohol or amine, hydrogen atom in hydrides) or chemical bond (double or triple bond in alkene and alkyne).

Figure 10.1 depicts some (A) electrophiles and (B) nucleophiles. The circled region comprises the electrophilic site and nucleophilic site, respectively.

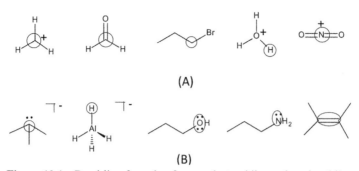

(A)

(B)

Figure 10.1   Bond line formula of some electrophiles and nucleophiles.

Some molecular groups can react as a nucleophile or an electrophile depending on the substituent effect and the other reactant. For example, aromatic compounds generally react as nucleophiles, but depending on the conditions mentioned above, they can react as electrophiles. Along the next chapters more information will be given about eletrophiles and nucleophiles, and electrophilic and nucleophilic sites. In the next book (a sequel to this very book), we will discuss nucleophilicity and electrophilicity deeply. *So far, one needs to know that reaction kinetics (i.e., the velocity of reaction) depends on nucleophilicity and electrophilicity. By increasing one of them or both, the velocity of the reaction increases.* In addition, the reaction kinetics is also related to the yield. By increasing the velocity of the reaction the yield increases.

Then, before a chemical reaction (which does not have radical mechanism), one needs to identify at first the electrophile, its electrophilic site, the nucleophile, and its nucleophilic site in order to understand how the reaction takes place. We will develop this reasoning in the next chapters.

GENERAL INFORMATION ABOUT MECHANISM

The mechanism indicates how the reactants approach to react by forming the transition state, yielding the products through the flux of electrons, bond breaking, and bond formation. The flux of electrons is indicated by arrows to represent the flux of electrons from the nucleophile towards the electrophile in order to pass by the transition state. *For simplicity, we will use transition state to refer to the transition structure or activated complex.*

A mechanism follows a reaction path, but one reaction can have more than one reaction path. One specific mechanism is associated with a specific potential energy surface. In the former it is represented the flux of electrons by arrows, bond breaking and/or forming by dashed lines, while the latter represents in a plot how the reaction energy changes from reactants to transition state and to products.

Mostly, the mechanisms are represented by bond-line formula or wedged-dashed line notation in cases of stereoselective reactions or one carbon center. In some cases drawing in perspective is crucial to understand the mechanism and the products formation.

One needs to identify the nucleophilic site of nucleophile and the electrophilic site of the electrophile. The arrow begins at the nucleophilic site and terminates at the eletrophilic site. In the transition state, dashed lines represent bonds being broken and/or formed. The reactions involving nucleophiles and electrophiles are called polar reactions and the arrow represents the displacement of a pair of electrons. Radical reactions involve the displacement of a single electron represented by a semi arrow.

Let us draw the mechanism for the reaction:

$$NaBr + CH_3Cl \rightarrow NaCl + CH_3Br$$

which was already discussed in chapter eight.

The bromide anion is the nucleophile (and nucleophilic site) and chloromethane is the electrophile (whose electrophilic site is the carbon atom). There is an arrow from the bromide anion pointing towards the carbon atom of chloromethane. Since carbon atom is tetravalent, its C-Cl bond has to be broken in order to form a new bond with bromide anion. Then, another arrow is used to represent the electron pair movement from C-Cl bond towards chlorine atom. Then, in the transition state, there appears a distorted trigonal bipyramidal geometry with one bond being formed (C-Br) and one bond being broken (C-Cl). Both of them are represented by dashed lines (Fig. 10.2). In the transition state, both chlorine and bromine atoms have partial atomic charge represented by δ (and not q) when its actual value is not known. The transition state is represented by square brackets and the symbol of transition state ($\neq$).

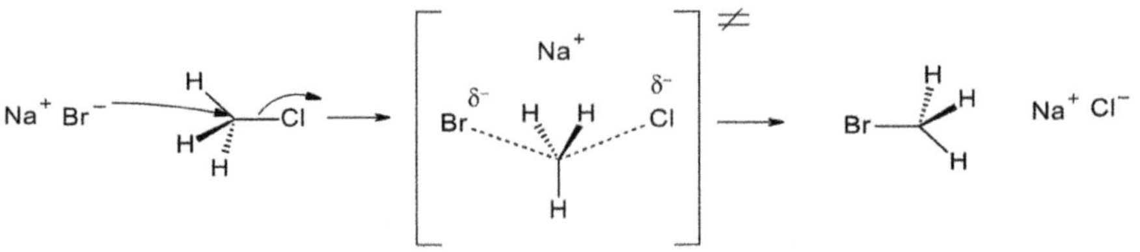

Figure 10.2   Mechanism of substitution reaction of sodium bromide and chloromethane.

The mechanism must represent as close as possible the actual geometries of reactants, intermediates (where it occurs), transition states, and products.

The mechanism of a reaction can have one step, two steps, three steps, or more. In this case, the term "step" means a part of the reaction. If a reaction has only one step, this step is the whole reaction. This is the case for the reaction of sodium bromide with chloromethane. It is a single-step reaction.

CHEMICAL KINETICS AND COLLISION THEORY

Chemical kinetics is the study of expressions to determine the rate of a reaction. The central idea of the collision theory is the requirement of appropriate collision between reactants. In that theory, the rate of a reaction is dependent upon the number of effective collisions between reactants. Effective collisions result in the formation of products from reactants. The minimum energy for an effective collision is given by the energy barrier or activation energy of the rate determining step. Besides minimum energy, effective collision also requires the reactants to be oriented in the correct position in order to promote the reaction. Then, *effective collision requires reactants to approach each other at the correct orientation and with the minimum energy*. More information about collision theory can be found elsewhere (Peters 2017).

HAMMOND´S POSTULATE

There are three energy profiles of energy barrier in a chemical reaction, which are named according to the similarities between transition state and reactant or transition state and intermediate/product. They are: (1) early transition state where the transition state and reactant have nearly similar energy and geometry; (2) late transition state where the transition state and intermediate or product have similar energy and geometry; and (3) here named symmetric transition state where the transition state is very different in energy and geometry with respect to the reactant, intermediate and product. In 1955, Hammond postulated that when a transition state and its subsequent intermediate have nearly the same energy content, their interconversion is a consequence of a small geometric reorganization. Then, any factor (for example, the type and number of substituents attached to a trivalent carbon in a carbocation) which affects the stability or instability of the intermediate in a late transition state also affects the stability or instability of the precedent transition state. Similarly, for an early transition state, the same reasoning is applied to the reactant and its subsequent transition state.

RATE LAWS AND REACTION RATE

Chemical kinetics is associated with the velocity of a given reaction which could be understood as the rate at which products are formed. For example, for the general reaction $R \rightarrow P$, the rate of reaction, r, is given by:

$$r = -\frac{d[R]}{dt} = \frac{d[P]}{dt}$$

Alternatively, the rate of the reaction can be written as:

$$r = k[R]$$

Where k is called the rate constant.

Both equations can be united in just one which is called the rate law.

$$\frac{d[P]}{dt} = k[R]$$

The rate law gives the information on how many molecules participate in the rate determining step of the reaction. *The rate law depends on the reactants that participate in the rate determining step of a reaction. The rate determining step has the highest energy barrier (when there are two or more energy barriers).*

The number of steps of a reaction depends on the number of energy barriers it has. If a reaction has one reaction barrier, it has only one step. In this case, the rate determining step is the only step in this reaction, obviously. If a reaction has two energy barriers, it has two steps. If a reaction has three energy barriers, it has three steps, and so on. In a multi-step reaction, the rate determining step can be the first, last, or in between. Moreover, one reaction can have a step without an energy barrier when this step occurs at a very fast velocity.

The number of molecules that participate in the transition state of the rate determining step defines its molecularity. It can be unimolecular where only one molecule is in the transition state of the highest energy barrier (which does not mean necessarily that only one molecule reacts to form a product or products), bimolecular where two molecules participate on the transition state of the rate determining step and termolecular where three molecules participate on the transition state of the rate determining step.

Let us suppose the reaction between *tert*-butanol and hydrogen chloride.

$$(CH_3)_3C - OH + HCl \rightarrow (CH_3)_3C - Cl + H_2O$$

This reaction has three steps:

1. $(CH_3)_3C - OH + HCl \rightarrow (CH_3)_3C - O^+H_2Cl^-$
2. $(CH_3)_3C - O^+H_2 \rightarrow (CH_3)_3C^+ + H_2O$
3. $(CH_3)_3C^+ + Cl^- \rightarrow (CH_3)_3C - Cl$

The second step is the rate determining step. Then, for this reaction the rate law is:

$$r = k[(CH_3)_3C - OH]$$

Its rate law is first order for *tert*-butanol and zero order for hydrogen chloride and it is known as first order kinetics. Figure 10.3 shows a partially pictorial representation of its potential energy surface (in terms of enthalpy). All minima in PES were calculated at $\omega B97XD/6-31G++(d,p)$ level of theory. Maxima in PES were not calculated, which explains the term "partially pictorial". In this PES one can see that the first step is very fast and there is no energy barrier for this reaction and the second step has the highest energy barrier which determines the velocity (the kinetics) of this reaction. It is important to say that at this moment it is not our objective to provide further details of the corresponding mechanism.

For the reaction described in the last section and in chapter eight:

$$NaBr + CH_3Cl \rightarrow NaCl + CH_3Br$$

Where both NaBr and CH_3Cl participate in the transition state of the rate determining step, its rate law is:

$$r = k[CH_3Cl][NaBr]$$

The rate law is first order for both NaBr and CH_3Cl, and it is known as second order kinetics.

By varying the concentration of reactants and evaluating the velocity of the reaction it is possible to obtain the rate law of a reaction. Usually, one measures the velocity of the reaction experimentally by means of the reaction time. In this case, there is an inverse relation between reaction rate and reaction time, i.e., by doubling or increasing the reaction rate four-fold, the reaction time decreases by a half or a quarter, respectively.

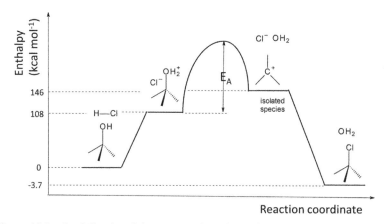

Figure 10.3   Partially pictorial representation of PES from the reaction between *tert*-butanol and hydrogen chloride, where all minima were calculated from ωB97XD/6-31G++(*d*,*p*) level of theory.

Let us consider a general reaction, $A+B \rightarrow C+D$, and we wonder whether this reaction is first order, second order, or something else. Firstly, one must begin the reaction with a reference concentration for the reactants, let's suppose 1 M each. Let us also suppose that for this case, this reaction finishes at a given time, t. This will be called the reference reaction (experiment #1 in Table 10.1). If the concentration of A is doubled to 2 M and the reaction time decreases by half the reference reaction, $t/2$, then reaction is first order for reactant A (experiment #2). If the reaction time decreases by a quarter of the reference reaction, $t/4$, then the reaction is second order for reactant A (experiment #3). If the reaction time does not change in comparison with the reference reaction, then the reaction is zero order for reactant A (experiment #4). The same reasoning is applied to reactant B, keeping reactant A with its concentration of the reference reaction, i.e., 1 M (experiments #5, #6, and #7).

Then, in Table 10.1, we show these experiments and some hypothetical rate law of a general reaction $A+B \rightarrow C+D$. Let us take some examples depicted in Table 10.1 The rate law $v=k[A][B]$ can be observed from experiments #2 and #5, where the concentration of A or B is doubled while the other (B or A) is kept constant, the reaction time diminishes by half (and velocity is doubled) in both cases. The rate law $r=k[A][B]^2$ can be observed from experiments #2 and #6 where the reaction time decreases by half in experiment #2 and decreases by a quarter in experiment #6. The rate law $r=k[A]$ is observed from experiments #2 and #7. In experiment #7, the reaction time does not change by doubling the concentration of B.

Table 10.1 Some hypothetical order and rate law for the general reaction $A+B \rightarrow C+D$.

Experiment	[A]/M	[B]/M	Reaction time	Order	Rate law
#1 (reference)	1	1	t	-	-
#2	2	1	$t/2$	1st for A	-
#3	2	1	$t/4$	2nd for A	-
#4	2	1	t	0th for A	-
#5	1	2	$t/2$	1st for B	-
#6	1	2	$t/4$	2nd for B	-
#7	1	2	t	0th for B	-
#2 and #5	-	-	-	2nd for reaction	$r=k[A][B]$
#2 and #6	-	-	-	3rd for reaction	$r=k[A][B]^2$
#2 and #7	-	-	-	1st for reaction	$r=k[A]$
#3 and #5	-	-	-	3rd for reaction	$r=k[A]^2[B]$
#3 and #6	-	-	-	4th for reaction	$r=k[A]^2[B]^2$
#3 and #6	-	-	-	2nd for reaction	$r=k[A]^2$
#4 and #5	-	-	-	1st for reaction	$r=k[B]$
#4 and #6	-	-	-	2nd for reaction	$r=k[B]^2$

ARRHENIUS EQUATION AND TRANSITION STATE THEORY

In 1884 van't Hoff found an expression for the temperature dependence of the equilibrium constant which greatly influenced Arrhenius to discover another expression for temperature dependence of the rate constant in 1889 (Laidler 1984). Hence, the transition state theory was started by Arrhenius's work about the temperature dependence of the rate constant of sugar isomerization according to the equation below (named after Arrhenius). Usually, by increasing the temperature, k also increases.

$$k = Ae^{(-E_A/RT)}$$

Where A is the pre-exponential factor and E_A is the activation energy (or the energy barrier of the rate determining step).

When $E_A \gg RT$, that is, when E_A is much higher than 0.60 kcal mol^{-1} at 300 K, it is assumed that the transition state (or activated complex) is in virtual equilibrium with the reactants, which is known as the transition state theory (Wright 2004).

$$RT = 8.31 \, JK^{-1}mol^{-1} \times 0.239 \times 10^{-3} \frac{kcal}{1J} \times 300 \, K = 0.60 \, kcal/mol$$

By taking the natural logarithm of the Arrhenius equation we have the linear equation:

$$\ln k = -\frac{E_A}{R}\left(\frac{1}{T}\right) + \ln A$$

Where the independent variable is $\frac{1}{T}$, the dependent variable is $\ln k$, the slope of the line is $-\frac{E_A}{R}$, and the intercept in y-axis is $\ln A$. By varying the reaction temperature and measuring the rate constant at each temperature, one can obtain A and E_A.

The influence of the temperature on the rate constant and reaction rate can be easily evaluated from the exponential part of the Arrhenius equation. Let us suppose activation energy of 20 kcal mol^{-1} at 300 K, then the exponential term is:

$$20\frac{kcal}{mol}\times\frac{4184J}{1kcal}\times\frac{1}{8.31JK^{-1}mol^{-1}}\times\frac{1}{300K}=-33.5$$

$$e^{(-33.5)}=2.82\times10^{-15}$$

The first important observation is that the exponential part of the Arrhenius equation is dimensionless and its value is very small. By increasing the temperature from 300 K to 310 K, the value of the exponential is more than three-fold higher, then, by increasing 10 K (or 10°C) from 300 K with energy barrier of 20 kcal mol^{-1} the reaction becomes three-fold faster.

$$\frac{10^4}{K^{-1}}\times\frac{1}{310K}=-32.4$$

$$e^{(-32.4)}=8.48\times10^{-15}$$

The unit of pre-exponential Arrhenius parameter, A, depends on the rate law. For first order reactions it is s^{-1} and for second-order reaction it is L mol^{-1} s^{-1} and each reaction has its specific value of A (Atkins 1998).

For a one-step or a multi-step reaction, the transition state theory, TST, assumes a surface dividing reactants and products where all molecules between reactants and transition state of the single barrier (in one-step reaction) or between reactants and transition state of the rate determining step, RDS, (the highest barrier of a multistep reaction) are in quasi-equilibrium. Moreover, the TST assumes that trajectories that pass over the highest barrier do not recross.

For a multi-step reaction, the barriers before the highest barrier do not interfere with the quasi-equilibrium condition in that "pool" between reactants and transition state of RDS. In addition, the steps before and after the RDS do not influence the overall rate of the multistep reaction.

The assumptions of quasi-equilibrium, QE, condition and rate determining-step for a multi-step reaction that follows TST are very important simplifications (when they are applicable) to derive rate laws for several reactions which could not be possible otherwise or it would be rather complicated (Peters 2017). For example, by only applying the pseudo-steady state approximation (a more general simplification than QE and RDS), the rate for aromatic nitration using HNO$_3$ and H$_2$SO$_4$ as reactants in a four-step reaction is given by the expression below (Peters 2017):

$$r_{ArNO_2}=\frac{\left(K_1K_2k_3/K_{H_2SO_4}\left[H_2O\right]\right)\left[HNO_3\right]\left[H^+\right]\left[ArH\right]}{1+\left(k_3/k_{-2}\left[H_2O\right]\right)\left[ArH\right]k_a}\therefore$$

$$k_a=\left\{1+\left(k_2/k_1\right)\left[1+\left[H^+\right]/K_{H_2SO_4}\right]/\left(\left[H_2SO_4\right]+\left[HSO_4^-\right]\right)\right\}$$

When using QE and RDS simplifications, the above equation becomes much more simple:

$$r_{ArNO_2} = k_3 [HNO_3][H^+][ArH]$$

Where the rate constant of the third step, k_3, is related to the rate determining step.

EYRING-POLANYI EQUATION TO THE TRANSITION STATE THEORY

In 1935, Eyring, Evans, and Polanyi improved the transition state theory, TST, independently (Eyring 1935, Evans and Polanyi 1935) by introducing partition functions in the reaction rate equation. Before Eyring, Evans, and Polanyi's works, Marcellin and Rice were the first to quantitatively describe the rate constant based on the statistical-mechanical treatment (Laidler and King 1983).

Both Eyring's work and Evans and Polanyi's work in 1935 introduced a rate constant equation expressed in terms of the partition functions, q, of the transition state, q_{TS}, and the reactants, q_i, similar to the equation below:

$$k = \kappa \frac{k_B T}{h} \frac{q_{TS}/V}{\prod_i (q_i/V)^{v_i}} e^{\left(-\Delta E^{\neq}/k_B T\right)}$$

Where V is the volume occupied by the reactants, $\Delta E^{\neq}$ is the energy barrier of the reaction, k_B is the Boltzmann constant, h is the Planck constant, and κ is the transmission coefficient representing the probability of activated complex of transition state (or simply the transition state) to form the product and not returning to reactants. In transition state theory, this probability is nearly 100% and κ is a unit.

One can also find this equation in the form (Laidler and King 1983):

$$k = \kappa \frac{k_B T}{h} \frac{Q_{TS}}{\prod_i (Q_i)} e^{\left(-\Delta E^{\neq}/RT\right)}$$

Where Q_{TS} and Q_i are the canonical partition functions of the transition state and reactants, i, and $\Delta E^{\neq}$ can be expressed in terms of the electronic energy.

In the same year, Eyring and Wynne-Jones developed another rate constant equation using thermodynamic parameters (Wynne-Jones and Eyring 1935). Before this work, other scientists also gave important contributions to the thermodynamic approach of the rate constant yielding similar expressions to that shown below (Laidler and King 1983).

$$k = \kappa \frac{k_B T}{h} e^{\left(-\Delta G^{\neq}/RT\right)}$$

This equation is very important to obtain activation enthalpy, $\Delta H^{\neq}$, and activation entropy, $\Delta S^{\neq}$, from the relation below using the plot $\ln[k\beta h]$ vs $1/T$:

$$\ln[k\beta h] = \Delta S^{\neq}/k_B - \beta \Delta H^{\neq}$$

Where the slope gives the activation enthalpy and intercept in ordinate gives the ratio $\Delta S^{\ne}/k_B$. This plot is known as the Eyring plot. More information about thermodynamic Eyring's equation can be found elsewhere (Carroll 2010).

Alternatively, the Eyring-WynneJones equation or simply the Eyring equation can be written as:

$$k = \frac{k_B T}{h} \frac{Q_{TS}/V}{\prod_i (Q_i/V)^{v_i}}$$

By knowing the equilibrium constant equations below (Peters 2017):

$$K = e^{(-\Delta G/RT)}$$

$$K = \frac{\prod_j (Q_j/V)^{v_j}}{\prod_i (Q_i/V)^{v_i}}$$

Where Q_j and Q_i are the canonical partition functions of products, j, and reactants, i.

The canonical partition function is given by:

$$Q = Q_{transl} Q_{rot} Q_{elect} Q_{vib} e^{\left(-\beta U_{vib(0)}\right)}$$
$$U_{vib(0)} = E_0 + ZPE \therefore \beta = 1/k_B T$$
$$ZPE = \tfrac{1}{2}\Sigma_i \hbar \omega_i$$

Where ω_i is the ith vibrational angular frequency, E_0 is the electronic energy, and ZPE is zero-point energy (chapter four). The canonical partition functions are given below (Peters 2017):

$$Q_{transl} = V/\lambda_T^3$$
$$Q_{rot(nonlinear)} = 8\pi^2 \Big/ \left\{\sigma_{sym} \prod_{k=1}^3 \varphi_T(I_k)\right\}$$
$$Q_{rot(linear)} = 4\pi \Big/ \left\{\sigma_{sym} \varphi_T(I_k)^2\right\}$$
$$Q_{vib} = \prod_i \left(1 - e^{-\beta \hbar \omega_i}\right)^{-1}$$
$$Q_{U(rotor)} = 2\sqrt{\pi}\Big/\varphi_T(I)$$

Where λ_T is de Broglie wavelength, φ_T is the thermal angle, and σ_{sym} is the symmetry number (an integer number according to the molecular point group).

Some reference values and useful combinations are given below (Peters 2017):

$$k_B T/h = 6.25 \times 10^{12}\ s^{-1} \therefore T = 300\,K$$
$$k_B T = 2.4945\ kJ \cdot mol^{-1} \therefore T = 300\,K$$
$$\hbar \omega = 11.95\ kJ \cdot mol^{-1} \therefore \omega = 1000\,cm^{-1}$$
$$\lambda_T = h\big/\sqrt{2\pi m k_B T} = 1.004\ Ang \therefore T = 300K \therefore m = 1.0\,amu$$
$$\varphi_T = h\big/\sqrt{2\pi I k_B T} = 1.004 \therefore T = 300K \therefore m = 1.0 (\text{Å})^2\,amu$$
$$\beta \hbar \omega = 4.79 \therefore T = 300\,K \therefore \omega = 1000\,cm^{-1}$$

In his book, Peters explains why it is possible to exclude the imaginary frequency from the Q_{TS} for the calculation of the rate constant (Peters 2017).

MORE ABOUT THE QUANTUM STATISTICAL TRANSITION STATE THEORY

First Approach

By considering the general reaction below:

$$A + B \rightleftarrows [A \text{----} B]^{\neq} \xrightarrow{\text{v}_1} C + D$$

The rate constant can be written as:

$$k = \kappa \frac{k_B T}{h} N_A \frac{q'^{\neq*}}{q'_A q'_B} e^{\left(-\Delta E^{\neq}/RT\right)}$$

$$\frac{k_B T}{h} = 10^{13}\,\text{s}^{-1}/\text{molecule} \therefore \text{at 300K}$$

$$A = \kappa \frac{k_B T}{h} N_A \frac{q'^{\neq*}}{q'_A q'_B}$$

Where $q'^{\neq}$, q'_A, and q'_B are the partition functions of the transition state, $[A\text{---}B]^{\neq}$, and reactants A and B per unit volume, respectively (Fogler 2011), and $q'^{\neq*}$ is the partition function of the transition state without one vibrational term associated with the imaginary frequency.

As already discussed in chapter four, partition function involves electronic, translational (related with volume of the container holding the molecule and its molecular mass), rotational (related with moments of inertia, I, and rotational symmetry numbers, S_y), and vibrational (related with fundamental frequencies of vibration, v) terms.

$$q = q_E q_T q_V q_R$$
$$k_B = 1.38 \times 10^{-23}\,\text{JK}^{-1}/\text{molecule}(\text{kg} \cdot \text{m} \cdot \text{s}^{-2} \cdot \text{K}^{-1}/\text{molecule})$$
$$h = 6.626 \times 10^{-34}\,\text{Js}(\text{kg} \cdot \text{m}^2 \cdot \text{s}^{-1})$$

The electronic partition function, q_E, is usually close to one. The molar partition function, q_m, is obtained from the ratio between partition function and number of moles, n. The partition function per unit volume is obtained by the ratio of molar partition function and molar volume, V_m, in dm^3/mol. From the partition function per unit volume, q', the unique term that is used per unit volume is the translational partition function, q'_T.

$$q_m = \frac{q}{n}$$

$$q' = \frac{q_m}{V_m} = q_E q'_T q_V q_R$$

In the translational partition function per unit volume, q'_T, each translation is in the order of 10^{30} m^{-3}. In the rotational partition function each rotation contributes to 10 to 1000, and in the vibrational partition function each vibration has from 1 to 10 (Wright 2004).

$$q'_T = \frac{(2\pi m k_B T)^{3/2}}{h^3}$$

$$q_V = \left(\frac{1}{1-e^{\left(-\frac{h\nu}{k_B T}\right)}}\right) \cong \frac{k_B T}{h\nu}$$

$$q_R = \frac{8\pi^2 I k_B T}{S_y h^2}$$

For example, in a reaction between two atoms to form one diatomic activated complex, the ratio of the partition functions is proportional to:

$$\frac{q'^{\neq}}{q'_A q'_B} \propto \frac{3transl, 2rot, 1vib}{3transl, 3transl} = \frac{2rot, 1vib}{3transl}$$

Where each atom has three translational movements, no vibrational movement, and no rotational movement, and a diatomic activated complex has three translational movements, two rotational movements, and one vibrational movement with imaginary frequency.

In Fig. 10.4, we see how kinetics is affected by temperature. By increasing the temperature, the population of reactant particles with minimum kinetic energy, KE_m, to surpass the energy barrier increases.

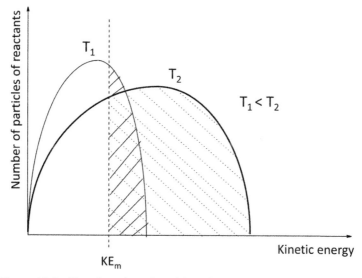

Figure 10.4 Plot of number of particles of reactants versus kinetic energy.

Second Approach

There is an alternative way to rationalize the rate of the reaction using partition functions. Let us suppose again the general reaction:

$$A+B \rightleftarrows [A----B]^{\neq} \xrightarrow{\;V_I\;} C+D$$

The rate of the reaction can be associated with the rate of crossing the energy barrier. In the transition state, there is one imaginary frequency (see chapter eight), v_I, associated with the reaction coordinate. The rate of crossing the energy barrier or the rate of the reaction is given by:

$$r = v_I C_{[A\cdots B]^{\neq}}$$

By knowing the equilibrium constant, $K^{\neq}$, for the equilibrium, $A+B \rightleftarrows [A\text{--}B]^{\neq}$, is given by:

$$K^{\neq} = \frac{C_{[A\cdots B]}^{\neq}}{C_A C_B}$$

Then, the rate of the reaction is:

$$r = v_I K^{\neq} C_A C_B$$

$$K^{\neq} = \frac{C_{[A\cdots B]}^{\neq}}{C_A C_B} = e^{\left(-\Delta E^{\neq}/RT\right)} N_A \frac{q'^{\neq}}{q'_A q'_B}$$

Where q' is the partition function per unit volume. Then, the rate of the reaction is:

$$r = v_I e^{\left(-\Delta E^{\neq}/RT\right)} N_A \frac{q'^{\neq}}{q'_A q'_B} C_A C_B$$

Let us expand the vibrational partition function of the transition state in order to factor out the term associated with the imaginary frequency.

$$q_V = \frac{k_B T}{h v}$$

$$q_V^{\neq} = q_V^{\neq *} q_{VI} = q_V^{\neq *} \frac{k_B T}{h v_I}$$

$$q'^{\neq} = q_E^{\neq} \cdot q_R^{\neq} \cdot q_V^{\neq} \cdot q_T'^{\neq}$$

$$q'^{\neq} = q_E^{\neq} \cdot q_R^{\neq} \cdot q_V^{\neq *} \cdot q_{VI}^{\neq} \cdot q_T'^{\neq}$$

$$q'^{\neq} = q_E^{\neq} \cdot q_R^{\neq} \cdot q_V^{\neq *} \cdot q_T'^{\neq} \frac{k_B T}{h v_I}$$

$$q_E^{\neq} \cdot q_R^{\neq} \cdot q_V^{\neq *} \cdot q_T'^{\neq} = q'^{\neq *}$$

Then
$$q'^{\neq} = q'^{\neq *} \frac{k_B T}{h v_I}$$

The last equation is the partition function for the transition state where the imaginary frequency has been factored out. By replacing the last equation of the partition function for the transition state in the reaction rate equation:

$$r = v_I N_A \frac{q'^{\neq}}{q'_A q'_B} e^{\left(-\Delta E^*/RT\right)} C_A C_B$$

We have the following equation:

$$r = \frac{k_B T}{h} N_A \frac{q'^{\neq*}}{q'_A q'_B} e^{\left(-\Delta E^*/RT\right)} C_A C_B$$

Be aware of the fact that the slight difference between the vibrational partition function for the transition state, $q^{\neq}$, and the vibrational partition function for the transition state without the imaginary frequency, $q^{\neq*}$, provides a significant difference in the reaction rate equation.

The ratio $k_B T/h$ is:

$$\frac{k_B T}{h} = \frac{1.38 \times 10^{-23}}{6.626 \times 10^{-34}} \times \frac{JK^{-1} molecule^{-1}}{Js} \times xK$$

$$\frac{k_B T}{h} = x \cdot 2.08 \times 10^{10} s^{-1} molecule^{-1}$$

The rate constant is:

$$k = \frac{k_B T}{h} N_A \frac{q'^{\neq*}}{q'_A q'_B} e^{\left(-\Delta E^*/RT\right)}$$

$$k = v_I N_A \frac{q'^{\neq}}{q'_A q'_B} e^{\left(-\Delta E^*/RT\right)}$$

Since the exponential part of the Eyring equation is dimensionless, the rate constant unit is:

$$units[k] = \left(\frac{1}{s \cdot molecule}\right) \times \left(\frac{\left(\frac{1}{dm^3}\right)}{\left(\frac{1}{dm^3}\right)\left(\frac{1}{dm^3}\right)}\right) \times \frac{molecule}{mol} = dm^3 mol^{-1} s^{-1}$$

The unit of rate constant is the same for both equations of reaction rate since the unit of v_I is $s^{-1} molecule^{-1}$, equivalent to that from the ratio $k_B T/h$.

See in Appendix 3 the relation between statistical thermodynamics and equilibrium constant to the transition state $K^{\neq}$.

FUNDAMENTALS OF HETEROGENEOUS CATALYSIS

Study Case: Hydrogenation of Alkenes

The catalyst is a molecule (usually in homogeneous catalysis), a macromolecule (in case of enzymes), or a solid material (in heterogeneous catalysis) that interacts or reacts with the reactants, promoting the decrease of the energy barrier of the rate determining step in comparison with the non-catalyzed reaction. *In most cases, catalysts increase the nucleophilicity or electrophilicity of the reactant, which lowers the energy barrier and increases the reaction rate.*

Some metals (nickel, copper, platinum, palladium) are used as catalysts in hydrogenation of alkenes and alkynes. These reactions are called heterogeneous catalysis, since there appears two phases: a gas or liquid phase and a solid phase. There are two ways in which the hydrogen is absorbed in metal surface: (1) molecular or nondissociative adsorption or physisorption (where hydrogen is absorbed as a molecule); and (2) dissociative absorption or chemisorption (where H-H bond in broken and hydrogen atoms are absorbed in the metal surface). In the former there is a van der Waals interaction between the gas and metal, and in the latter there is a direct chemical bond between the atom and metal (Fig. 10.5). Nondissociative adsorption of hydrogen occurs with copper and dissociative adsorption of hydrogen occurs with palladium (Lapujoulade and Perreau 1983, Fogler 2011). In the case of dissociative adsorption of hydrogen in palladium, it is not clear whether it has an unactivated path (without energy barrier) or an activated path with very low barrier or both (Johansson et al. 2010). *The physisorption or chemisorption is the first step of the catalyzed hydrogenation reaction whose rate determining step has a lower barrier (not bold path in Fig. 10.5(C)) than that from non-catalyzed reaction (bold path in Fig. 10.5(C)).* The metal increases the electrophilicity of the hydrogen.

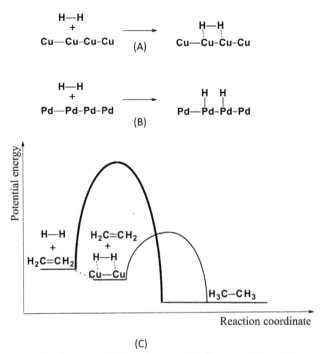

Figure 10.5 (A) Physiorption of H_2 on cooper; (B) chemisorption of H_2 on palladium; and (C) potential energy surface of catalyzed (thin line) and noncatalyzed (line in bold) hydrogenation of ethene.

FUNDAMENTALS OF HOMOGENEOUS CATALYSIS

Study Case: Bronsted-Lowry Acid-base Catalysis

An important role in catalysis: acids increase the electrophilicity of the electrophilic reactant and bases increase the nucleophilicity of nucleophilic reactant. The mode catalysis occurs (i.e., acceleration of reaction rate) by increasing nucleophilicity or electrophilicity of the reactant by means of Bronsted-Lowry acid or base. In most cases, the acid and base used as homogeneous catalysts (where there is only one phase) are inorganic.

Acid or base catalysis can occur in two different ways: (1) specific acid or base catalysis and (2) general acid or base catalysis. *Specific acid catalysis occurs when the rate determining step is not the proton transfer towards the organic reactant. This occurs when the proton is transferred to oxygen atoms in carbonyl compounds, for example. General acid catalysis occurs when the proton transfer to the organic reactant is the rate determining step. This is the case of proton transfer to carbon atoms* (Ault 2007).

In the specific acid catalysis when protic polar solvent is used, there are two fast reaction steps before the rate determining step: (1) proton transfer from the inorganic acid to the protic solvent and (2) proton transfer from the hydronium ion (when the solvent is water) to the organic reactant (mainly to its heteroatom). The third step is the rate determining step. In Fig. 10.6, there is an example of specific acid catalysis. The first step is the proton transfer from sulfuric acid to water. followed by the proton transfer from the hydronium ion to the formaldehyde forming the protonated formaldehyde, whose carbonyl carbon atom is more electrophilic than that from neutral formaldehyde. In other words, the hydronium ion increased the electrophilicity of the organic substrate (formaldehyde) and increased the reaction rate. The third is the rate determining step where water molecule (from its oxygen atom as the nucleophilic site) attacks the protonated formaldehyde (in the carbon atom as the electrophilic site).

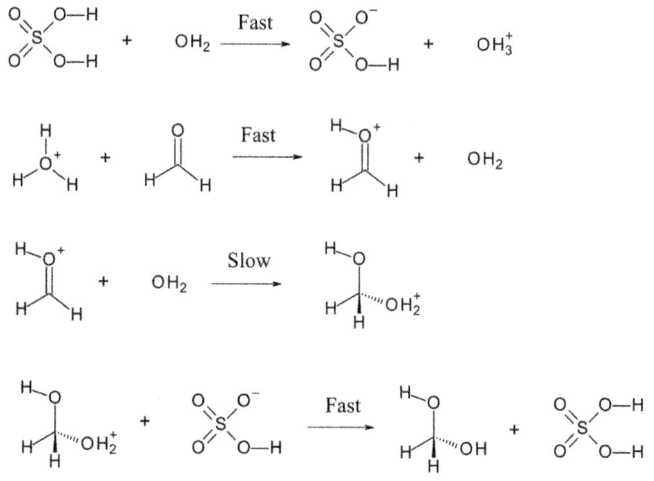

Figure 10.6   Steps of the specific acid hydrolysis of formaldehyde.

In the general acid catalysis there are two different reaction paths that take place simultaneously. In the first path, the first step (the proton transfer from inorganic acid to water) is fast and the second step (the proton transfer from hydronium ion to organic reactant) is the rate determining step. In the second path, the first step (the proton transfer from inorganic acid to the organic reactant directly) is the rate determining step. Figure 10.7 shows one example of general acid catalysis – the hydration of an alkene catalyzed by acid. In this case, the reaction is first order for the acid (hydronium ion and/or the inorganic acid directly) and alkene (Anslyn and Dougherty 2006).

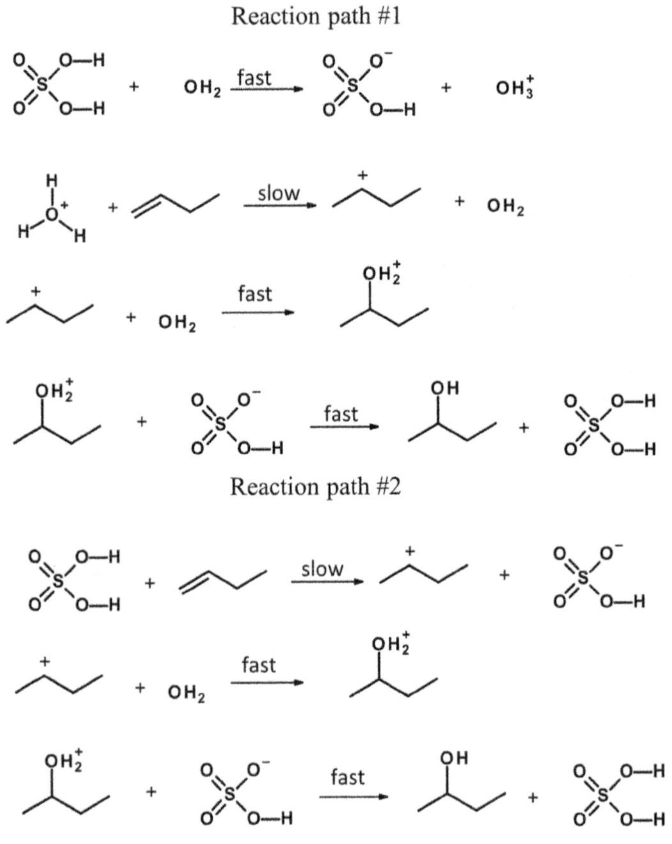

Reaction path #1

Reaction path #2

Figure 10.7   Steps of acidic hydrolysis of but-1-ene.

EXCEPTIONS AND LIMITATIONS OF THE TRANSITION STATE THEORY

As Peters mentioned in his book: "Eyring plots are best to interpret activation parameters for elementary steps. When applied to pseudo-rate constants that emerge from multistep mechanisms, the activation parameters from Eyring plot may inadvertently inherit the characteristics of various adsorptions constants and equilibrium constants in the overall mechanism (...) which can lead to a negative activation enthalpy. (...) Even elementary rate constants involve tunneling and recrossing effects that are not included in the TST expression" (Peters 2017).

Some reactions do not follow the linearity of Eyring plot. One of these exceptions is the reaction where tunneling effect takes place. Tunneling effect occurs when reactants jump the energy barrier avoiding its maximum critical point. Tunneling effect is more common for proton or hydrogen atom transfer. Not all proton or hydrogen transfer have tunneling effect, for example, proton transfer in acid hydrolysis of alkenes. The kinetics of reactions with tunneling effect do not follow transition state theory. The rate constant equation is rather different from the Arrhenius equation (Anslyn and Dougherty 2006).

$$k = \left[\frac{e^{E/RT}}{\beta - E/RT} \left(\beta e^{-E/RT} - (E/RT) e^{-\beta} \right) \right] A e^{-E/RT}$$

$$\beta = 2a\pi^2 (2mE)^{1/2} / h$$

Where A is the Arrhenius pre-exponential factor and a is half the barrier width.

Another type of exception to the transition state theory is the reaction that is (partly, at least) dictated by dynamical behavior (Anslyn and Dougherty 2006). Further details are not within the scope of this book, and can be found in Peters's book.

EXERCISES

1. Give the rate of reaction at 300 K involving two non-linear molecules with three atoms each (A and B) (whose $DOF_{(v)} = 3N-6$) forming one non-linear activated complex (whose $DOF_{(v)} = 3N_A + 3N_B - 6$) where each translational, vibrational, and rotational partition functions are $10^{30}\,m^{-3}$, 10 and 100, respectively. Each molecule has three translational movements and three rotational movements. The concentration of each reactant is 1 M, the energy barrier is 20 kcal mol^{-1}, and the imaginary vibrational frequency of the transition state is 750 cm^{-1}. Data: $R = 8.31\,J\,K^{-1}\,mol^{-1}$. Use both equations for rate of reaction:

$$r = v_I N_A \frac{q'^{\ne}}{q'_A q'_B} e^{\left(-\Delta E^{\ne}/RT \right)} C_A C_B$$

$$r = \frac{k_B T}{h} N_A \frac{q'^{\ne *}}{q'_A q'_B} e^{\left(-\Delta E^{\ne}/RT \right)} C_A C_B$$

Notes

(a) remember that speed of light, c, is the product of frequency, in Hz, and wavelength, in m.

$$\lambda v = c$$

In chemistry, it is usual to represent frequencies in cm^{-1} which is the wavenumber. Then to convert a frequency in cm^{-1} to Hz (or s^{-1}), one needs to multiply the former by the speed of light ($2.998 \times 10^{10}\,cm\,s^{-1}$).

$$\tilde{v}(\text{cm}^{-1}) = \frac{1}{\lambda} = \frac{v}{c}$$

(b)
$$k_B = 1.38 \times 10^{23} \, \text{kg} \cdot \text{m} \cdot \text{s}^{-2} \cdot \text{K}^{-1} / \text{molecule}$$
$$h = 6.626 \times 10^{-34} \, \text{kg} \cdot \text{m}^2 \cdot \text{s}^{-1}$$

(c) $N_A = 6.02 \times 10^{23}$ molecules/mol
(d) $1\,\text{kcal} = 4184\,\text{J}$
(e) $-\Delta E_0(\text{kcal mol}^{-1})/RT$

$$-\frac{x \times 4184\,\text{J}}{\text{mol}} \times \frac{1}{8.31\,\text{JK}^{-1}\text{mol}^{-1}\,y\text{K}} = -x(\text{kcal/mol}) \times \frac{503.49}{y(\text{K})}$$

(f) By factoring out the imaginary frequency, the number of vibrational partition functions for the transition state becomes: $3N_A + 3N_B - 7$

Answers

$$\frac{q'^{\neq}}{q'_A q'_B} = 0.148 \times 10^{-36} \, \text{dm}^3$$

$$v_I N_A = 13.53 \times 10^{36} \, \text{s}^{-1}\text{mol}^{-1}$$

$$\exp(-E_A/RT) = 2.82 \times 10^{-15}$$

$$r = 5.65 \times 10^{-15} \, \text{mol} \cdot \text{s}^{-1} \cdot \text{dm}^3$$

$$\frac{k_B T}{h} N_A = 3.76 \times 10^{36} \, \text{s}^{-1}\text{mol}^{-1}$$

$$\frac{q'^{\neq *}}{q'_A q'_B} = 0.135 \times 10^{-36} \, \text{dm}^3$$

$$r = 0.51 \times 10^{-15} \, \text{mol} \cdot \text{s}^{-1} \cdot \text{dm}^3$$

The difference between the values of the reaction rate is due to the approximation used in the calculation of the partition functions.

2. Give the rate constant at 330 K involving one atom and one linear diatomic molecule to form one non-linear activated complex where each translational, vibrational, and rotational partition functions are $10^{27} \, \text{dm}^{-3}$, 1 and 10, respectively. The energy barrier is 40 kJ mol^{-1} and the imaginary vibrational frequency of the transition state is 500 cm^{-1}. Use both equations for the rate constant.

3. Give the rate constant for the $H_2 + H$ reaction passing through [H---H---H] transition state at T = 1000 K by using the equation below (Peters 2017):

$$k = \frac{k_B T}{h} \frac{Q_{TS}/V}{\prod_i (Q_i/V)^{v_i}}$$

$$k = \frac{k_B T}{h} \sigma_{H_2} \frac{\lambda_H^3 \lambda_{H_2}^3}{\lambda_{TS}^3} \frac{\varphi_{H_2}^2}{\varphi_{TS}^2} \prod_i^{TS} \frac{\left(1 - e^{-\beta\hbar\omega_i}\right)^{-1}}{\left(1 - e^{-\beta\hbar\omega_{H2}}\right)^{-1}} e^{-\beta(\Delta E_{vib(0)})}$$

Data

	H	H_2	$[H\text{---}H\text{---}H]$
σ_{el}	2	1	2
σ_{sym}	1	2	2
E_0 (kJ mol^{-1})	-	0.0	41
ω (cm$^{-1)}$	-	4395	2058, 909, 909, 1511i
$d_{H\text{-}H}$ (Å)	-	0.74	0.93, 0.93

For the TS, the center of mass is at the central H, and then $I_{TS}=2(0.74/2)^2$ 1.007 amu $= 1.742$ Å^{2}amu. For H_2, $I_{H2}=0.276$Å^{2}amu. The thermal de Broglie angle φ_T is 1.004 (for $I=1.0$Å^{2}amu and $T=300$K). The thermal angle for TS, φ_{TS}, is 0.643 at 420 K and for H_2 is 1.615. The thermal de Broglie wavelengths for TS, λ_{TS}, is 0.488 Å, for H_2 is 0.598 Å, and for H is 0.846 Å at 420 K. The vibrational energy difference at 0 K ($\Delta E_{vib(o)}$) is 37.9 kJ mol^{-1} and transition state symmetry number is 1.

Answer: $k = 2.63 \times 10^9$ Å/reaction/s (Peters 2017).

REFERENCES CITED

Anslyn E.V. and Dougherty, D.A. 2006. Modern Physical Organic Chemistry. University Science Books, Sausalito.

Atkins, P.W. 1998. Physical Chemistry. Oxford University Press, Melbourne.

Ault, A. 2007. General acid and general base catalysis. J. Chem. Educ. 84: 38-39.

Carroll, F.A. 2010. Perspectives on Structure and Mechanism in Organic Chemistry. John Wiley & Sons Inc. Hoboken.

Evans, M.G. and Polanyi, M. 1935. Some applications of transition state method to the calculation of reaction velocities, especially in solution. Trans. Faraday Soc. 31: 875-894.

Eyring, H. 1935. The activated complex in chemical reactions. J. Chem. Phys. 3: 107-115.

Fogler, H.S. 2011. Essentials of Chemical Reaction Engineering. Prentice Hall, Pearson Education, Inc. Westford.

Lapujoulade, J. and Perreau, J. 1983. The diffraction of molecular hydrogen from copper surfaces. Physica Scripta 1983: 138.

Laidler K.J. and King, M.C. 1983. The development of transition state theory. J. Phys. Chem. 87: 2657-2664.

Laidler, K.J. 1984. The development of the Arrhenius equation. J. Chem. Educ. 61: 494-498.

Johansson, M.K., Skúlason, E., Nielsen, G., Murphy, S., Nielsen, R.M. and Chorkendorff, I.B. 2010. Hydrogen adsorption on palladium and palladium hydride at 1 bar. Surface Sci. 604: 718-729.

Peters, B. 2017. Reaction Rate Theory and Rare Events. Elsevier. Amsterdam.

Wright, M.R. 2004. An Introduction to Chemical Kinetics. John Wiley & sons, Ltd. The Atrium.

Wynne-Jones, W.F.K. and Eyring, H. 1935. The absolute rate of reactions in condensed phases. J. Chem. Phys. 3: 492-502.

Chapter Eleven

Intermolecular Interactions

DIPOLE MOMENT

Dipole moment, μ, is a vector physical quantity (i.e., it needs direction and magnitude to describe it) and a unimolecular/microscopic property (i.e., it needs only one molecule to obtain its value theoretically). Its vector points towards the negative charge from the positive charge (Fig. 11.1(A)) and it is the product of displacement vector, r, and charge, q. The dipole moment vector arises from the electronegativity difference between the bonding atoms. The more electronegative atom has the partial negative charge and the less electronegative atom has the partial positive charge.

$$\vec{\mu} = \vec{r} \cdot q$$

The unit for dipole moment is Debye, which is historically defined as 10^{-10} esu. Å, where esu means electrostatic unit of charge. The atomic (or elementary) charge is 1.602×10^{-19} Coulomb or 4.8×10^{-10} esu (or statcoulombs). For example, for a chemical bond having 0.2 esu of positive and negative charges separated by 1.54 Å:

$$\vec{\mu} = (0.2 \times 4.8 \times 10^{-10}) \times (1.54) = 1.48 \, D$$

In SI units the dipole moment is given by C m, where $1 D = 3.336 \times 10^{-19}$ C m. Then, the dipole moment of NaCl (each ion having a unit charge) separated at distance 1.874 Å (where $1 \, Å = 10^{-10}$ m) is:

$$\mu = \frac{1.602 \times 10^{-19} \times 1.874 \times 10^{-10}}{3.336 \times 10^{-30}} = 9 \, D$$

For heteronuclear diatomic molecules, there is a single dipole moment vector, for example, in NO, CO, HBr, etc. Figure 11.1(B) shows the electron charge concentrated in fluorine atom (having negative partial charge) in HF. For homonuclear diatomic molecules, there is no dipole moment vector because both nuclei are equal and then there is no difference of electronegativity between then, for example, in H_2, N_2, O_2, Cl_2, etc. Figure 11.1(C) shows that the electron charge is equally distributed between both fluorine atoms in F_2, having no dipole moment.

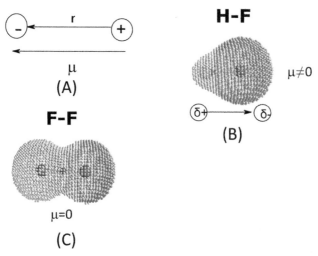

Figure 11.1 (A) Schematic representation of dipole moment vector; (B) dipole moment vector in HF and its corresponding electron density; (C) electron density of F_2 molecule.

In poliatomic molecules (with three or more heteronuclear atoms), there are two or more vectors of dipole moment components whose resultant vector is the dipole moment of the molecule. This resultant vector is obtained from the vector sum of the vector components.

$$\vec{\mu} = \vec{\mu_1} + \vec{\mu_2} + ... + \vec{\mu_i}$$

The resultant vector is obtained by the use of parallelogram law (Fig. 11.2(A)). For the simplest case, a triatomic molecule, H_2O molecule, for example, there are two component vectors of dipole moment pointing from each hydrogen atom towards the oxygen atom. By using parallelogram law, the resultant vector points from mid space between both hydrogen atoms towards the oxygen atom (Fig. 11.2(B)). In some cases, it is possible to have non-zero dipole moment components and zero dipole moment resultant, as in the case of tetrachloromethane and fumaric acid (Fig. 11.2(C)).

Sometimes when comparing dipole moments, atomic charge prevails over the interatomic distance for determining the dipole moment. That is the case when comparing dipole moment of ketone or aldehyde with that from alcohol with similar molecular mass. For example, in methanol, the C-O(H) bond length is higher than C=O bond length in formaldehyde, but dipole moment of formaldehyde is higher than that from methanol because sp^2 oxygen (in formaldehyde) is more electronegative than sp^3 oxygen (in methanol). Then, partial charge in formaldehyde oxygen is higher than that in methanol oxygen, which explains the higher dipole moment of formaldehyde with respect to methanol (Fig. 11.3).

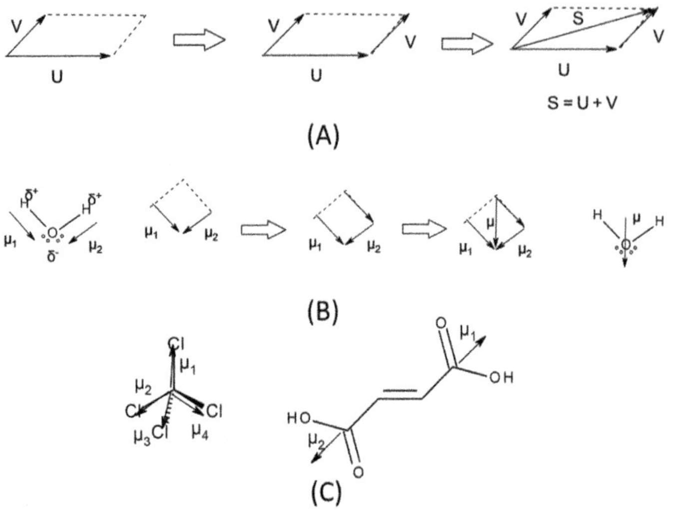

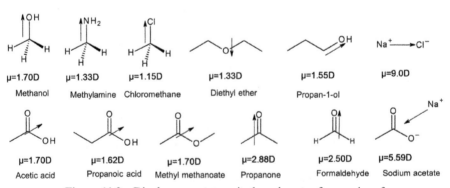

Figure 11.2 (A) Parallelogram law for general vectors U and V; (B) parallelogram law in water molecule for its dipole moment components; (C) dipole moment components of tetrachloromethane and fumaric acid.

Carboxylic acid and ester have smaller dipole moments than aldehyde and ketone because dipole moment components in carboxylic acid and ester have somewhat opposite directions decreasing the dipole moment resultant in comparison with that from aldehyde and ketone. Compare propanoic acid and methyl methanoate with propanone in Fig. 11.3 (although propanone has slightly smaller molecular mass, this fact does not influence dipole moment decisively).

μ=1.70D	μ=1.33D	μ=1.15D	μ=1.33D	μ=1.55D	μ=9.0D
Methanol	Methylamine	Chloromethane	Diethyl ether	Propan-1-ol	
μ=1.70D	μ=1.62D	μ=1.70D	μ=2.88D	μ=2.50D	μ=5.59D
Acetic acid	Propanoic acid	Methyl methanoate	Propanone	Formaldehyde	Sodium acetate

Figure 11.3 Dipole moment magnitude and vector for a series of organic molecules plus NaCl.

Inorganic salts or even organic salts always have higher dipole moments than corresponding neutral organic molecules, because in the former there is a whole charge in cation and anion, while in neutral organic molecules there is only a partial charge (smaller than a whole charge). In addition, the distance between cation and anion is higher than that in covalent bond length in neutral organic molecules. In

Fig. 11.3 one can compare the dipole moments of acetic acid (1.70 D) and sodium acetate (5.59 D).

The electronegativity order oxygen > nitrogen > chlorine gives the same order of their partial charge (considering that they are bonded to same atom, in that case, the carbon atom), which is responsible for the dipole moment order: methanol > methylamine > chloromethane (see in Fig. 11.3).

The electrostatic potential map gives the information of the charge distribution represented by a colored surface over the molecule backbone. The red area (a low potential) shows the abundance of electrons (or the nucleophilic site). The blue area (a high potential) depicts the charge dispersion (or the electrophilic site), which indicates the regions of (partial) negative charge and (partial) positive charge, respectively, i.e., both charged poles of a molecule. See the example of acetone (propanone) in Fig. 11.4(A), where the carbon atom is the positive pole and oxygen atom is the negative pole. The electrostatic potential map can also provide the quadrupole of molecule. For example, benzene has no dipole moment but it has a quadrupole moment with two negative poles below and above the aromatic ring and two positive poles within the aromatic ring (Fig. 11.4(B)).

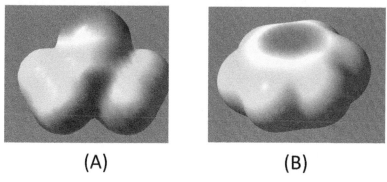

(A) **(B)**

Figure 11.4   Electrostatic potential map of (A) acetone and (B) benzene.
Color version at the end of the book

INTERMOLECULAR INTERACTIONS

Intermolecular interaction is a macroscopic property (i.e., it needs a bulk of molecules to determine its magnitude). However, *a handy, effective model to interpret the intermolecular interactions is the bimolecular model where a complex of two molecules interact with each other*. Intermolecular interactions play a decisive role in several areas of science and technology. Some physical quantities of substances are directly or partly related to their corresponding intermolecular interactions, for instance, solubility, viscosity, melting point, boiling point, surface tension, etc. As for the industrial sector, intermolecular interaction is crucial for the pharmaceutic industry (interaction enzyme-pharmacophore), dye industry (interaction surface-dye/pigment), perfume industry (interaction skin-fixative), cosmetic industry (interaction oil-surfactant), flotation process, etc.

Intermolecular interactions depend on the electrostatic interaction between opposed charges of neighbor molecules. The types of intermolecular interactions are: (1) dipole-dipole interaction; (2) induced dipole-induced dipole interaction; (3) dipole-induced dipole interaction; (4) ion-dipole interaction; (5) ion-induced dipole interaction; (6) hydrogen bond; (7) hydrogen-hydrogen bonding; (8) π-stacking; (9) hydrophobic interaction; (10) cation-π bond interaction. The term dipole can be replaced by multipole because the latter is more comprehensive. For example, the benzene molecule has no dipole but it has non-zero quadrupole and it is able to interact with molecules and ions through its π electrons.

The van der Waals interactions encompass several intermolecular interactions, but not all. They are: (1) London dispersion (or only dispersion) associated with induced multipole-induced multipole interaction; (2) Debye interaction (or polarization or induction) associated with the permanent multipole-induced multipole interaction; (3) electrostatic or Coulombic interaction; and (4) repulsion from Pauli exclusion principle. Mistakenly, some of these terms are associated with the word "force" which is not correct because in nature there are only four fundamental forces and all of these interactions have the same electrostatic nature (see chapter three). Nonetheless, the molecular energy can be decomposed in these terms. For example, the Symmetry Adapted Perturbation Theory (SAPT) can be used to decompose the binding energy of binary molecular systems into electrostatics (or Coulombic), exchange, induction, and dispersion components (Szalewicz 2011).

$$E_{int}^{SAPT0} = E_{coulomb}^{(1)} + E_{exch}^{(1)} + E_{ind}^{(2)} + E_{disp}^{(2)}$$

The exchange component or exchange term in the energy partitioning (or decomposition) is due to the antissymmetric feature of the fermions when two particles are exchanged. Unlike all other energy terms from SAPT, the exchange component has no force associated with it. Sometimes, the exchange component is called Pauli repulsion.

DIPOLE-DIPOLE INTERACTION

The dipole-dipole interaction occurs between neighboring polar molecules (i.e., molecules with dipole moment vector). Each polar molecule has a negative pole (with negative partial charge) and a positive pole (with positive partial charge).

By using the bimolecular model, before optimization, both chloromethane molecules are arbitrarily positioned with the same poles on the same side in parallel direction. After optimization, the dipoles stand in opposing directions with opposite poles on the same side which indicates (through optimization) the maximum attractive interaction (Fig. 11.5(A)).

Figure. 11.5(B) shows the optimized geometries of chloromethane chloro-methane and fluoromethane-fluoromethane dipole-dipole interaction from ωB97XD /6-311G++(2d,p) level of theory. The average distance between opposite poles from vicinal molecules is 3.80 and 3.10 Å for chloromethane-chloromethane and fluoromethane-fluoromethane in parallel directions, respectively.

The energy of dipole-dipole interaction can be calculated from the bimolecular model using enthalpy formation of dipole-dipole complex. Remember the inverse relation between potential energy and force. Then, the smallest (or more negative) interaction energy (or formation enthalpy of dipole-dipole complex) indicates the strongest dipole-dipole interaction.

$$\Delta H_{f(complex)} = H_{(complex)} - 2H_{(isolated\text{-}molecule)}$$

The dipole-dipole interaction increases as the dipole moment of the involved molecules increases. For example, Fig. 11.5(B) depicts the formation enthalpy of the dipole-dipole complex, $\Delta H_{f(dipole\text{-}dipole\ complex)}$, of a couple of methyl halides (according to the bimolecular model), where we see that dipole-dipole interaction energy of chloromethane ($\mu = 1.99\,D$) < fluoromethane ($\mu = 1.87\,D$). Then, the dipole-dipole interaction, $F_{dipole\text{-}dipole}$ (as well as dipole moment) follows an opposite trend: $F_{dipole\text{-}dipole(chlorometane)} > F_{dipole\text{-}dipole(fluorometane)}$.

The dipole-dipole interaction is ruled by the electrostatic force between opposing partial charges. The energy of this dipole-dipole interaction can be directly calculated from electrostatic force ($E = F.r$). If the dipoles are in parallel directions Fig. 11.5(C), the electrostatic force and the corresponding energy of dipole-dipole interaction are:

$$F_{dipole\text{-}dipole(parallel)} = K\frac{\delta_A^+\delta_B^-}{r_{\delta_A^+\delta_B^-}^2} + K\frac{\delta_A^-\delta_B^+}{r_{\delta_A^-\delta_B^+}^2}$$

$$E_{dipole\text{-}dipole(parallel)} = K\frac{\delta_A^+\delta_B^-}{r_{\delta_A^+\delta_B^-}} + K\frac{\delta_A^-\delta_B^+}{r_{\delta_A^-\delta_B^+}}$$

Where A and B are vicinal molecules (from same or different substances).

Alternative equations for dipole-dipole interaction force and energy (in parallel directions) are given below by using dipole moment instead of charge. Similar equations can be found elsewhere (Isaacs 1995, Anslyn and Dougherty 2006).

$$F_{dipole\text{-}dipole(parallel)} = 2K\frac{(\mu)(\mu')}{r^4} \therefore \mu = q.r$$

$$E_{dipole\text{-}dipole(parallel)} = 2K\frac{(\mu)(\mu')}{r^3}$$

When the dipoles of chloromethane complex are nearly aligned in the same axis, the $\Delta H_{f(dipole\text{-}dipole\ complex)}$ is higher (i.e., its dipole-dipole interaction is smaller) than that in parallel dipoles (Figs. 11.5(B) and (C)). Then, the dipole-dipole interaction for aligned dipoles is smaller (its electrostatic force is weaker) than that in parallel dipoles, since in the latter there is a sum of two electrostatic interactions and in the aligned direction there is only one electrostatic interaction.

$$F_{dipole\text{-}dipole(aligned)} = K\frac{\delta_A^+\delta_B^-}{r_{\delta_A^+\delta_B^-}^2} \quad \text{or} \quad F_{dipole\text{-}dipole(aligned)} = K\frac{(\mu)(\mu')}{r^4}$$

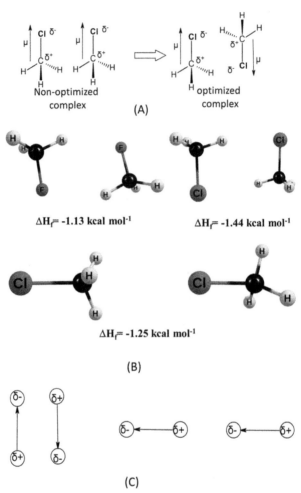

Figure 11.5 (A) Schematic representation of initial (and arbitrary) and final geometries of chloromethane-chloromethane in the optimization process; (B) optimized chloromethane-chloromethane (in parallel and aligned in the same axis) and fluoromethane-fluoromethane complexes along with their corresponding formation enthalpy of dipole-dipole complex, $\Delta H_{f(dipole\text{-}dipole\ complex)}$ calculated at ωB97XD/6-311G++(2d,p) level of theory; (C) schematic representation of dipole-dipole interaction in parallel and aligned directions.

INDUCED DIPOLE-INDUCED DIPOLE INTERACTION

The induced dipole-induced dipole interaction (or London interaction) arises from an instantaneous dipole in an apolar molecule due to the random fluctuation of the electron density (which can be understood from resonance theory) inducing vicinal apolar molecules to adopt instantaneous dipoles as well. Then there appears an electrostatic attractive interaction between these instantaneous dipoles. The instantaneous dipole moment is smaller than a permanent dipole moment due to the difficulty to create partial charge in an atom bonded to other atom with the same electronegativity.

The induced dipole-induced dipole interaction is an interaction between apolar non-benzenoid molecules (H_2, Cl_2, N_2, O_2, noble gas, etc.). We have already shown one example of induced dipole-induced dipole interaction in chapter six for He-He interaction (see Fig. 6.4(B)).

The induced dipole-induced dipole interaction occurs when one apolar molecule adopts partial charges during a very short lifetime, which induces its vicinal apolar molecules to adopt opposite instantaneous partial charges (see Fig. 11.6(A)) This instantaneous dipole fluctuates in synchronicity. It can be rationalized from resonance theory, where the coefficients for the polarized resonance structures are non-zero but very close to zero.

In Fig. 11.6(B), the optimized interaction between two hydrogen molecules and the corresponding molecular graph is shown. The distance between the H_2 molecules is $3.710\,\text{Å}$ and from a front-view perspective, both molecules have a cross-shaped arrangement. The charge density of the bond critical point, ρ_b, between hydrogen atoms of vicinal hydrogen molecules is $0.0007\,\text{au}$. which is more than ten times smaller than that from dipole-dipole interaction and less than 1% of the ρ_b from a covalent H-H bond. The formation enthalpy of induced dipole-induced dipole complex from hydrogen molecules (according to bimolecular model) is $0.011\,\text{kcal mol}^{-1}$ at ωB97XD/6-311G level of theory. Since its interaction energy is higher than that in dipole-dipole interaction (see Fig. 11.5(B)), the strength of induced dipole-induced dipole interaction is weaker than that from dipole-dipole interaction.

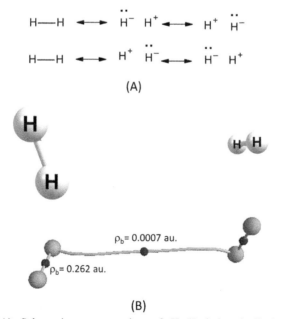

(B)

Figure 11.6 (A) Schematic representation of H_2-H_2 induced dipole-induced dipole interaction by using resonance theory; (B) optimized geometry and corresponding molecular graph of H_2-H_2 induced dipole-induced dipole interaction at ωB97XD/6-311G level of theory.

DIPOLE-INDUCED DIPOLE INTERACTION

The dipole-induced dipole interaction (or Debye interaction) arises from one polar molecule which induces instantaneous dipoles in vicinal apolar molecules. One example is the interaction between HCl and an argonium atom. Figure 11.7 shows the optimized geometry of Ar-HCl-Ar and its corresponding molecular graph in a nearly aligned direction where the enthalpy formation for this interaction is -0.87 kcal mol^{-1}. The values of charge density of bond critical point between the interacting atoms Ar-Cl and H-Ar are 0.0014 and 0.0028 au., respectively, which are higher than that from induced dipole-induced-dipole and smaller than that from dipole-dipole interaction (H---Cl, ρ_b= 0.0060 and H---F, ρ_b= 0.0086 au., from chloromethane-chloromethane and fluoromethane-fluoromethane interactions, respectively. See Fig. 11.7(B)).

Then, the dipole-induced dipole interaction is stronger than induced dipole-induced dipole interaction (it has a smaller interaction energy) but it is weaker than dipole-dipole interaction (it has a higher interaction energy than that from dipole-dipole interaction). As a consequence, the order of the van der Waals interactions is: dipole-dipole > dipole-induced dipole > induced dipole-induced dipole.

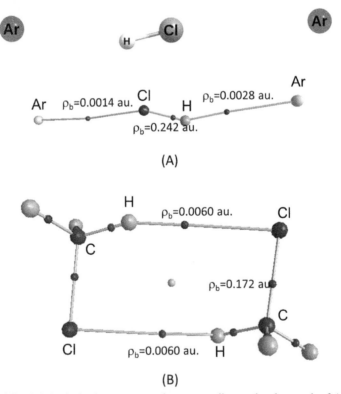

(A)

(B)

Figure 11.7 (A) Optimized geometry and corresponding molecular graph of Ar-HCl-Ar complex from ωB97XD/6-311G++(2d,p) level of theory; (B) molecular graph of chloromethane-chloromethane interaction along with ρ_b values.

ION-DIPOLE INTERACTION

The ion-dipole interaction involves an ionic compound and a polar covalent compound as represented in Fig. 11.8(A). It is a directional interaction as observed from QTAIM results (Fig. 11.8(B). The anion interacts with the positive partial charge of the polar molecule and the cation interacts with the negative partial charge of the polar molecule.

Figure 11.8(C) depicts the optimized geometry of the interaction between each pair – chloride anion-chloromethane, sodium cation-chloromethane, and sodium chloride-chloromethane. Chloride anion in chloride anion-chloromethane pair is closer to the methyl carbon (whose interatomic distance is 3.19 Å) being placed in the same axis than the C-Cl bond from chlromethane. Sodium cation in sodium cation-chloromethane pair is closer to the chlorine atom of chloromethane (whose interatomic distance is 2.71 Å). As for the NaCl-CH$_3$Cl interaction, the interatomic distance Na$^+$ Cl$^-$ in isolated optimized NaCl increases slightly from 2.39 Å to 2.43 Å with respect to the NaCl interacting with chloromethane. The sodium-chlorine (from chloromethane) interatomic distance in sodium chloride-chloromethane increases slightly to 2.80 Å with respect to that from the sodium cation-chloromethane pair. As for the chlorine anion interacting with CH$_3$(Cl), the chlorine atom is positioned beside one hydrogen atom of chloromethane (with interatomic distance 2.54 Å) and not in the same axis of (H$_3$)C-Cl bond, as it occurs in chloride anion-chloromethane pair.

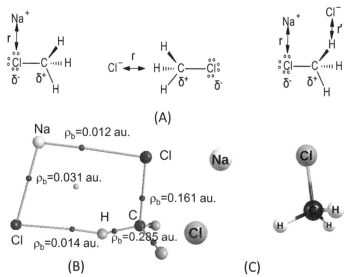

Figure 11.8 (A) Schematic representation of isolated ion-dipole interaction and ionic salt-dipole interaction; (B) QTAIM molecular graph along with charge density of the bond critical point, ρ_b; and (C) corresponding optimized geometry of sodium chloride-chloromethane.

Figure 11.8(B) shows the charge density of the bond critical point, ρ_b, of sodium chloride-chloromethane. The ρ_b of intermolecular interactions is two to three times smaller than that from NaCl ionic bond, which is much smaller than those from covalent bonds. The ρ_b of Na-Cl(CH$_3$) and Cl-H(CH$_2$Cl) interatomic interaction are 0.012 and 0.014 au., which is two times higher than the H-Cl interatomic interaction from chloromethane-chloromethane complex—a dipole-dipole interaction (Fig. 11.7(B)).

The interaction energies, according to formation enthalpy of the ion-dipole complex, $\Delta H_{f(ion\text{-}dipole\ complex)}$, of the isolated sodium cation-chloromethane, isolated chloride anion-chloromethane and sodium chloride-chloromethane are −15.06, −445.39 and −11.29 kcal mol^{-1}, respectively, from ωB97XD/6-311G++(2d,p) level of theory. The very high value of interaction energy from isolated chloride anion-chloromethane is probably because of its particular geometry where chloride anion is positioned at the same axis of (H$_3$)C-Cl in chloromethane which becomes the complex for a substitution reaction. The isolated cation and anion interacting with a polar molecule is an unreal situation which has been analyzed only for comparison. The real system is the cation-anion salt interacting with a polar molecule, such as the sodium chloride-chloromethane pair. In this case, the ion-dipole interaction energy ($\Delta H_f = -11.29$ kcal mol^{-1} for NaCl-CH$_3$Cl) is much smaller than that from dipole-dipole interaction energy of chloromethane-chloromethane with $\Delta H_f = -1.44$ kcal mol^{-1} in parallel direction. That is because the electrostatic force between ion and partial charge is greater than the electrostatic force between opposite partial charges, i.e., $F_{(ion\text{-}dipole)} > F_{(dipole\text{-}dipole)}$. As the last analysis, a charge in an ion, Q^+ or Q^-, has a unit value or more, which is generally much more than a partial charge, δ^+ or δ^-, in a polar molecule, i.e., $|Q| > |\delta|$.

$$F_{isolated\text{-}ion\text{-}dipole} = K\frac{Q\delta}{r_{Q\delta}^2}$$

$$F_{cation\text{-}anion\text{-}dipole} = K\frac{Q^-\delta_A^+}{r_{Q^-\delta_A^+}^2} + K\frac{Q^+\delta_B^-}{r_{Q^+\delta_B^-}^2}$$

HYDROGEN BOND

The hydrogen bond is an electrostatic, through-space directional intermolecular/intramolecular interaction between hydrogen atom (bonded to a heteroatom, which is called donator or Dn) and heteroatom from another bond (from a vicinal molecule for intermolecular interaction or in the same molecule for intramolecular interaction), which is called acceptor or Ac. As depicted in Fig. 11.9(A), the hydrogen bond can be rationalized from a formation process where prior to the hydrogen bond formation there is an interchange of charge density from covalent H-heteroatom (donator atom) bond and another heteroatom (acceptor atom), forming an intermolecular/intramolecular interaction between Ac and H(−Dn). The electrostatic force of a hydrogen bond, $F_{H\text{-}bond}$, can be rationalized by means of two distinguished equations: one using hydrogen and acceptor partial charges (δ) and the other using charge density in hydrogen bond, $\rho_{H\text{-}bond}$, and Z_{eff} of hydrogen and acceptor. The first equation is easier to use because it depends only on the electronegativity of the atoms. For the second equation, the charge density in hydrogen bond which can be obtained from QTAIM using the charge density of the bond critical point of the hydrogen density, $\rho_{b(H\text{-}bond)}$ is needed to know.

$$F_{H\text{-}bond} = K \frac{\delta_H^+ \delta_{Ac}^-}{r_{\delta_H^+ \delta_{Ac}^-}^2}$$

$$F_{H\text{-}bond} = K \frac{\rho_e Z_{eff\,(C)}}{r_{C1}^2} + K \frac{\rho_e Z_{eff\,(H)}}{r_{H1}^2}$$

Figure 11.9(B) depicts the molecular graph from optimized geometries of H_2O-H_2O, NH_3-H_2O, and NH_3-NH_3 complexes along with their corresponding formation enthalpy of hydrogen bond complexes, $\Delta H_{f(H\text{-}bond\ complex)}$ and the charge density of the bond critical point in hydrogen bond The order of interaction energy of hydrogen bond is: H_2O-H_2O < NH_3-NH_3 < NH_3-H_2O, which gives an opposite order for the $F_{H\text{-}bond}$. Except for the NH_3-H_2O complex, the order of $F_{H\text{-}bond}$ can be understood by the electronegativity order: O > N, which yields the same order for their negative partial charge and for the hydrogen bonded to them (the more electronegative the heteroatom bonded to the hydrogen the more positive the partial charge in hydrogen).

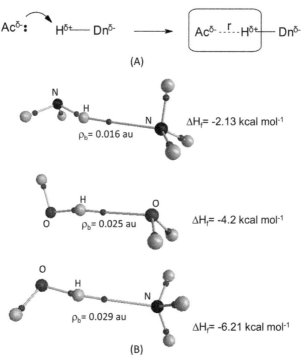

$Ac^{\delta-}:$ $H^{\delta+}$ — $Dn^{\delta-}$ → $Ac^{\delta-}$ ---^{r}--- $H^{\delta+}$ — $Dn^{\delta-}$

(A)

ρ_b = 0.016 au ΔH_f = -2.13 kcal mol^{-1}

ρ_b = 0.025 au ΔH_f = -4.2 kcal mol^{-1}

ρ_b = 0.029 au ΔH_f = -6.21 kcal mol^{-1}

(B)

Figure 11.9 (A) Schematic representation of hydrogen bond formation process; (B) molecular graph of the optimized geometry of water-water, ammonia-water, and ammonia-ammonia hydrogen complexes, along with their corresponding formation enthalpy of hydrogen bond complexes, $\Delta H_{f(H\text{-}bond\ complex)}$ and the charge density of the bond critical point in hydrogen bond, $\rho_{b(H\text{-}bond)}$, from ωB97XD/6-311G++(2d,p) level of theory.

$$F_{H\text{-}bond(H_2O\text{-}H_2O)} = K \frac{\delta_{H(O)}^+ \delta_O^-}{r_{\delta_{H(O)}^+ \delta_O^-}^2} \quad \therefore F_{H\text{-}bond(NH_3\text{-}NH_3)} = K \frac{\delta_{H(N)}^+ \delta_N^-}{r_{\delta_{H(N)}^+ \delta_N^-}^2}$$

$$\delta_O^- > \delta_N^- \quad \therefore \delta_{H(O)}^+ > \delta_{H(N)}^+$$

Then
$$F_{H\text{-}bond(H_2O\text{-}H_2O)} > F_{H\text{-}bond(NH_3\text{-}NH_3)}$$

In order to understand the whole order, it is necessary to use the equation of electrostatic force according to charge density of hydrogen bond, $\rho_{b(H\text{-}bond)}$, and Z_{eff} of the bonded atoms. From Fig. 11.9(B), one can see that the order of $\rho_{b(H\text{-}bond)}$ is $NH_3\text{-}H_2O > H_2O\text{-}H_2O > NH_3\text{-}NH_3$ which explains the interaction energy order (regarding the inverse relation between energy and force). The higher the $\rho_{b(H\text{-}bond)}$, the stronger the hydrogen bond and the smaller the corresponding hydrogen bond interaction energy. In addition, these ρ_b values (from 0.016 to 0.029 au.) are higher than those from the ion-dipole interaction. Then, the hydrogen bond is stronger than the ion-dipole interaction.

When considering the experimental data of Table 11.1, the strength of hydrogen bond $O\text{-}H\text{-}\text{-}\text{-}OR_2 > O\text{-}H\text{-}\text{-}\text{-}SR_2 > O\text{-}H\text{-}\text{-}\text{-}SeR_2$ can be rationalized from the electronegativity order $O > S > Se$, which gives the same order of their partial charge in modulus. By using the electrostatic equation with the partial charge of hydrogen and acceptor atom, it is easy to observe that the order of the hydrogen bond strength depends on the partial charge of acceptor ($O > S > Se$) and the distance hydrogen-acceptor which increases as the atomic volume of the acceptor increases. Both factors explain the data given in Table 11.1.

Likewise, the following orders of hydrogen bond strength: $HF > HCl > HBr > HI$ and $H_2O > NH_3 > H_2S > H_3P$ can be reasoned from the electrostatic force based on partial charges (equation below), where the more electronegative atom ($F > Cl > Br > I$ and $O > N > S > P$) leads to the more partial charge in modulus in acceptor and hydrogen atoms, giving higher hydrogen bond in terms of electrostatic force.

$$F_{H\text{-}bond} = K \frac{\delta_H^+ \delta_{Ac}^-}{r_{\delta_H^+ \delta_{Ac}^-}^2}$$

Table 11.1 Experimental dissociation energy, in kcal mol^{-1}, of the hydrogen bond between phenol (as hydrogen donor) and dioxane/dibutyl sulfide/dibutyl selenide in carbon tetrachloride.*

Hydrogen bond	Involved compounds	Dissociation energy (kcal mol^{-1})
O-H---OR$_2$	Phenol/dioxane	−5.0
O-H---SR$_2$	Phenol/dibutyl sulfide	−4.2
O-H---SeR$_2$	Phenol/dibutyl selenide	−3.7

*Data from Anslyn and Dougherty 2006.

The order of formation enthalpy of complexes with the intermolecular interaction is ion-dipole interaction ($\Delta H_f = -11.29$ kcal mol^{-1} for NaCl-chloromethane) < hydrogen bond ($\Delta H_f = -6.21$ kcal mol^{-1} for $H_2O\text{-}NH_3$) < dipole-dipole interaction

($\Delta H_f = -1.44$ kcal mol^{-1} for chloromethane-chloromethane) < dipole-induced dipole interaction ($\Delta H_f = -0.87$ kcal mol^{-1} for argonium-hydrogen chloride-argonium) < induced dipole-induced dipole interaction ($\Delta H_f = 1.00$ kcal mol^{-1} for H_2-H_2). The strength of these interactions has an opposite order: ion-dipole interaction > hydrogen bond > dipole-dipole interaction > dipole-induced dipole interaction > induced dipole-induced dipole interaction.

HYDROGEN-HYDROGEN BOND(ING)

Hydrogen-hydrogen bond(ing), H-H bond, is a new type of inter/intramolecular interaction, which was discovered in 2003 using QTAIM (Matta et al. 2003) for analyzing the intramolecular interaction in some aromatic compounds where they found that H-H bond decreases the energy of a molecule up to 10 kcal mol^{-1}. After that some scientists questioned the H-H bond stabilizing effect in planar biphenyl based on an arbitrary set of molecular orbitals (which we have shown in chapter two; they are not univocal). This criticism was properly responded to by Bader later on (Bader 2006). In 2014, we published a paper that we believe cleared any doubt about hydrogen-hydrogen bond as a stabilizing effect (Monteiro and Firme 2014). We have studied alkane complexes and we have found a very good correlation between the number of H-H bonds in alkane complexes and alkane's boiling point (Fig. 11.10(A)). Besides, we have shown that during the optimization of a hydrogen atom not belonging to a H-H bond, its atomic energy (from QTAIM) has an erratic behavior and finally its energy increases after optimization of the alkane complex (Fig. 11.10(B)), while for another hydrogen atom (from the same molecule in the complex) that participates in a H-H bond, its atomic energy decreases by 1.6 kcal mol^{-1} at the end of the optimization process (Fig. 11.10(C)).

Hydrogen-hydrogen bond is a through-space inter/intramolecular interaction between hydrogen atoms from different C-H bonds, whose atomic charges are nearly zero. Hydrogen-hydrogen bond is the main intermolecular interaction in alkanes and some alkenes, and it is a secondary intermolecular interaction in alcohols, ethers, and amines. Any molecule with saturated long hydrocarbon chain has hydrogen-hydrogen bond. See the hydrogen-hydrogen bonds, H-H bonds, in propane-propane complex represented by bond paths indicated in Fig. 11.11(A).

A broader view of the intermolecular interactions is given by the NCI theory. While QTAIM shows only bond paths, NCI shows the whole surface of intermolecular interactions (Contreras-García, J. et al. 2011). See the details about NCI theory in the end of chapter two. Figure 11.11(B) shows the surface of NCI in propane-propane complex. In this last figure, one can see the dark gray surface (corresponding to dark green in NCI coloring scheme that is a stronger van der Waals interaction) representing H-H bonds and light gray surface (which corresponds to light green surface in NCI coloring scheme) representing a weaker van der Waals interaction. The latter corresponds to a secondary interaction in alkanes, besides H-H bond (see more details in chapter fourteen).

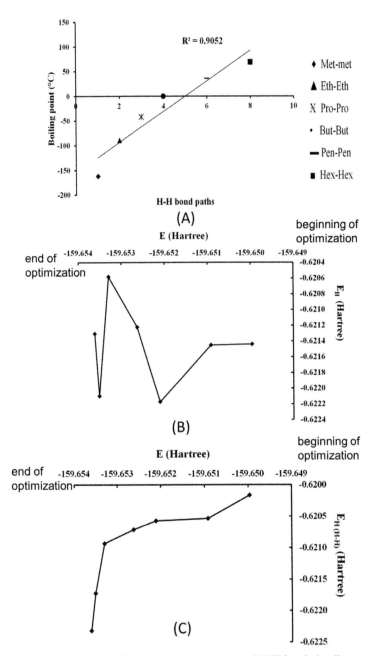

Figure 11.10 (A) Plot of boiling point versus number of H-H bonds in alkane complexes from methane to hexane; (B) and (C) Plots of electronic energy of ethane-ethane complex at the beginning of the optimization process (with an arbitrary initial geometry) versus atomic energy of the hydrogen atom (according to QTAIM) not belonging to a H-H bond and a hydrogen atom belonging to a H-H bond, respectively. Courtesy of American Chemical Society (see the Acknowledgment section).

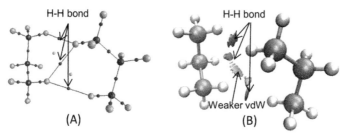

Figure 11.11 (A) Molecular graph of optimized propane-propane complex from ωB97XD/6-311G++(d,p) level of theory; (B) NCI isosurface with sign(l_2)ρ coloring scheme of the propane-propane complex where dark and light green are represented by dark and light gray, respectively, here.

π-STACKING INTERACTION

The π-stacking interaction occurs between two benzene molecules or any pair of phenyl ring which are displaced from one another so that one carbon of a phenyl ring interacts with the other carbon of the other phenyl ring, i.e., it is a C-C interaction between displaced phenyl rings. When both phenyl rings are in parallel, there is a repulsion instead of an attraction interaction.

The π-stacking can be analyzed by QTAIM. However, to observe the repulsion from parallel phenyl rings, one has to use the Non-Covalent Interaction, NCI, method.

A very interesting case to study the π-stacking interaction is in cyclophanes, which are a closed circuit of phenyl rings and alyphatic bridges (Firme and Araújo 2018). In some cyclophanes there is an attractive π-stacking interaction (when the phenyl rings are displaced from one another) and in other cyclophanes there is a huge repulsion region between phenyl rings (when they are in parallel). See in Fig. 11.12(A) the multi-layered metacyclophane containing two π-stacking interaction besides some H-H bonds, according to QTAIM. In Fig. 11.12(B), the [2.2.2.2.2.2](1,2,3,4,5,6)cyclophane, a superphane, has parallel phenyl rings and huge repulsion area, which has a red surface in the NCI coloring scheme (represented here as a dark gray surface).

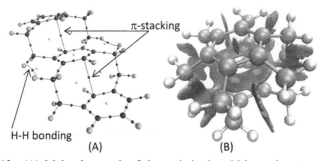

Figure 11.12 (A) Molecular graph of the optimized multi-layered metacyclophane, and (B) NCI isosurface with sign(l_2)ρ coloring scheme of the superphane where the red surface is represented by dark gray here.

EXERCISES

1. Explain the following boiling point order: propane $(-42.1°C)$ < ethylamine $(16.6°C)$ < ethanol $(78.5°C)$.
2. Using bimolecular model, explain why 2-methyl propane, 2-methyl butane, and 2-methyl pentane have smaller boiling points than their corresponding unbranched alkanes. Give the boiling point order for these branched alkanes.
3. Why is solid maleic acid highly soluble in water, while its isomer solid fumaric acid is poorly soluble in water?

Acknowledgment

Figure 11.10 was reprinted with permission from Monteiroand Firme (2014). Hydrogen-hydrogen bonds in highly branched alkanes and in alkane complexes: a DFT, *ab initio*, QTAIM and ELF study. J. Phys. Chem. A 118: 1730-1740. Copyright 2014 American Chemical Society.

REFERENCES CITED

Anslyn, E.V. and Dougherty, D.A. 2006. Modern Physical Organic Chemistry. University Science Books, Sausalito.

Bader, R.F.W. 2006. Pauli repulsions exist only in the eye of the beholder. Chem. Eur. J. 12: 2896-2901.

Contreras-García, J., Johnson, E.R., Keinan, S., Chaudret, R., Piquemal, J.-P., Beratan, D.N. and Yang, W. 2011. NCIPLOT: A program for plotting non-covalent interactions. J. Chem. Theory Comput. 7: 625-632.

Isaacs, N.S. 1995. Physical Organic Chemistry. Longman Group UK Limited, Harlow.

Firme, C.L. and Araújo, D.M. 2018. Revisiting electronic nature and geometric parameters of cyclophanes and their relation with stability - DFT, QTAIM and NCI study. Comp. Theor. Chem. 1135: 18-27.

Matta, C.F., Hernández-Trujillo, J., Tang, T.-H., and Bader, R.F.W. 2003. Hydrogen-hydrogen bonding: a stabilizing interaction in molecules. Chem. Eur. J. 9: 1940-1951.

Monteiro, N.K.V. and Firme, C.L. 2014. Hydrogen-hydrogen bonds in highly branched alkanes and in alkane complexes: a DFT, ab initio, QTAIM and ELF study. J. Phys. Chem. A 118: 1730-1740.

Szalewicz, K. 2011. Symmetry-adapted perturbation theory of intermolecular forces. WIREs Comput. Mol. Sci. 2: 254-272.

Chapter Twelve

Carbocations

DEFINITION AND CLASSIFICATION

Carbocations are alkyl ions containing one trivalent, positively charged carbon atom and they are the intermediate of electrophilic addition to alkenes. Carbocations are electrophiles whose electrophilic center is the trivalent carbon atom. However, the positive charge is delocalized among neighbor hydrogen and carbon atoms by inductive effect. The trivalent carbon atom is sp^2 and it has a trigonal planar geometry, i.e., the nucleophile might attack the carbocation upwards or downwards.

Carbocations can be defined as classical ion (or carbenium ion) and non-classical ion (or carbonium ion). In the former, there is a formal charge localized in just one carbon (although, actually the positive charge is delocalized among the neighbor hydrogen and carbon atoms by means of inductive effect). In the latter, the positive charge is delocalized to other two carbon atoms by a through-space assistance of a π-bond or σ-bond, forming a formal $3c$-$2e$ or $4c$-$2e$ multicenter bonding where some of them have homoaromatic properties.

Carbocations can also be classified according to the number of alkyl groups bonded to the trivalent carbon: (1) methyl cation where only hydrogen atoms are bonded to the trivalent carbon; (2) primary carbocation where one alkyl group is bonded to the trivalent carbon; (3) secondary carbocation where two alkyl groups are bonded to the trivalent carbon; and (4) tertiary carbocation where three alkyl groups are bonded to the trivalent carbon. All represented carbocations in Fig. 12.1 (ethyl cation, 2-propyl cation and tert-butyl cation) are carbenium ions. Each alkyl group bonded to the trivalent carbon atom in all carbocations in Fig. 12.1 is a methyl group.

Alkyl group is a hydrocarbon fragment which is bonded to a hydrocarbon chain or an electrophilic/nucleophilic center. The alkyl group is an alkane with a missing hydrogen atom. It can be a methyl (CH_3-), ethyl (CH_3CH_2-), propyl ($CH_3CH_2CH_2$-), 2-propyl (($CH_3)_2CH$-), butyl ($CH_3CH_2CH_2CH_2$-), *tert*-butyl (($CH_3)_3C$-), and so on. The dash (-) represents the place where the alkyl group will be (or is) linked to the rest of the molecule. They are generally represented by R, R', R'' or R_1, R_2, R_3. Each carbocation can have, at most, three alkyl groups (see more information in chapter fourteen).

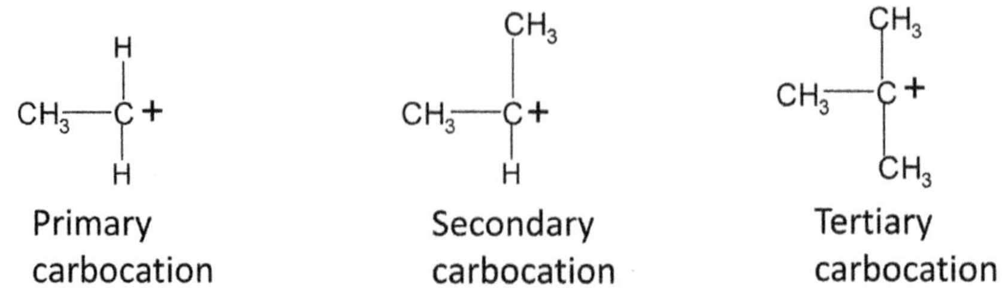

Figure 12.1 Representation of the simplest primary, secondary, and tertiary carbocations where each alkyl group is a methyl.

The characterization of the first stable alkyl cation (*tert*-butyl cation) was done by George Olah and collaborators in 1963 (Olah et al. 1963). This isolation and characterization was possible due to the discovery of the super acids (or magic acids) – a mixture of antimony pentafluoride (SbF_5) and fluorosulfonic acid (FSO_3H) – in 1960. The magic acid– an acid with acidity greater than pure sulfuric acid – can protonate methane to yield *tert*-butyl cation (Olah and Schlosberg 1968).

$$CH_4 + H^+ \rightarrow CH_5^+$$
$$CH_5^+ \rightarrow CH_3^+ + H_2$$
$$CH_3^+ + 3CH_4 \rightarrow (CH_3)_3 C^+ + 3H_2$$

INDUCTIVE EFFECT OF ALKYL GROUPS

Inductive effect is a σ-bonded electronic communication between one substituent (for example, one alkyl group) and an electrophilic or nucleophilic center in the same molecule. The inductive effect implies the charge delocalization (from the substituent towards the electrophilic/nucleophilic center or vice-versa) by means of the 'bridge' of a single or multiple σ bonds (two or three σ bonds) that link them.

The inductive effect decreases exponentially as the distance between the substituent and electrophilic/nucleophilic site increases. It is maximum at one σ-bond distance and it is minimum at three σ-bond distance, i.e., null at higher distances (more than four σ-bond distance).

Figure 12.2 shows the optimized geometries of the derivatives of *n*-silanyl-2-pentyl cation, *n*=2,3,4,5, along with their corresponding absolute enthalpy value (in Hartree) from ωB97XD/6-311G++(2*d,p*) level of theory. Roughly speaking, as the silanyl group (H_3Si-) is placed far away from the trivalent carbon, that is, from *n*=2 to *n*=5, the stability of the derivative decreases, although there is one particularity: both 3-silanyl-2-pentyl cation and 4-silanyl-2-pentyl cation have the same absolute enthalpy. Both have 3*c*-2*e* multicenter bonding involving the silanyl group. In 3-silanyl-2-pentyl cation, there is a partial rearrangement (see the next section) of the silanyl group from C3 towards C2 (Fig. 12.2(B)). In 4-silanyl-2-pentyl cation, there is firstly a hydrogen migration (or 1,2-hydride shift) from C3 to C2 followed by partial rearrangement of the silanyl group from C4 towards C3, forming a *3c-2e* multicenter bonding very similar to that from 3-silanyl-2-pentyl cation (Fig. 12.3(C)). This similarity in the multicenter bonding leads to similar geometries and then similar absolute energy.

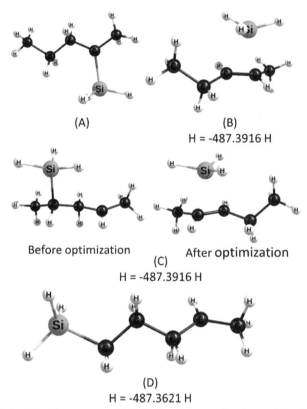

(A)   (B)

H = -487.3916 H

Before optimization   After optimization

(C)

H = -487.3916 H

(D)

H = -487.3621 H

Figure 12.2 Optimized geometry of (A) 2-silanyl-2-pentyl cation and its (B) 3-silanyl-, (C) 4-silanyl-, and (D) 5-silanyl isomers, along with their absolute enthalpies from ωB97XD/6-311G++(2d,p) level of theory.

Like alkyl groups, substituent group (or just substituent) is a part of the molecule. In fact, alkyl group is a type of substituent. The substituent can be classified as: (1) electron withdrawing group, EWG, by inductive effect (e.g., Br-, Cl-, F-, F_3C-, etc.); (2) electron withdrawing group by resonance effect (e.g., O_2N-); (3) electron donating group, EDG, by inductive effect (e.g., Li-, H_3Si-) and; (4) electron donating group by resonance effect (e.g., HO-, CH_3O-). More information about substituent groups is given in chapter twenty.

The alkyl group has ambiguous behavior as substituent group, that is, it can be moderate EWG or moderate EDG by inductive effect depending on the electron deficiency or electron accumulation of the electrophilic site or the nucleophilic site, respectively. In case the alkyl group is linked to an electrophilic site (for example, carbonyl group or trivalent, positive carbon atom), it behaves as a moderate EDG transferring charge density to the electrophilic center, which leads to the delocalization of the (partial) positive charge and its consequent stabilization. On the other hand, in case the alkyl group is linked to a nucleophilic site (for example, C=C or C≡C bond in alkenes or alkynes, respectively, or negatively charged carbon atom in carbanions), then it becomes a EWG, removing charge density from the nucleophilic site so that the electron density is more delocalized. In both cases, the alkyl group leads to a higher stabilization by means of the charge density delocalization (as stated before, the more delocalized the unitary charge or charge density, the more stable is the molecular system).

STABILITY OF CARBENIUM IONS

Figure 12.3(A) shows the optimized geometries of methyl cation, ethyl cation (a primary carbocation), 2-propyl cation (a secondary carbocation), and tert-butyl cation (a tertiary carbocation) from G4 method (Curtis et al. 2007). When going from methyl cation to ethyl cation, from ethyl cation to 2-propyl cation, and then to *tert*-butyl cation, one methyl group is linked to the positive, trivalent carbon atom. Then, methyl cation, ethyl cation, 2-propyl cation, and *tert*-butyl cation have 0, 1, 2, and 3 methyl groups linked to positive, trivalent carbon atom, respectively. Figure 12.3(A) also gives the reference enthalpy change according to the reaction:

$$CR_3^+ + NaH \rightarrow CR_3H + Na^+ \quad \therefore R = H, CH_3$$

This enthalpy change provides a measure of the stability of the carbocations where the reaction between methyl cation and sodium hydride is taken as the reference. The more negative the reference enthalpy change, the more stable the carbocation. Hence, we have the following order of the stabilization: methyl cation (0 alkyl group) < ethyl cation (1 alkyl group) < 2-propyl cation (2 alkyl group) < *tert*-butyl cation (3 alkyl group). This order is a natural consequence of EDG inductive effect of the methyl group which delocalize the positive charge and stabilizes the cation. The more alkyl groups bonded to the positive, trivalent carbon atom, the more stable the corresponding carbocation. Therefore, we have the following law: *primary carbocation is less stable than secondary carbocation (by 18 kcal mol^{-1}), and secondary carbocation is less stable than tertiary carbocation (by 15 kcal mol^{-1}).*

An important information to add is the possibility of the primary carbocation to form 3c-2e bonding system (becoming an carbonium ion, as in ethyl cation in Fig. 12.3(A)) or to rearrange to more secondary carbocation (for example, the propyl cation before geometry optimization becomes 2-propyl cation, a secondary carbocation, after the geometry optimization as depicted in Fig. 12.3(B)).

The partial atomic charge can be used to rationalize the charge delocalization. In Fig. 12.3(A), we also show the QTAIM atomic charges in trivalent carbon atom of methyl cation, ethyl cation, 2-propyl cation, and *tert*-butyl cation. We can see the following decreasing order of partial atomic charge in carbon atom: methyl cation > 2-propyl cation > *tert*-butyl cation. This order is in accordance with the positive charge delocalization of the alkyl group acting as EDG in the trivalent carbon atom, which leads to its smaller positive partial charge. Although partial atomic charge of trivalent carbon in ethyl cation is smaller than that from methylcation, it does not follow the expected trend when compared to 2-propyl cation and *tert*-butyl cation, probably because of its unusual multicenter bonding.

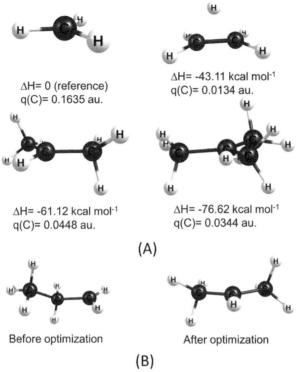

$\Delta H= 0$ (reference)
$q(C)= 0.1635$ au.

$\Delta H= -43.11$ kcal mol^{-1}
$q(C)= 0.0134$ au.

$\Delta H= -61.12$ kcal mol^{-1}
$q(C)= 0.0448$ au.

$\Delta H= -76.62$ kcal mol^{-1}
$q(C)= 0.0344$ au.

(A)

Before optimization   After optimization

(B)

Figure 12.3 (A) Optimized geometry of methyl cation, ethyl cation, 2-propyl cation, and *tert*-butyl cation along with their reference enthalpy change from G4 method and QTAIM partial atomic charge in trivalent carbon atom; (B) arbitrary initial geometry of propyl cation prior to optimization process and its final, optimized geometry forming 2-propyl cation from G4 method.

REARRANGEMENT OF CARBENIUM IONS

Rearrangement process is the complete or partial migration of a hydrogen, methyl, silanyl, or phenyl group from one carbon atom to its vicinal carbon atom in order to stabilize the carbocation (see Fig. 12.4). When the migration is complete, there is a complete rearrangement which changes the type of carbocation (or not). When the migration is partial it forms a multicenter bonding.

Primary carbocations tend to rearrange to secondary or tertiary carbocation (a complete rearrangement process) or tend to form a multicenter bonding (passing from classical cation to non-classical cation). In some cases, secondary carbocations also tend to form multicenter bonding – a partial rearrangement. Then, complete or partial rearrangements are processes to transform less stable carbocations into more stable carbocation (by complete or partial change of carbocation type). For example, the ethyl cation is not a classical primary carbocation since it has a 3c-2e multicenter bonding as a consequence of partial rearrangement of hydrogen atom (partial 1,2-hydride shift). Another example: propyl cation has a complete rearrangement to 2-propyl cation, passing from primary carbocation to secondary carbocation (a complete rearrangement of its hydrogen atom known as 1,2-hydride shift). In the case of 3- and 4-silanyl-2-pentyl cations, silanyl partial rearrangement happens, which forms 3c-2e multicenter bonding involving silanyl group instead of hydrogen as in ethyl cation. In this last case, the positive charge is delocalized in the 3c-2e bonding system, which is more stable than a formal localized positive charge. Figure 12.4 shows the schematic representation of the partial rearrangement in ethyl cation and 3-silanyl-pentyl cation and the complete rearrangement in propyl cation. Moreover, it is also possible to occur complete rearrangement without changing the type of carbocation. For example, 2-pentyl cation can rearrange to 3-pentyl cation (both of them secondary carbocations) by 1,2-hydride shift because the barrier of hydrogen migration is low.

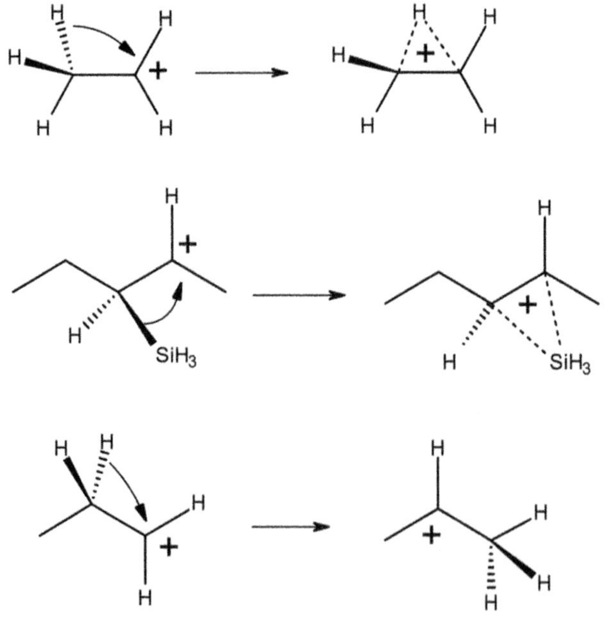

Figure 12.4 Schematic representation of partial rearrangement of hydrogen in ethyl cation and 3-silanyl pentyl cation and complete rearrangement in propyl cation to 2-propyl cation through hydrogen migration.

The rearrangement is a concerted process and not a stepwise process, i.e., it has no intermediates. We can observe this fact from the optimization process of the ethyl cation, propyl cation, and 3- and 4-silanyl pentyl cations, in Figs. 12.3 and 12.2, respectively. The initial arbitrary geometry was a classical primary carbocation for ethyl and propyl cation and after optimization process, they changed into non-classical ion and secondary carbocation, respectively. In *n*-silanyl pentyl cations, they are secondary carbocations which undergo partial rearrangement during optimization process as well. These are clear examples of the above-mentioned statement of the direct process of the rearrangement.

The concerted rearrangement instead of stepwise rearrangement can also be observed in chemical reactions as well. For example, in the elimination reaction of neo-pentanol by dilute sulfuric acid. After the protonation of the alcohol, the departure of the leaving group, water, assisted by other water molecule, from the protonated neo-pentanol does not occur without further assistance of complete migration of the methyl group (a rearrangement process). Figure 12.5 shows the first and second step of this reaction. This fact can be proved by the lack of transition state for the departure of the leaving group without simultaneous methyl rearrangement. Figure 12.6(A) shows one transition state which is not associated with the reaction coordinate, as one can observe by the arrows indicating the movement of the atoms according to its imaginary frequency. In this case, the movements are not associated with the reaction coordinate. On the other hand, there is a transition state where assisted methyl migration happens, as depicted in Fig. 12.6(B), which shows the expected movement of atoms in the imaginary frequency of the transition state associated with the reaction coordinate.

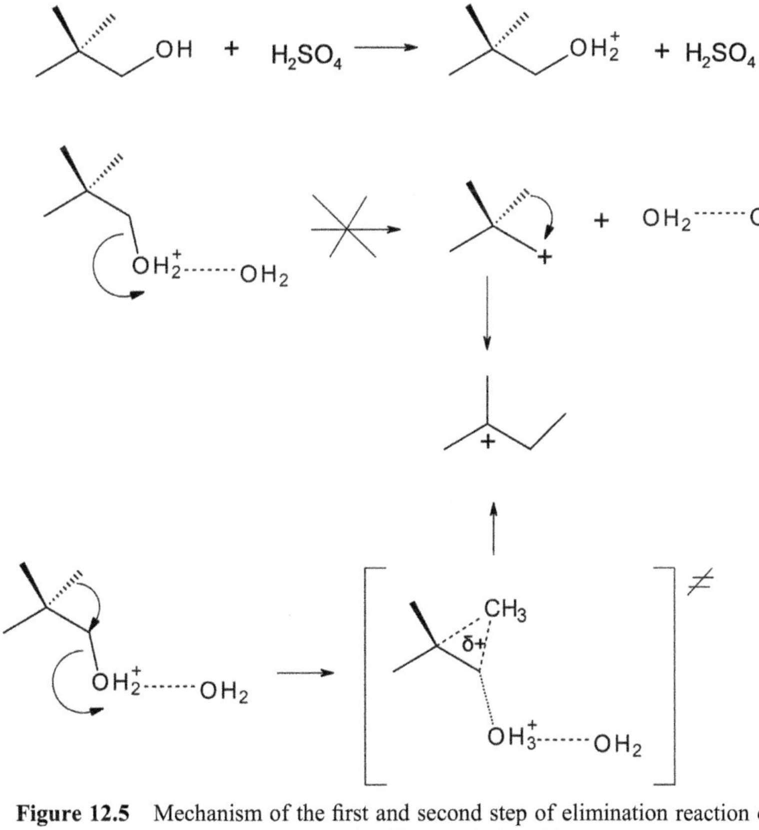

Figure 12.5   Mechanism of the first and second step of elimination reaction of neo-pentanol in dilute sulfuric acid.

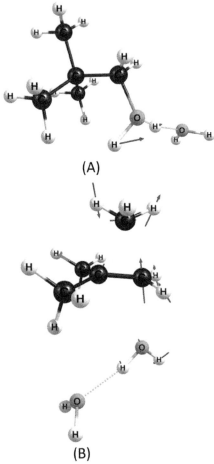

Figure 12.6 (A) Optimized geometry of the transition state of the elimination reaction of protonated neo-pentanol assisted by water, whose arrows correspond to the vibrations of its imaginary frequency; (B) Optimized geometry of the transition state of the elimination reaction of protonated neo-pentanol assisted by water and methyl migration, whose arrows correspond to the vibrations of its imaginary frequency.

CARBONIUM IONS

In mid-20th century, Winstein and Trifan were the first to postulate the anchimeric assistance of σ electrons from C6-C1 bond (according to IUPAC numbering rule for bicyclic compounds) to accelerate the reaction of 2-*exo*-norbornyl-brosilate by forming the 2-norbornyl cation as an intermediate (see Fig. 12.7) in comparison with its isomer 2-*endo* (Winstein and Trifan 1949, 1952). This first step is the rate determining step of this solvolysis reaction. *The anchimeric assistance plays a central role in the formation of the carbonium ion. It is a through-space σ-electron or π-electron donation from a vicinal bond towards an electrophilic center promoting acceleration of departure of the leaving group and the formation of the multicenter bonding.* This multicenter bonding leads to the delocalization of the positive charge within three atoms instead of only one, and gives a higher stabilization of the carbocation in comparison with classical carbocations.

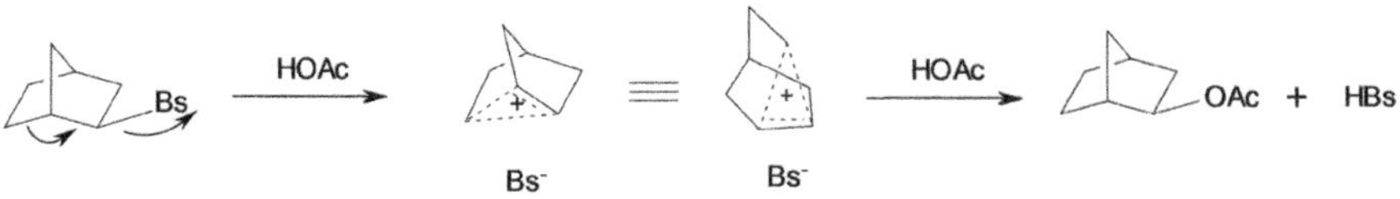

Figure 12.7   Schematic representation of the anchimeric assistance of C6-C1 σ bond in 2-*exo*-norbornyl-brosilate to form 2-norbornyl cation.

Subsequent MO theoretical work indicated anchimeric assistance of C6-C1 σ bond to the departure of the leaving group located in *exo*-C2 of norbonyl derivative, but no equivalent assistance to the departure of leaving group in *endo*-C2 position. However, our work using QTAIM later proved that both *endo*- and *exo*-C2 leaving groups have anchimeric assistance from C6-C1 σ bond. In fact, our QTAIM results showed that anchimeric assistance was twice stronger towards *exo*-C2 leaving group than in *endo*-C2 (Firme 2012). In Fig. 12.8, the PES of the departure of water (as the leaving group of the protonated 2-*exo*- and 2-*endo*-norbornanol) appears from the norbornyl moiety, forming the 2-norbornyl cation which was the first studied carbonium ion in 1949-1952. One can see in Fig. 12.8 that the departure of leaving group from *exo* position has a much smaller energy barrier than that from *endo* position, which indicates that the anchimeric assistance is higher in *exo* position (for this case in particular) than in *endo* position.

The molecular graph of norbonyl cation (Fig. 12.9(A)) indicates a bond path from mid C1-C2 bond towards C6 atom – a clear topological indication of 3c-2e multicenter bonding, the main principal feature of a carbonium ion (Firme et al. 2008a).

The importance of the anchimeric assistance for the formation of carbonium ion can be observed by comparing the molecular graph of 2-norbornyl cation and 6,6-difluoro-2-norbornyl cation (Figs. 12.9(A) and (B)). The latter has a very distinguished electronic nature from the former: it has no bond path in the middle of C2-C1 bond and its DI (C1-C6) is four times smaller than that in the former. Remember that DI (delocalization index) is an amount of shared electrons between two atoms. Obviously the 3c-2e bonding system in the 6,6-difluoro-2-norbornyl cation is much weaker than that in the 2-norbornyl cation because of the influence of both fluorine atoms as EWGs, which prevents the σ-electron donation from C1-C6 towards trivalent C2 atom (Firme et al. 2008a).

The term "non-classical ion" was coined by Roberts and Mazur, who studied the carbonium ions from solvolysis of cyclopropylcarbinyl derivative, e.g., (chloromethyl)cyclopropane (Roberts and Mazur 1951), by forming an intricate non-classical ion which gives rise to three very distinguished products – an indirect experimental outcome of its existence (Fig. 12.10(A)). In 1970, Olah and collaborators experimentally proved the non-classical nature of norbornyl cation by NMR data (Olah et al. 1970). Figure 12.10(B) shows different types of anchimeric assistance (through σ bond and through π bond), which leads to three different types of carbonium ions (tris-homocyclopropenyl cation, 2-norbornyl cation, and 7-norbornenyl cation). Then, *carbonium ions are non-classical ions having multicenter bonding which are triggered by a prior reaction having an anchimeric assistance.*

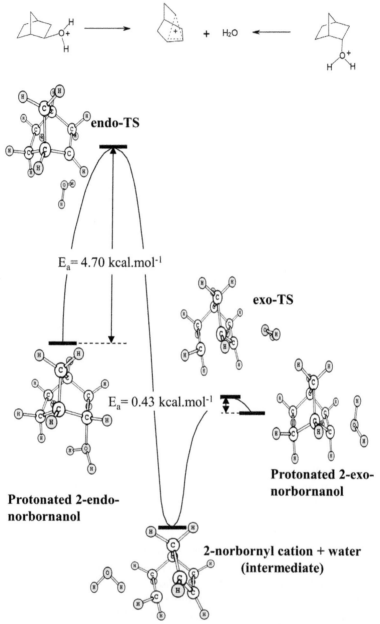

Figure 12.8 Schematic representation and potential energy surface of the departure of water leaving group in both protonated 2-*exo*- and 2-*endo*-norbornanol forming 2-norbornyl cation (see the Acknowledgment section).

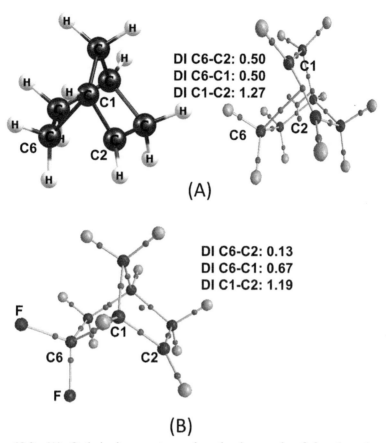

DI C6-C2: 0.50
DI C6-C1: 0.50
DI C1-C2: 1.27

(A)

DI C6-C2: 0.13
DI C6-C1: 0.67
DI C1-C2: 1.19

(B)

Figure 12.9 (A) Optimized geometry and molecular graph of 2-norbornyl cation; (B) molecular graph of 6,6-difluoro-2-norbornyl cation. See the Acknowledgment section for further information.

The molecular graphs of the 7-norbornenyl cation, tris-homocyclopropenyl cation, and 3-cyclopent-1-enyl cation are also shown in Figs. 12.11(A), (B) and (C), respectively, along with their corresponding silicon analogs. While 7-norbornenyl cation and tris-homocyclopropenyl cation are carbonium ions, 3-cyclopent-1-enyl cation is a cabernium ion. For the carbonium ions in Fig. 12.11, there is no bond path linking all three atoms of the multicenter bonding. When comparing to their silicon analogs, for the case of the silicon analog of 7-norbornenyl cation, there are bond paths linking all silicon atoms of the multicenter bonding, probably because the Si-Si π-bond is weaker than C-C π-bond and Si is more electropositive than carbon atom. These facts in silicon analog lead to a stronger multicenter bonding which can also be observed by higher DIs from Si2 and Si3 to Si7 for the silicon analog of 7-norbornenyl cation. The same reasoning is used to explain the existence of bond paths linking silicon atom of the multicenter bonding in the silicon analog of 3-cyclopent-1-enyl cation (Fig. 12.11), which is noticeably a non-classical ion, while 3-cyclopent-1-enyl cation is a classical ion.

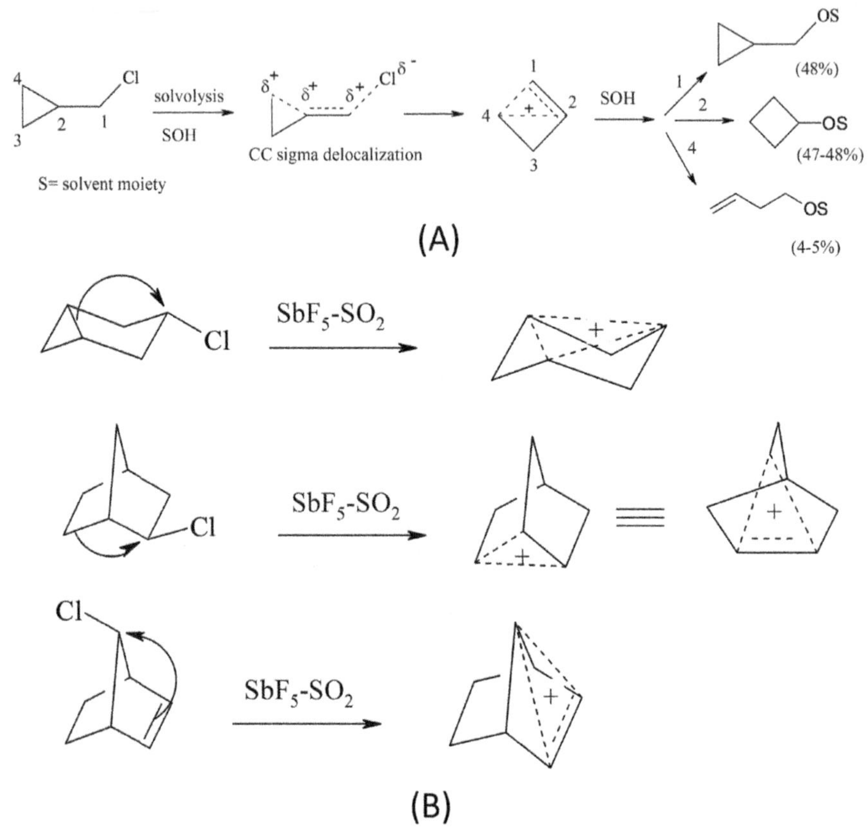

(A)

(B)

Figure 12.10 (A) Schematic representation of solvolysis of (chloromethyl)cyclopropane; (B) Schematic representation of solvolysis reactions giving tris-homocyclopropenyl cation, 2-norbornyl cation, and 7-norbornenyl cation.

All of these above-mentioned carbonium ions have $3c$-$2e$ bonding, which is a planar multicenter bonding. However, there are tri-dimensional multicenter bonding as well, for example, the $4c$-$2e$ bonding system in 1,3-dehydro-5,7-adamantanediyl dication and derivatives. Figure 12.12(A) shows the four resonance structures of this dication which leads to the real, hybrid molecule where two positive charges are delocalized among four carbon atoms. The 1,3-dehydro-5,7-adamantanediyl dication (Fig. 12.12(B)) is a stable species which can be obtained synthetically. However, the 5,7-adamantanediyl dication (Fig. 12.12(C)) has not been obtained so far and our theoretical results also indicate that it is more unstable than 1,3-dehydro-5,7-adamantanediyl dication (Firme et al. 2008b). The delocalization indexes involving the carbon atoms C1, C3, C5, and C7 in 5,7-adamantanediyl dication (or adamantyl dication) are smaller (twice or more smaller) and not uniform than those from 1,3-dehydro-5,7-adamantanediyl dication. Then, the multicenter bonding in 5,7-adamantanediyl dication is much smaller than that in 1,3-dehydro-5,7-adamantanediyl dication (Firme et al. 2008b), which might explain the instability of 5,7-adamantanediyl dication.

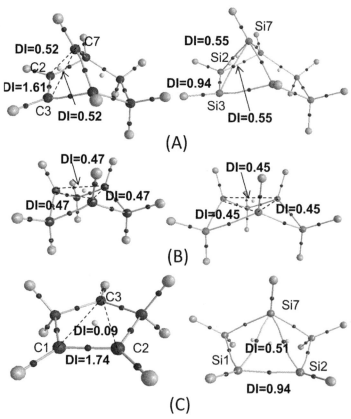

Figure 12.11   Molecular graph of (A) 7-norbornenyl cation, (B) tris-homocyclopropenyl
cation, and (C) 3-pent-1-enyl cation and their silicon analogs.
See the Acknowledgment section for further information.

Another example of tri-dimensional and 4c-2e bonding system of a carbonium
ion is the bisnoradamantenyl dication (Fig. 12.12(D)), which is one derivative of
bisnoradamantene. Bisnoradamantene (Fig. 12.12(F)) is a highly pyramidalized
and unstable caged alkene which was already obtained synthetically under special
conditions. One can note that the double CC bond of bisnoradamantene (and also
its cation and dication derivatives) is in the middle of its structure. Its dication and
cation bisnoradamantenyl (Fig.12.12(E)) are carbonium ions with 4c-2e and 3c-2e
multibonding system, respectively, using the π electrons of the double CC bond for
the multicenter bonding (Firme et al. 2013).

As for 1,3-dehydro-5,7-adamantanediyl, its 4c-2e multicenter bond is done by
the use of σ bond between C1 and C3. The 5,7-adamantanediyl dication does not
have this σ-bond in C1-C3 for the multicenter bond, which accounts for the lower
values of DI's, involving C1, C3, C5, and C7 in comparison with those from its
parent 1,3-desidro.

It is important to add that all carbonium ions are homoaromatic species, that
is, they have aromatic behavior/properties which will be studied in chapter nineteen.

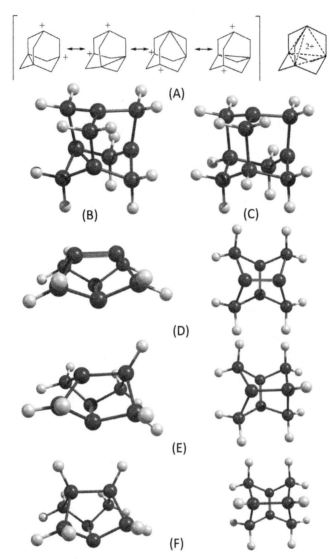

Figure 12.12 (A) Schematic representation of resonance structures and hybrid of 1,3-dehydro-5,7-adamantanediyl dication; optimized geometries of (B) 1,3-dehydro-5,7-adamantanediyl dication and (C) 5,7-adamantanediyl dication along with their C1,C3,C5, and C7 DI's. Optimized structures from side and upper views of (D) bisnoradamantenyl dication; (E) bisnoradamantenyl cation; and (F) bisnoradamantene.

IDENTIFICATION OF CARBONIUM ION AND CARBENIUM ION

The identification of carbonium and carbenium ions can be done experimentally or theoretically. Schleyer and collaborators developed an index based on ^{13}C NMR chemical shift difference between a carbocation and its corresponding neutral hydrocarbon (Schleyer et al. 1980). On the other hand, Gassman and Fentiman used ^{1}H NMR chemical shifts to identify carbonium ions and carbenium ions (Gassman and Fentiman 1970). However, Olah showed that these NMR analyses are limited when no structural changes occur in solvolysis (Olah et al. 1981). Then, in order to overcome these experimental limitations, we have developed a theoretical tool for this identification based on QTAIM (Firme et al. 2009). We have chosen the same set of carbocations used for Gassman and Fentiman: a set of 7-aryl-*p*-substituted-7-norborneyl cations) and other five-membered ring systems. We have used H-, CH_3O- (an EDG), and F_3C- (an EWG) as substituents in the phenyl ring attached to the norborneyl moiety. The EDG decreases the positive partial charge in C7 from the norborneyl moiety, which decreases its interaction with π-bond electrons and leads to the formation of a carbenium ion. On the other hand, the EWG increases the partial positive charge in norborneyl C7, favoring the interaction with the C2-C3 π-bond electrons, which yields a carbonium ion given by the notably higher proximity between C3/C3 and C7 atoms from norborneyl moiety. Moreover, we have selected some topological data that can easily distinguish these cations when no drastic geometric change occurs (Fig. 12.13). We can see that DI (C2-C3) decreases from CH_3O-, H-, and F_3C-, while DI (C2/C3-C7) increases in the same order, giving a clear evidence of increasing π-bond interaction with C7 atom. Then, we have extreme examples of carbenium and carbonium ions: 7-*p*-methoxy-phenyl-7-norborneyl cation (no multicenter bonding) and 7-*p*-trifluoromethyl-phenyl-7-norborneyl cation (with a 3c-2e bonding), respectively.

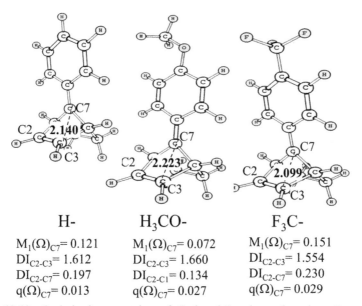

H-

$M_1(\Omega)_{C7}$= 0.121
DI_{C2-C3}= 1.612
DI_{C2-C7}= 0.197
$q(\Omega)_{C7}$= 0.013

H_3CO-

$M_1(\Omega)_{C7}$= 0.072
DI_{C2-C3}= 1.660
DI_{C2-C1}= 0.134
$q(\Omega)_{C7}$= 0.027

F_3C-

$M_1(\Omega)_{C7}$= 0.151
DI_{C2-C3}= 1.554
DI_{C2-C7}= 0.230
$q(\Omega)_{C7}$= 0.029

Figure 12.13 Optimized geometries of 7-phenyl-7-norborneyl cation, 7-*p*-methoxy-phenyl-7-norborneyl cation, and 7-*p*-trifluoromethyl-phenyl-7-norborneyl cation along with interatomic distances (in A) and some topological information. See the Acknowledgment section for further information.

FLUXIONAL CARBOCATIONS

Some hypercoordinate carbonium ions, e.g., protonated alkanes such as CH_5^+ or $C_2H_7^+$, have fluxional property, which is a dynamical instability of its geometry where the multicenter bonding interchanges among all hydrogen atoms, which leads to abrupt changes in its geometry and energy (see Fig. 12.14). Molecules without fluxional property have small changes in geometry and energy during the molecular dynamics.

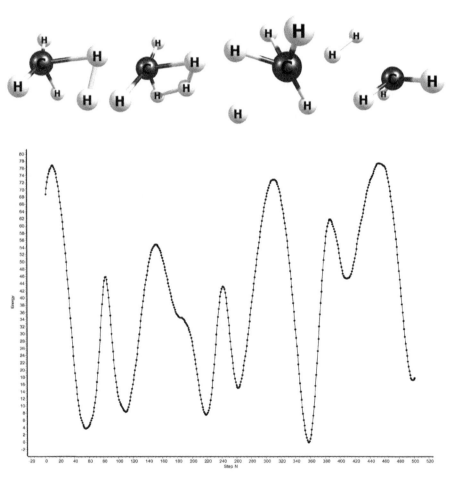

Figure 12.14   Some selected snapshots from the Bohr-Oppenheimer molecular dynamics of CH_5^+ along with the PES of the whole dynamics.

Based on experimental ^{13}C NMR data, cyclopentyl cation has fluxional property by means of 1,2-hydride shift, since its carbon chemical shifts give just one value indicating that all of its carbon atoms have, at a given time, a positive charge (Anslyn and Dougherty 2006).

EXERCISES

1. Do the rearrangement process which changes cyclohexyl cation into 1-methyl-1-cyclopentyl cation.
2. Explain why 2-norbornyl cation (a secondary carbocation) does not rearrange to 1-norbornyl cation. Actually, 2-norbornyl cation is a carbonium ion and not a carbenium ion. Give its real structure using bond-line notation.
3. Explain the increasing stability order of the following carbocations (carbenium ions): tert-butyl cation < 2-methyl-2-butyl cation < 2-methyl-2-pentyl cation < 3-methyl-3-pentyl cation (Data from Anslyn and Dougherty 2006).
4. Give the five resonance structures of cyclopentyl cation using 1,2-hydride shift.
5. Explain why protonated pentan-2-ol has two carbocations as intermediates after the departure of water as the leaving group.
6. Give the mechanism to explain the rearrangement of 2-cyclopentyl-2-propyl cation (the carbocation with formal positive charge at C2 of the isopropyl substituent from isopropyl cyclopentane) towards 1,2-dimethylcyclohexyl cation (where formal positive charge is placed at one of the carbon atoms of the ring bonded to a methyl group). Tip: One hydride shift happens, followed by a 1,3-methyl shift (from isopropyl group) along with 1-3 C-C bond formation simultaneously. See nomenclature rules in chapters thirteen and fourteen.

Acknowledgment

Figures 12.8 and 12.13 were published in our own article, whose copyrights belong to Sociedade Brasileira de Química, and we appreciate the publisher's permission to print these figures in this book. Figures 12.9 and 12.11 were reprinted with permission from Firme et al. 2008a. Electronic nature of carbonium ions and their silicon analogues. J. Phys. Chem. A. 112: 3165-3171. Copyright 2008 American Chemical Society.

REFERENCES CITED

Anslyn, E.V. and Dougherty, D.A. 2006. Modern Physical Organic Chemistry. University Science Books. Sausalito.

Curtis, L.A., Redfern, P.C. and Raghavachari, K. 2007. Gaussian-4 theory. J. Chem. Phys. 126: 084108.

Firme, C.L., Antunes, O.A.C. and Esteves, P.M. 2008a. Electronic nature of carbonium ions and their silicon analogues. J. Phys. Chem. A. 112: 3165-3171.

Firme, C.L., Antunes, O.A.C. and Esteves, P.M. 2008b. Electronic nature of the aromatic adamantanediyl ions and its analogues. J. Braz. Chem. Soc. 19: 140-149.

Firme, C.L., Antunes, O.A.C. and Esteves, P.M. 2009. Identification of carbonium and carbenium ions by QTAIM. J. Braz. Chem. Soc. 20: 543-548.

Firme, C.L. 2012. Topological study of the first step of nucleophilically unassisted solvolysis of protonated 2-endo/exo-norbornanol and protonated 2-endo/exo-oxabicycloheptanol. J. Braz. Chem. Soc. 23: 513-521.

Firme, C.L., da Costa, T.F., da Penha, E.T. and Esteves, P.M. 2013. Electronic structures of bisnoradamantenyl and bisnoradamantanyl dications and related species. J. Mol. Model. 19: 2485-2497.

Gassman, P.G. and Fentiman, A.F. 1970. Characteristics of abrupt change from participation to nonparticipation of a neighboring group. J. Am. Chem. Soc. 92: 2549

Olah, G.A., Tolgyesi, W.S., Kuhn, S.J., Moffatt, M.E. Bastien, I.J. and Baker, E.B. 1963. Stable carbonium ions. IV. 1a secondary and tertiary alkyl and aralkyl oxocarbonium hexafluoroantimonates. formation and identification of the trimethylcarbonium ion by decarbonylation of the tert-butyl oxocarbonium ion. J. Am. Chem. Soc. 85: 1328-1334.

Olah, G.A. and Schlosberg, R.H. 1968. Chemistry in super acids. I. Hydrogen exchange and polycondensation of methane and alkanes in FSO_3H-SbF_3 ('magic acid') solution. Protonation of alkanes and the intermediacy of CH_5^+ and related hydrocarbon ions. The high chemical reactivity of 'paraffins' in ionic solution reactions. J. Am. Chem. Soc. 90: 2726-2727.

Olah, G.A., White, A.M., DeMember, J.R., Commeyras, A., and Lui, C.Y. 1970. Stable carbonium ions .100. structure of the norbornyl cation. J. Am. Chem. Soc. 92: 4627-4640.

Olah, G.A., Berrier, A.L., Arvanaghi, M. and Prakash, G.K.S. 1981. Stable Carbocations .231. C-13 NMR Spectroscopic study of the application of the tool of increasing electron demand to the 7-aryl-7-norbornenyl, 7-aryl-7-norbornyl, 2-aryl-2-bicyclo[2.1.1]hexyl, 1-aryl-1-cyclobutyl, and 3-aryl-3-nortricyclyl cations. J. Am. Chem. Soc. 103: 1122-1128.

Roberts, J.D. and Mazur, R.H. 1951. The nature of the intermediate in carbonium ion-type interconversion Reactions of cyclobutyl, cyclopropylcarbinyl and allylcarbinyl derivatives. J. Am. Chem. Soc. 73: 3542-3543.

Schleyer, P.v.R., Lenoir, D., Mison, P., Liang, G., Prakash, G.K.S. and Olah, G.A. 1980. Chemistry of proto-adamantane. 7. rapidly equilibrating unsymmetrically bridged 1,3,5,7-tetramethyl-2-adamantyl and rapidly equilibrating trivalent 1,2,3,5,7-pentamethyl-2-adamantyl cations - additivity of C-13-NMR chemical-shifts relating the classical vs non-classical nature of carbocationsy. J. Am. Chem. Soc. 102: 683-691.

Winstein, S. and Trifan, D.S. 1949. The Structure of the bicyclo [2,2,1]2-heptyl (norbornyl) carbonium ion. J. Am. Chem. Soc. 71: 2953-2953.

Winstein, S. and Trifan, D. 1952. Neighboring carbon and hydrogen. X. Solvolysis of endo-norbornyl arylsulfonates. J. Am. Chem. Soc. 74: 1147-1154.

Chapter Thirteen

Isomerism

ISOMERISM AND TYPES OF ISOMERISM

Isomerism gives rise to isomers that are molecules with the same chemical formula but different structural parameters or different spacial structures (different type of branching or different type of functional group or different position of the same functional group or different arrangement of substituents or different absolute configuration of the asymmetric atom).

As for structural isomerism, there are three types: chain isomerism, position isomerism (or regioisomerism), and functional isomerism.

The chain isomerism is related to the different position of the branching in its main chain. For example, butane and 2-methyl-propane are isomers (C_4H_{10}); pentane, 2-methyl-butane and 2,2-dimethyl-propane are isomers (C_5H_{12}); but-1-ene and 2-methyl propene are isomers (C_4H_8) as well (see Fig. 13.1(A)).

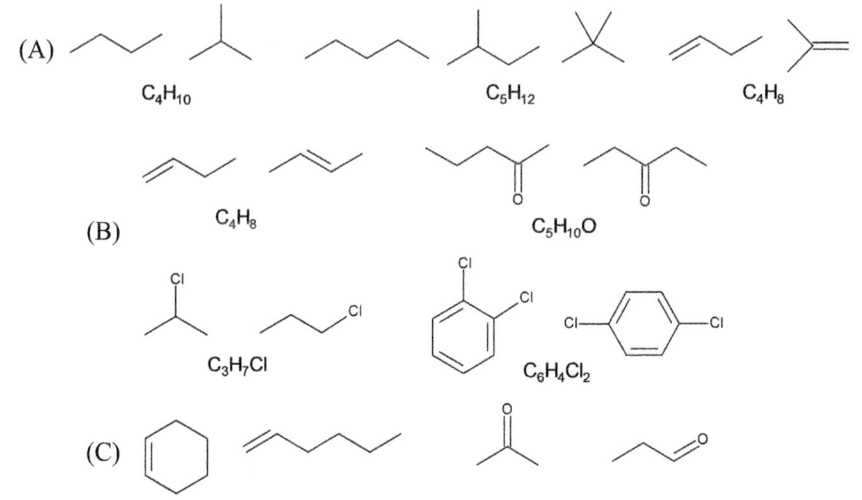

Figure 13.1 Bond line formula of (A) chain isomers, (B) regioisomers, and (C) functional isomers.

The position isomerism (or regioisomerism) is related to a different position of the substituent group or functional group in the molecule. For example, but-1-ene and but-2-ene are isomers (C_4H_8), pentan-2-one and pentan-3-one are isomers ($C_5H_{10}O$), 2-chloro-propane and 1-chloro-propane are isomers (C_3H_7Cl), *orto*-dichlorobenzene and *para*-dichlorobenzene are isomers ($C_6H_4Cl_2$) as well (see Fig. 13.1(B)).

Functional isomerism is related to different functional groups with the same molecular formula. For example, propanone and propanal are isomers (C_3H_6O), hexan-1-ene and cyclohexane (C_6H_{12}) are isomers as well (see Fig. 13.1(C)).

GEOMETRIC STEREOISOMERISM

Stereoisomerism is related to specific arrangements of substituents where two isomers are differentiated by their spacial disposition. Stereo means spacial.

Geometric stereoisomerm occurs in alkenes or derivatives and in substituted cycloalkanes. They generate two isomers called *cis* and *trans* or *E* and *Z*.

Geometric stereoisomerism in alkenes is more restricted than in cycloalkanes. In alkenes, two larger substituents (which can be same or different substituents) have to be in each sp^2 carbon atom. When they are placed in the same side of the double bond it is called *cis* or *Z* , e.g., *cis*-but-2-ene, and when they are placed in opposite sides of the double bond, e.g., *trans*-but-2-ene, it is called *trans* or *E*. In both *cis*- and *trans*-but-2-ene, the largest substituents are methyl group. It is possible to use different substituents, then, only the largest ones are taken into account. For example, *cis*-2,3-dichloro-but-2-ene and *trans*-2,3-dichloro-but-2-ene (see Fig. 13.2(A)). It is possible to have four different substituents and only the largest of each sp^2 carbon atom is taken into account, e.g., *cis*-/*trans*-2-bromo-3-chloro-pent-2-ene (see Fig. 13.2(A)).

There are some situations in which there is no geometric stereoisomerism. Firstly, if they are placed in the same sp^2 carbon atom: for example, 2-methylpropene. In that case, both methyl groups are at C2 carbon. Secondly, if there is only one substituent (hereafter, we consider hydrogen as a "neutral" substituent) at one of the sp^2 carbon atoms, e.g., in propene. Thirdly, if there are three or four equal substituents at sp^2 carbon atoms. For example, 2-methyl-but-2-ene and 2,3-dimethyl-but-2-ene, where three or four methyl groups are placed at sp^2 carbon atoms (Fig. 13.2(B)). A similar situation occurs for 2-ethyl-but-1-ene, but-1-ene, 3-ethyl-hex-3-ene, and 2,3-diethyl-hex-3-ene, respectively. See Fig. 13.2(C) for the bond-line notation of all these molecules.

Regarding polyenes or derivatives, each double bond will have *E* or *Z* configuration, which must be indicated in their nomenclature. For example, in (2*E*,4*E*,6*Z*,8*E*)-3,7-dimethyl-9-(2,6,6-trimethyl-1-cyclohenenyl)nona-2,4,6,8-tetraenoic acid, there are four CC double bonds and each one has *E* or *Z* configuration (Fig. 13.2(D)).

Substituted cycloalkanes with two or more substituents in different carbon atoms of the ring also have geometric stereoisomerism. Any n, m-disubstituted cycloalkanes have *cis* and *trans* isomers, for example, *cis*-(or *trans*-)1,2-dimethylcyclopropane, *cis*-(or *trans*-)1,2-(or 1,3-) dimethylcyclobutane, *cis*-(or *trans*-)1,2-(or 1,3-or 1,4-) dimethylcyclopentane. Any other substituent larger than methyl can also be used (see Fig.13.2(E)). Any n,m,o-trisubstituted cycloalkane have *cis-cis, trans-trans, cis-trans* configuration. For example, *cis-cis*-1,2,3-trimethyl cyclohexane, *trans-cis*-1,2,4-trimethyl cyclohexane.

The *cis-trans* stereoisomerism or geometric isomerism is a type of diastereoisomerism, that is, they are easily isolated, they have different physical properties, and they have no optical activity (i.e., they are not non-superimposable mirror images).

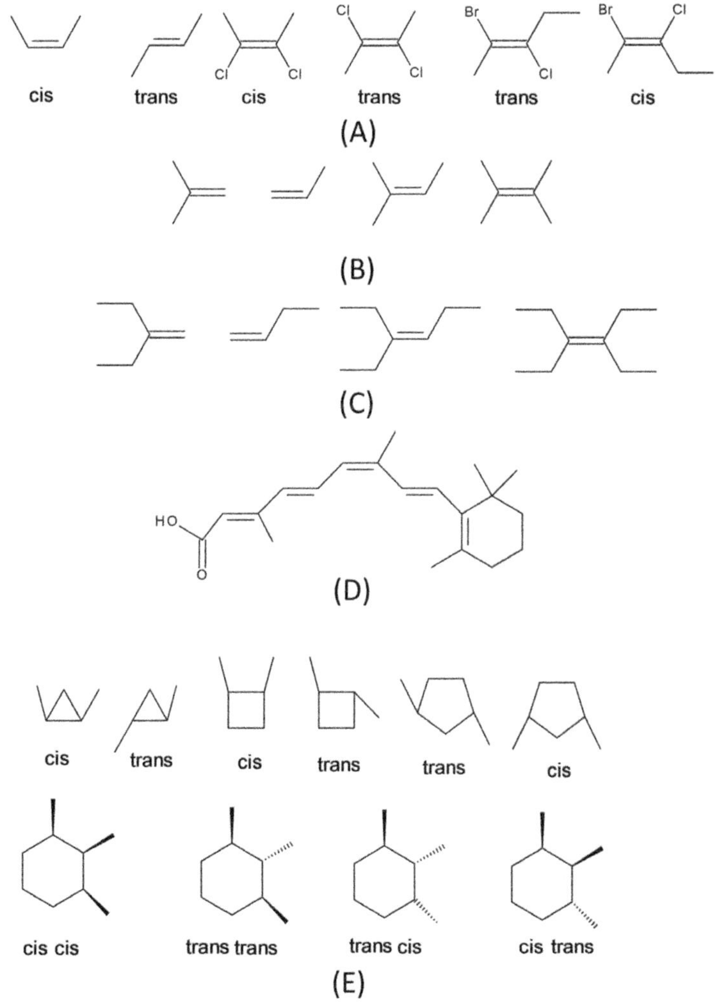

Figure 13.2 (A) *cis*-(or *trans*-) alkenes and derivatives; (B) and (C) alkenes with no geometric isomerism; (D) (2E,4E,6Z,8E)-3,7-dimethyl-9-(2,6,6-trimethyl-1-cyclohenenyl) nona-2,4,6,8-tetraenoic acid; (E) *cis*-(or *trans*-) substituted cycloalkanes.

Geometric stereoisomerism exists because C-C σ bond in cycloalkanes cannot rotate (because C-C rotation increases the ring strain in the cycloalkane- see chapter fifteen) and C=C double bonds in alkenes cannot rotate because of the π bond repulsion with respect to the substituents of each sp^2 carbon. (See chapter sixteen).

OPTICAL STEREOISOMERISM

Optical stereoisomerism is related to the pair of isomers which have distinguished optical properties (or which are optically active), although any other property is the same for both isomers. Optically active molecules are a pair of molecules which are non-superimposable mirror images and they are called enantiomers, i.e., they are chiral molecules or they have chirality (more details in the next subsection). The enantiomers deviate the polarized light at certain wavelength to opposing directions but same magnitude. This is experimentally observed by the use of polarimeter, which measures the angle of rotation when the polarized light passes through an optically active molecule.

Optically active molecules or chiral molecules can have one of the three options: (1) stereogenic/chirality/asymmetric center or stereocenter which is a tetrahedral atom (carbon, nitrogen, silicon, etc.) with four different substituents; (2) axial chirality which is a spatial arrangement involving a set of atoms in one axis that gives rise to a pair of chiral molecules (non-superimposable mirror images); (3) planar chirality which is a pair of chiral molecules having non-coplanar rings which are asymmetric from one another and cannot easily rotate about one another due to a steric hindrance.

General cases of axial chirality are substituted allenes or substituted spirobicycloalkanes. In allenes, there are two double bonds in the same sp carbon atom bonded to two sp^2 carbon atoms at each side. In order to avoid electronic repulsion of both π bonds in the same plane, there is a right angle between them which makes the substituents in each sp^2 carbon atom adopt a right angle with respect to the substituents of the other sp^2 carbon atom. If there are, at least, one substituent different from hydrogen in each sp^2 carbon atom, there appears chirality in the molecule, e.g., 1,3-dichloro propadiene and 4-chloropenta-2,3-dien-2-ol (Fig. 13.3(A)) have enantiomerism. As for spirobicycloalkanes, there is one and only one carbon uniting two saturated rings (called spirocarbon). Since the spirocarbon is a tetrahedron, one ring is placed in the plane of the paper and the other ring is placed at a right angle to the plane of the paper. The substituents in each opposing carbon atom with respect to the spirocarbon adopt opposing orientations of the whole ring it belongs in order to respect the fact that in a tetrahedron, two bonds are in the plane and two other bonds are out of the plane. If there are, at least, one substituent different from hydrogen in each of these carbons, there is chirality in the molecule. For example, 2,6-dichloro-spiro[3,3]heptane has enantiomerism (Fig. 13.3(B)).

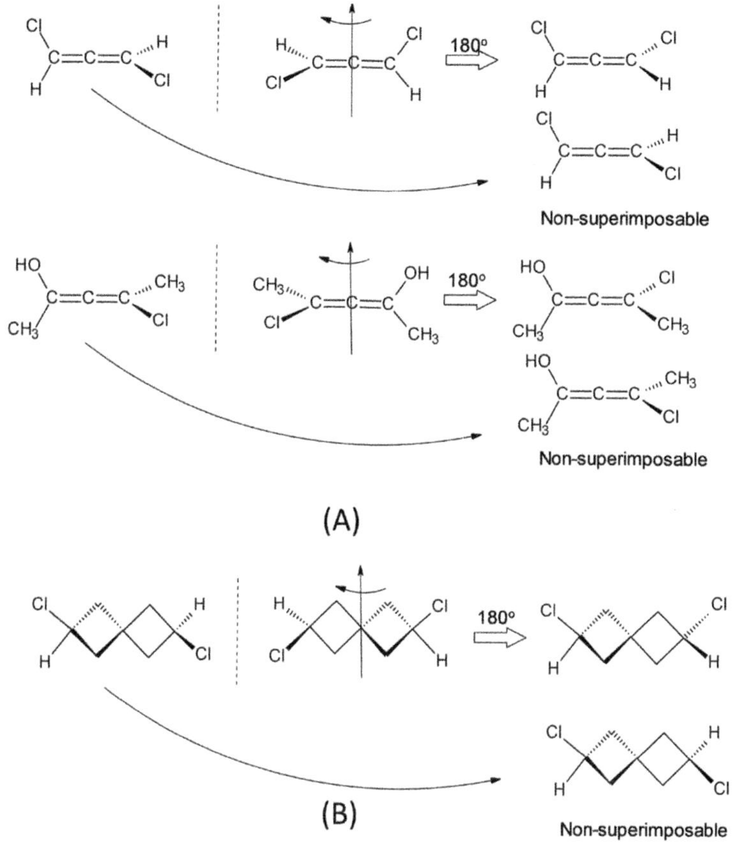

Figure 13.3 (A) (R)- and (S)-1,3-dichloro propadiene and (R)- and (S)-4-chloropenta-2,3-dien-2-ol; (B) (R)- and (S)- 2,6-dichloro-spiro[3,3]heptane.

One iconic example of planar chirality is the 1,1′-bi(2-naphthol) (see Fig. 13.4).

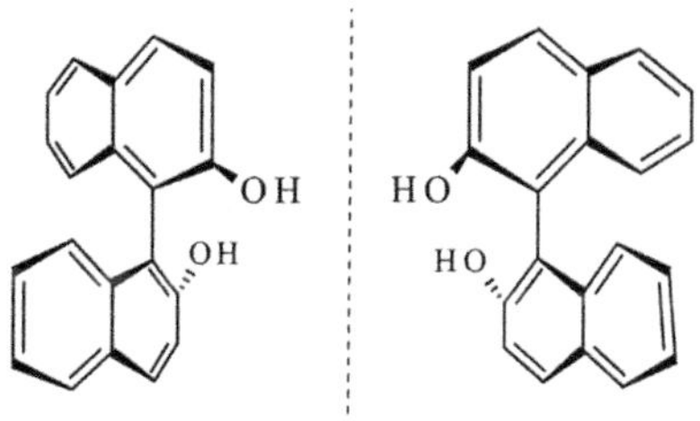

Figure 13.4 (R)- and (S)- 1,1′-bi(2-naphthol).

CHIRALITY

Chirality comes from Greek, meaning "hand" which is the iconic example of chirality. An object is chiral when its mirror image is non superimposable to it, i.e., the object and mirror image are distinguished, different entities, so-called entantiomorphs, a Greek word for "opposite forms". On the other hand, achiral object and its mirror image are superimposable, being exactly the same entity. Besides hands, other chiral objects are feet, glasses, gloves, socks, shoes, and tablet arm chair, for example. Achiral objects are an ordinary chair, pencil, mirror, cell phone, fork, plate, spoon, etc.

In chemistry, enantiomorphs are called entantiomers. The types of enantiomers were mentioned in the previous subsection. Except for enantiomers, all other molecules are achiral. Some achiral molecules might have stereoisomerism, for example, diastereoisomers and mesoisomers (see further subsections). Other achiral molecules have no type of stereoisomerism.

Most building blocks (monomers) for biomacromolecules in living beings (animals and plants) are chiral. For example, cytidine (a nucleoside molecule made up of cytosine and a ribose ring) for building DNA has four stereocenters (Fig. 13.5(A)). Except for glycine, all other alfa-amino acids, for building proteins and enzymes, have a C_α chiral center. As for monosaccharides, D-glyceraldehyde (which has one chiral center) is the building block for constructing other larger monosaccharides (aldoses) and then polysaccharides (Fig. 13.5(B)).

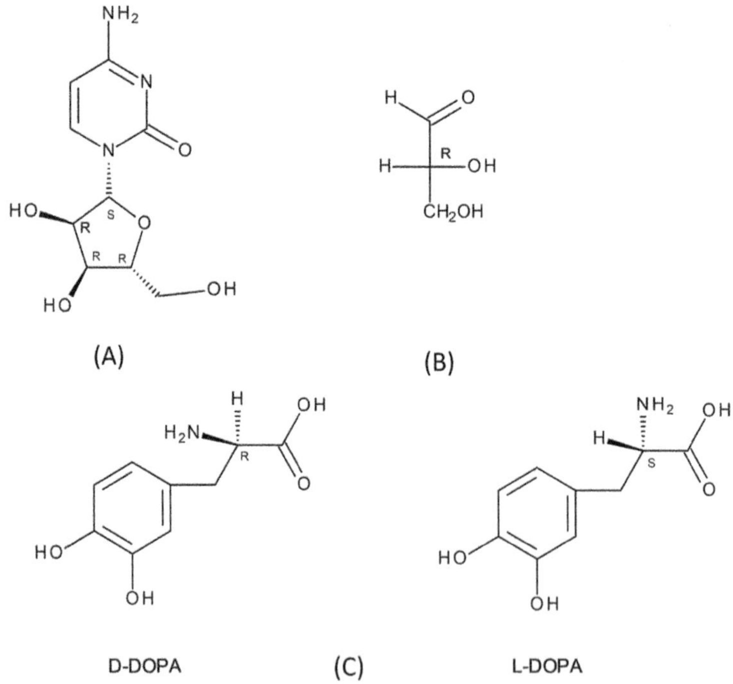

(A)   (B)

D-DOPA   (C)   L-DOPA

Figure 13.5   (A) cytidine, (B) D-glyceraldehyde, (C) L-DOPA, and D-DOPA along with the absolute configuration of stereo centers.

Moreover, all reactions in living beings are catalyzed by enzymes which can be enantioselective. Then, the effect of each enantiomer in animals can be completely different. In case of drugs, 50% of marked drugs are chiral, in which 90% are mixtures of enantiomers (racemates) rather than a pure single enantiomer (Nguyen et al. 2006). McConathy and Owens state that "two enantiomers of a chiral drug may differ significantly in their bioavailability, rate of metabolism, metabolites, excretion, potency, and selectivity for receptors, transporters and/or enzymes, and toxicity. The use of single-enantiomer drugs can potentially lead to simpler and more selective profiles, improved therapeutic indices, simple pharmacokinetics due to different rates of metabolism of the different enantiomers, and decreased drug interactions. For example, one enantiomer may be responsible for the therapeutic effects of a drug, whereas the other enantiomer is inactive and/or contributes to undesirable effects" (McConathy and Owens 2003). For example, L-DOPA is a drug for the treatment of Parkinson's disease, while D-DOPA is biologically inactive (Fig. 13.5(C)).

ENANTIOMERS FROM STEREOGENIC CENTRES AND SYMMETRY NOTION

Enantiomers from stereogenic centers have one (or more) tetrahedral sp^3 hybridized atom (carbon, silicon, germanium, positively tetracoordinated nitrogen) with four distinguished substituents attached to this atom. That is, the stereogenic/chiral center has no symmetry element, which is also called asymmetrical center. A molecule with one asymmetrical center has a mirror image which is not superimposible and is its enantiomer (see Fig.13.6 (A)).

The symmetry element is a geometrical entity used for a symmetry operation in which a transformation occurs in this molecule giving the exact same molecule after this operation. For example, one can use the plane as a symmetry element to apply two symmetry operations in a water molecule (with an angular geometry) passing through all three atoms, a horizontal plane (σ_h), and passing through oxygen atom forming a right angle with the former, a vertical plane (σ_v). One can also use a line or axis as a symmetry element passing through oxygen atom in the vertical direction and rotate the water molecule about this axis by 180° giving the exact water molecule with an inverted position of hydrogen atoms, which is called a C_2 operation of symmetry, where 2 is the ratio 360/180 (Fig. 13.6(B)). A further example is dibromomethane which also has σ_h, σ_v, and C_2 symmetry operations as water although it has a different geometry, a tetrahedral geometry (Fig. 13.6(B)). The center of symmetry is the third element of symmetry, where there is a central point from which two lines are crossed and each of them passes through two atoms or groups in the molecule. For example, in *trans*-dibromoethene, one line passes through two bromine atoms and the other line passes through two hydrogen atoms. These lines cross in the middle of the C=C double bond. When a symmetry operation called inversion operation (i) is applied in the center of symmetry (inversion center), both bromine atoms are inverted in position and both hydrogen atoms are inverted in positions, giving exactly the same molecule.

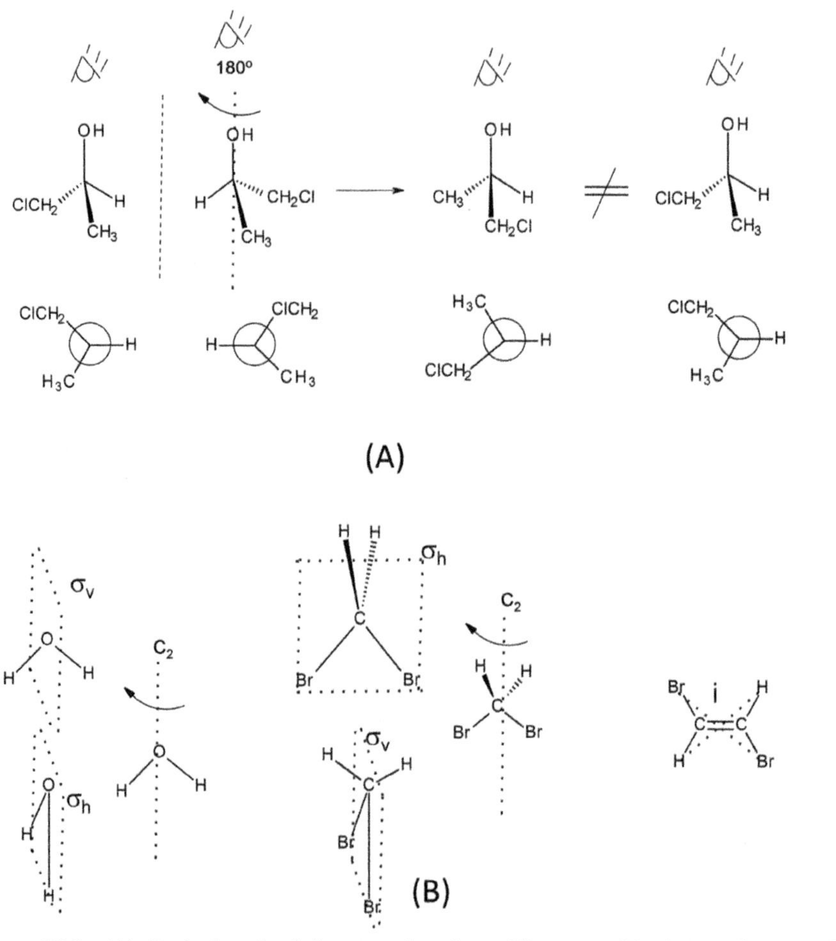

(A)

(B)

Figure 13.6 (A) Dashed-wedged line notation from C2 atom of both enantiomers of 3-chloro-propan-2-ol along with their corresponding planarized view from C-OH axis; (B) symmetry elements used for symmetry operations in water, dibromomethane, and *trans*-dibromoethene.

A symmetrical tetrahedral atom is not a stereogenic or chiral center, i.e., its mirror image is superimposable. A symmetrical tetrahedral atom has, at least, two equal substituents in the molecule (for example, methyl carbon and methylene carbon). Chloromethane, for example, has three equal substituents (three hydrogen atoms) and it has three σ_v and three C_6 symmetry elements from C-Cl axis (Fig. 13.7(A)). Dichloromethane (with two pairs of equal substituents) has two σ_v and two C_2 symmetry elements from the vertical axis passing through C atom (Fig. 13.7(B)). Dichlorobromomethane (with three different substituents) has only σ_h symmetry elements (Fig. 13.7(C)). Consequently, tetrahedral atoms with one type of substituent (e.g., hydrogen atom in methane), two types of substituents (e.g., chloromethane and dichloromethane) or three types of substituents (e.g., dichlorobromomethane) have symmetry elements and this tetrahedral atom is not an asymmetrical/chiral atom. On the other hand, a tetrahedral atom with four different types of substituents, for example, in bromochloromethanol (Fig. 13.7(D)) has no symmetry element and it is an asymmetrical or chiral atom. In other words it has a non-superimposable mirror image which is its enatiomeric form.

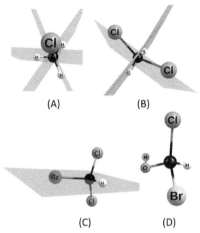

(A)   (B)

(C)   (D)

Figure 13.7 Optimized geometries along with symmetry elements of (A) chloromethane, (B) dichloromethane, (C) dichlorobromomethane, and (D) bromochloromethanol.

Important rule: except for a very few cases, all molecules with inversion center of symmertry and/or plane of symmetry are always achiral, although some molecules with other symmetry elements can also be achiral.

Hints: (1) Use free quantum chemistry visualization programs, such as Avogadro to draw the pair of enantiomers and observe more realistically that they are not superimposable. Do the same for a achiral molecule and draw its mirror image and realize they are superimposable.

(2) Draw four equilateral triangles with borders on a hard paper in order to build a tetrahedral pyramid and paint each vertex with different colors. Make another pyramid changing the colors in two vertices and observe that they are not superimposable.

OPTICAL PROPERTIES FROM POLARIZED LIGHT

Light travels through space as a sine wave of electric field synchronously with a sine wave of magnetic field at a right angle, represented by the following wave function:

$$\Psi = \vec{E} + i\vec{B}$$

The peak-to-peak amplitude of each field is represented by a double arrow. A non-polarized light has infinite double arrow electric and magnetic field when viewed through a transverse section. A polarizing filter is a polymer stretched out in one direction. When the light passes through this filter, the electric fields in all directions are blocked except the one in the direction of the stretched polymer. As a consequence, a polarized light (in which the electric field propagates in only one direction) is formed. If there is a second polarizing filter at a right angle with the first, then the polarized light is completely blocked and no light is observed after the second filter (Fig. 13.8(A)). However, it is important to emphasize that the polarized light has small oscillations in both directions (clockwise and counter-clockwise), providing a chiral environment to detect chiral molecules.

A chiral molecule is able to deviate the polarized light in a way that the second polarizing filter has to be placed at $90+\alpha$ or $90-\alpha$ in order to block out the polarized light because the chiral molecule had deviated the polarized light by $+\alpha$ or $-\alpha$ angle (Fig. 13.8(B)).

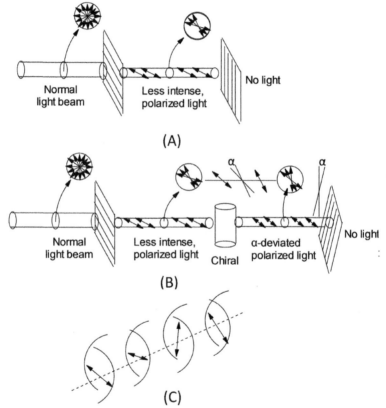

Figure 13.8 (A) Propagation of normal light through polarizing filter and then polarized light blocked-out in the second polarizing filter at right angle with the first; (B) propagation of the polarized light through chiral sample deviating the light; (C) circularly polarized light varying the angle of polarized light along with the propagation direction.

The deviation of the polarized light by a chiral matter is only possible due to the small oscillations of the polarized light in clockwise and counter-clockwise directions. When light passes through a transparent or even opaque (chiral or achiral) material (solid, liquid, and even gas) scattering of light occurs. The scattering of the light is the successive electron excitation by photons and re-emission of them in each molecule of the matter when light passes through that material (a gaseous, liquid, or solid matter). This re-emission of the photons can be in random directions (when matter is a gas), giving rise to the Rayleigh scattering (for example, the sunlight scattering in the Earth's atmosphere resulting in blue sky during the day and red sky during sunfall and sunrise), or in an ordered manner (when matter is a liquid or solid), giving rise to refraction phenomenon (a well-defined direction of light deviation when travelling a solid or liquid matter). When light passes through a liquid or solid, the speed of light decreases (as a consequence of successive photon excitation and re-emission) and this velocity decay is dependent on the refractive index of the matter. When a polarized light passes through a chiral matter, it provides different velocities for both oscillations of polarized light, which leads to the deviation of the polarized light (Anslyn and Dougherty 2006). Quite recently, Jenkins' group used QTAIM descriptors (ellipticity and stress tensor) to obtain their values along the rotation of a bond (change in the dihedral angle) involving the asymmetric/symmetric center. They showed that each enantiomer has a helictical behaviour of the stress tensor descriptor which opposite to that from its enantiomer. Each helictical behavior indicated the absolute configuration of each enantiomer according to the clockwise or counter-clockwise direction of their heliticities. On the other hand, an achiral molecule does not have this helictical behaviour (Xu et al. 2019). This seems to be the first rationalization of a electronic property of a chiral molecule associated with its optical activity.

In a circularly polarized light, the polarized light is rotated over 360° along with its propagation direction at a constant magnitude (Fig. 13.8(C)). In circular dichroism (CD) spectroscopy, two circularly polarized lights (CPL) are generated: left-handed CPL, L-CPL, and right-handed CPL, R-CPL. When both CPL lights pass through a chiral sample, a delta absorbance (difference between absorbance of left circularly polarized and absorbance of right circularly polarized light) occurs. One state of the circularly polarized light will be absorbed to a greater or lesser extent than the other, giving a negative or positive signal. The CD spectroscopy of a simple chiral molecule gives several positive and negative absorption bands.

Vibrational circularly dichroism (VCD) spectroscopy measures the differential response between L-CPL and R-CPL of a polarized infrared radiation during vibrational transition of a chiral sample. The VCD can be used to determine the absolute configuration of an enantiomer, since each enantiomer exhibits a symmetrically different absorption spectrum. For example, (2S)-butan-2-ol has a symmetrically opposing VCD spectrum (Fig. 13.9(A)) to (2R)-butan-2-ol (Fig. 13.9(B)).

Enantiomers have the same physical properties (boiling point, melting point, viscosity, density, solubility, surface stress, dipole moment, refractivity from normal light, etc.), except for the optical properties (deviation of polarized light and differential absorption of circularly polarized light). Even other spectroscopic methods other than circular dichroism give similar results for both enantiomers when using achiral solvents for the chiral sample. However, chiral molecules from the same pair of enantiomers interact differently with chiral molecules, such as a chiral solvent or chiral catalyst (a chiral organometallic complex or a chiral enzyme or even another chiral reactant), giving rise to differentiated results of reaction (stereoselectivity) or separation of enantiomers or different spectroscopic results. These topics will be discussed in the next book (see notes at the beginning of this book).

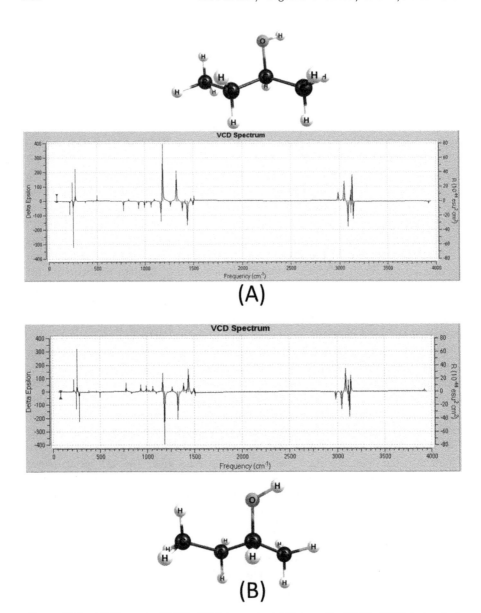

Figure 13.9 (A) Theoretical VCD of (2S)-butan-2-ol; (B) theoretical VCD of (2R)-butan-2-ol.

The polarimeter is an equipment which gives the observed optical rotation, α, which is linearly dependent upon the chiral sample concentration. The higher the concentration, the higher the observed optical rotation in modulus. This linear rotation can be used to find the unknown concentration of the same chiral substance. Firstly, one needs to calibrate the polarimeter with pure achiral solvent, for example, water. Next, different concentrations of the same chiral substance are measured in the polarimeter. Table 13.1 shows the values of α for each concentration of sucrose.

Table 13.1 Observed optical rotation, in degrees, for each concentration, in g mL^{-1}, of sucrose at 25°C.

Sucrose solution concentration/g mL^{-1}	0.0	0.02	0.04	0.1
Observed optical rotation/°	0.0	1.1	2.4	7.0

From a linear regression, one finds a linear expression which can be used for obtaining the concentration of an unknown sample (of the same chiral substance).

In order to normalize one specific value of the optical rotation from a determined chiral substance, the conditions for measuring the optical rotation in the polarimeter must be standardized (temperature, usually at 25°C; light wave length, λ, usually in 589.3 nm; concentration, c, in g mL^{-1}; length of the sample container, l, in dcm) and one expression is used to provide the specific rotation:

$$[\alpha]_\lambda^{t^o} = \frac{\alpha}{l \cdot c} \therefore \frac{[\text{deg}]}{\left[10 \cdot \text{g} \cdot \text{cm}^2\right]}$$

The specific rotation has an awkward unit, so it is usual to represent the value of specific rotation without units.

CAHN-INGOLD-PRELOG RULES

Cahn-Ingold-Prelog rules or CIP rules are used to assign the absolute configuration of a stereogenic/chiral center, named R (from Rectus) or S (from Sinister), from a molecule represented by dashed-wedged line notation or stick-and-ball model. Experimentally, the assignment of a chiral center can be done by X-ray crystallography, optical rotatory dispersion (where specific rotation of a chiral sample varies with the change of the wavelength of the polarized light) or vibrational circular dichroism.

Firstly, one has to find the stereocenters or chiral centers in the molecule, that is, asymmetric tetrahedral carbon atom (or other sp^3 atom) with four different substituents. Afterwards, one has to assign the priorities of these susbtituents attached to the asymmetric carbon atom. There are some rules that have to be followed to assign the priorities correctly. They are:

1. Each first atom from each substituent bonded to the chiral tetrahedral atom is given a number from one to four. The number one is given to the atom with the highest atomic number (highest priority) and number four is given to the atom with the lowest atomic number (lowest priority).

2. If two or more of these first atoms of these substituents have the same atomic number, then we analyze the atomic number of all atoms bonded to the first atom of each substituent and we sum all of them. The highest sum will have the highest priority. For example, (C,C,H) and (C,H,H) are the atoms bonded to the first atom of each substitutent. The vicinal atoms (C,C,H) has the highest sum of Z and the highest priority. In case the sum of atomic number of the atoms bonded to the first atom of each substituent is equal, for example, (C,C,H) and (C,C,H), we analyze the atomic numbers (and its sum) from the atoms bonded to the second atom in the chain of the substituent. This process goes further until there are no more ties to assign the priority to.

3. The substituent of lowest priority (number four) has to be placed behind the viewer and then the sequence of 1-2-3 numbers has a clockwise or counter-clockwise direction.
4. If it is clockwise, the chiral carbon is assigned R configuration. If it is counter-clockwise, the chiral carbon is assigned S configuration.
5. Its enantiomer obviously has the opposite assignment which can be done by exchanging any pair of the substituents bonded to the chiral atom.
6. In case the lowest substituent is placed in front of the viewer (and not in plane of the paper), then analysis of item (4) is the opposite to that when the lowest subtituent is placed behind the viewer.
7. The atomic number of an atom in a double or triple bond is counted twice or thrice, respectively.

Example Determine the priority order (according to CIP rules) for: methyl, bromomethyl, ethyl, hydrogen. Afterwards, draw one of the possible enantiomers from dashed-wedged line notation from the perspective of the asymmetric carbon bonded to these substituents and next in the perspective of the main chain. Do the same for the other enantiomer.

Answer Firstly, use the condensed formula (not totally condensed) to draw each substituent and give the priority number. Having a tie for the first atom in each substituent, put between brackets the vicinal atoms bonded to each first atom. Then, we have:

As a consequence: -CH$_3$ (H,H,H); -CH$_2$Br (H,H,Br), -CH$_2$CH$_3$ (H,H,C); -H
-CH$_2$Br (1) > -CH$_2$CH$_3$ (2) > -CH$_3$ (3) > -H (4)

Now, draw the dashed-wedged line notation of a tetrahedral carbon with these substituents where you draw the substituent with lowest priority backwards and the other substituents at a random position, and hence you can determine the absolute configuration of the asymmetric carbon (Fig. 13.10(A)). Next, change any pair of substituents and then you have its enantiomer (Fig. 13.10(B)). Afterwards, you need to find the main chain from that structure by counting the carbon atoms in a sequence that gives the largest amount of C atoms. Finally, draw the corresponding dashed-wedged line notation to this structure from the perspective of the main chain (Fig. 13.10(C)). Then, you change the position of the two substituents out of the main chain and you obtain the other enantiomer from this perspective as well (Fig. 13.10(C)).

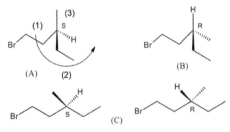

Figure 13.10 Dashed-wedged line notation from the perspective of the tetrahedral asymmetric carbon atom of (A) (3S)-1-bromo-3-methyl pentane; and of (B) (3R)-1-bromo-3-methyl pentane; and (C) their corresponding dashed-wedged line notation from the perspective of the main chain.

Hint   There are free softwares such as Accelrys Draw (previous name for Biovia Draw), where you can draw the molecule and obtain the configuration of asymmetric carbon, in case you are in doubt about the correct configuration.

DIASTEREOISOMERISM

Diastereoisomerism is another type of stereoisomerism different from enantiomerism, where the diastereoisomers are not mirror images and have different physical properties. Diastereoisomers have two or more stereogenic/chiral centers where at least one (but not all) of them have different configurations. When diastereoisomers have different configuration in just one asymmetric atom they are called epimers, for example, α and β anomers of cyclic monosaccharides.

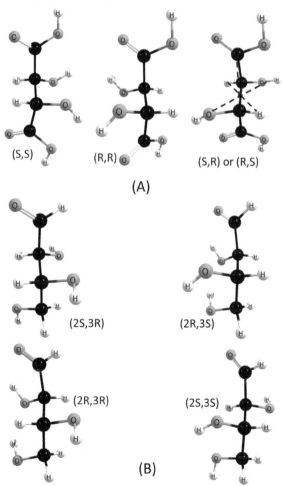

Figure 13.11   (A) Optimized structures of (2*R*,3*R*)-, (2*R*,3*S*)-, and (2*S*,3*S*)-tartaric acid; (B) Optimized structures of (2*R*,3*R*)-, (2*S*,3*S*)-, (2*S*,3*R*)-, and (2*R*,3*S*)-aldotetrose or so-called D-Erythrose, L-Threose, D-Threose, and L-Erythrose, respectively.

If a molecule has two stereocenters, there are up to four stereoisomers, (R,R), (S,S), (R,S), and (S,R). Sometimes, (R,S) and (S,R) are mesoisomers (see the next subsection) and there are only three stereoisomers. For example, tartaric acid has three stereoisomers: (R,R), (S,S), and (R,S) or (S,R) because $(2R,3S)$- and $(2S,3R)$-tartaric acid have an inversion center of symmetry and they are the same molecule (a meso isomer). In Fig. 13.11(A), the optimized structures of the stereoisomers of tartaric acid are shown. In that case, (R,R) and (S,S) are enantiomers, but (S,S) and (R,S) or (R,R) and (R,S) are epimers. On the other hand, for aldotetroses, there are four stereoisomers: (R,R), (S,S), (R,S), and (S,R), where (R,S) and (S,R) are enantiomers and not the same molecule because there is no symmetry element for these molecules (Fig. 13.11(B)).

When a molecule has three stereocenters, it might have up to eight stereoisomers: (R,R,R), (S,S,S), which are enantiomers, (R,S,R), (S,R,S), which are enantiomers, (R,R,S), (S,S,R) which are enantiomers, and (S,R,R) and (R,S,S) which are enantiomers as well. When you compare (R,R,R) to (R,S,R) or (S,R,S) or (R,R,S) or (S,S,R) or (S,R,R) or (R,S,S), these pairs are all diastereoisomers. A similar result occurs when comparing molecules with one or two stereocenters to opposing absolute configuration.

The total amount of stereoisomers is 2^n where n is the amount of stereocenters in the molecule. For example, aldopentose, has eight stereoisomers: D-Ribose, L-Ribose, D-Arabinose, L-Arabinose, D-Xylose, L-Xylose, D-Lyxose, and L-Lyxose (Fig. 13.12). The most convenient and simplest model to represent these monosaccharides is Fischer projection (chapter nine).

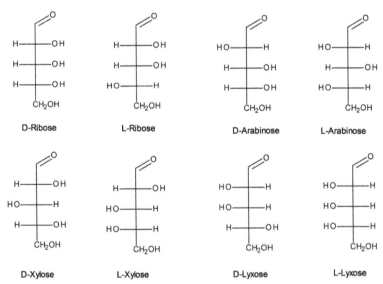

Figure 13.12   Fischer projection of D-Ribose, L-Ribose, D-Arabinose, L-Arabinose, D-Xylose, L-Xylose, D-Lyxose, and L-Lyxose.

Fischer projection can also be used to assign the configuration of the asymmetric atoms (or stereocenters) of a molecule with two or more stereocenters. We took as an example the four stereoisomers of threonine, one of the building blocks of proteins (Fig. 13.13). For each asymmetric carbon, one has to place the hydrogen (the substituent with lowest priority in each asymmetric carbon) at the vertical position of the Fischer projection by using one of the movements in Fischer projection which does not change the configuration of the stereocenter. In that case, we use the so-called "one fixed, three switched" movement (see Fig. 9.7(F) in chapter nine). Afterwards, we give the priority numbers for each substituent bonded to each stereocenter in order to assign their absolute configuration.

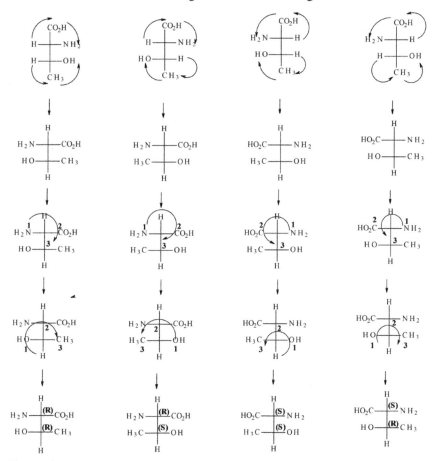

Figure 13.13   Fischer projection and procedure to assign the absolute configuration by using Fischer projection of estereoisomers of threonine.

MESO ISOMER

A meso isomer (please do not confuse with mesomerism) or meso compound is a molecule with two stereocenters, but its mirror image is equivalent to the molecule, i.e., an object and its mirror image are superimposable and represent the same molecule. As a consequence, meso isomer is a non-optically active (achiral) molecule although its stereoisomers are optically active. A meso compound has a plane of symmetry and/or inversion center of symmetry. As mentioned above tartaric acid has three stereoisomers: (R,R), (S,S), and (R,S) or (S,R) where $(2R,3S)$- or $(2S,3R)$-tartaric acid is the same molecule (a meso compound) because they have an inversion center of symmetry (Fig. 13.11(A)).

Other types of meso compounds can be found in some disubstituted cycloalkanes. Similarly, *cis*-dimethylcyclopropane has two stereocenters (R,S), but it is also a meso compound due to its plane of symmetry (Fig. 13.14(A)). From a two-dimensional view of *trans*-dimethylcyclopropane, one might think it has an inversion center of symmetry, but it does not. Then, *trans*-dimethylcyclopropane has a pair of enantiomers (Doering and Kirmse 1960). The same reasoning is applied to *cis*- and *trans*-1,3-dimethylcyclohexane and *cis*- and *trans*-1,2-dimethylcyclobutane, giving the same result as that from *trans*-dimethylcyclopropane (Fig. 13.14(B)). Then, *any trans-disubstituted cycloalkane (not having a plane of symmetry) has two enantiomers.*

However, *cis*- and *trans*-1,3-dimethylcyclobutane are both meso compounds, because both have a plane of symmetry. Like $(2R,3S)$- or $(2S,3R)$-tartaric acid, 1,2-dibromobutane has an inversion center of symmetry and it is a meso compound as well (Fig. 13.14(B)).

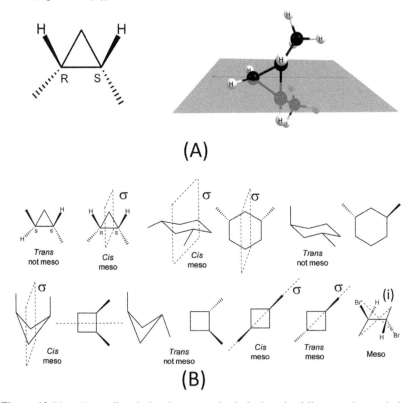

Figure 13.14 (A) *cis*-dimethylcyclopronane in dashed-wedged line notation optimized stick-and-ball model; (B) some meso and not-meso compounds.

CONFORMATIONAL ISOMERISM

Conformers (or rotamers) are isomers of the same molecule generated by the rotation of one single bond. Then, conformational isomerism is a type of stereoisomerism. Each single bond may give rise to two, four, or six conformers. In each pair, one conformer is a minimal in the potential energy surface and the other conformer is the maximum in PES (that is, it is a transition state). Conformers exist in any molecule with single bond: alkanes, amines, alcohols, ethers, and even alkenes and alkines with, at least, two saturated carbon atoms. *It is very important to mention that, in vast majority of the situations, the most stable conformer gives rise to all properties of the molecule.* The conformers can be easily obtained using Newman projection (see the example in Fig. 9.5 in chapter nine). At each 60° rotation of a molecule in Newman projection, it is possible to find its rotamer or the same rotamer. *After obtaining all the conformers of a molecule, one may plot a potential energy surface with these rotamers, by previously knowing the energy difference between then. This procedure is called conformational analysis.* See chapter fourteen for more details. *Conformational analysis is always related to two vicinal atoms (two carbon atoms, or one oxygen/nitrogen and one carbon atom), where one of them is kept unmoved and the other is rotated from 60 to 60°.* Performing conformation analysis is extremely important to obtain the most stable conformer of a molecule that might have several conformers in order to use the most stable conformer for subsequent calculations and property analysis. In Fig. 13.15(A), there are examples of the optimized geometry of the most stable conformer (after conformational analysis) of some alcohols and amines.

The systematic nomenclature of conformers is based on the Klyne-Prelog system, where a circle (the same used to represent the first carbon from Newmann projection) is divided in two parts. The upper part (where the reference group is located) is called "syn" and the lower part is called "anti". The vertical line that crosses the circle is called "periplanar", and diagonal lines 60° before and after the periplanar line are called "clinal". See the scheme in Fig. 13.15(B). Let us consider conformers of an alkane where the methyl is the reference in the first carbon (Fig. 13.5(B)). If the most voluminous group of the second carbon (so-called R) is on the same side (upper part) and in the periplanar line, the conformer is called synperiplanar. If R is at the upper part and in the clinal line, the conformer is called synclinal. If R is at the lower part (opposite to the reference group in the first carbon) in the periplanar line, the conformer is called antiperiplanar. Finally, if R is at the lower part in the clinal line, the conformer is called anticlinal.

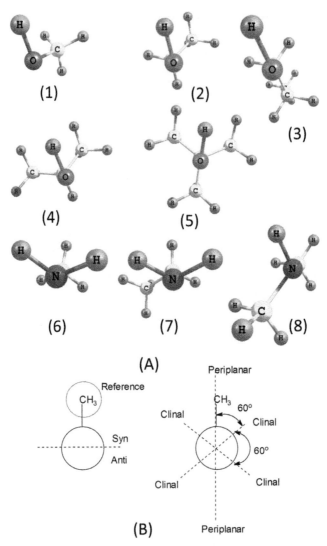

Figure 13.15 (A) optimized geometry of most stable conformer of (1) methanol, (2) ethanol, (3) propan-1-ol, (4) propan-2-ol, (5) tert-butanol, (6) methylamine, (7) ethylamine, (8) dimethylamine; (B) schematic representation for the nomenclature of conformers.

EXERCISES

1. Determine the priority order (according to CIP rules) for the substituents in the items below. Afterwards, draw one of the possible enantiomers from dashed-wedged line notation from the perspective of the asymmetric carbon bonded to these substituents, and next from the perspective of the main chain. Do the same for the other enantiomer. Finally, assign the absolute configuration for each structure.

   A) Butyl; 1-methyl-propyl; hydrogen; 2-methyl-propyl
   B) 2-methyl-propyl; 1-methyl-ethyl; cyclohexyl; 1,1-dimethyl-ethyl
   C) Ethyl; 1-chloroethyl; 1-bromoethyl; 2-bromoethyl

2. Find the chiral carbon and assign its absolute configuration for the molecules in Fig. 13.16.

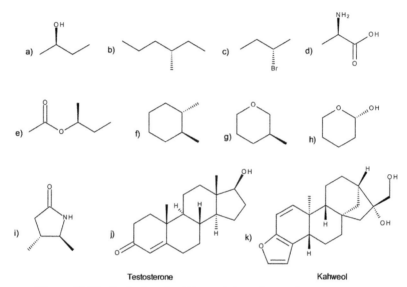

Figure 13.16 Dashed-wedged line notation of several structures.

3. The structures in Fig. 13.11 are in stick-and-ball model of the optimized structures in zig-zag carbon chain. Change these structures into Fischer projection and assign the absolute configuration of their stereocenter using Fischer projection as well. See an example in Fig. 13.17.

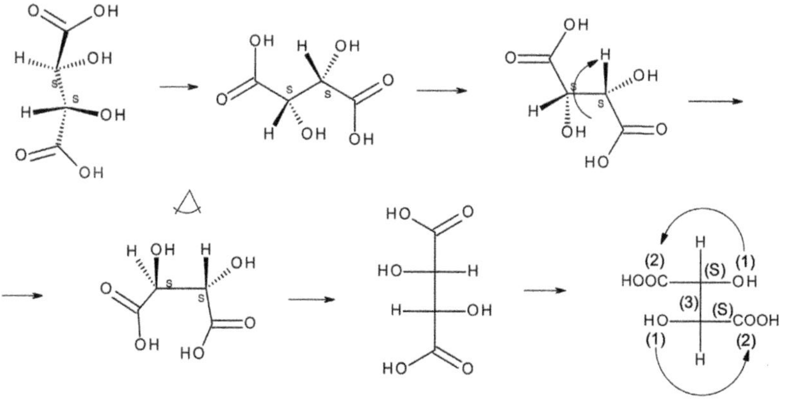

Figure 13.17 (2S,3S)-aspartic acid in dashed-wedged line notation and in Fischer projection.

REFERENCES CITED

Anslyn, E.V. and Dougherty, D. 2006. Modern Physical Organic Chemistry. University Science Books. Sausalito.

Doering, W. von E. and Kirmse, W. 1960. Absolute configuration of trans-1,2-dimethylcyclopropane. Tetrahedron 11: 272-275.

McConathy, J. and Owens, M.J. 2003. Stereochemistry in drug action. Prim. Care Companion J. Clin. Psychiatry 5: 70-73.

Nguyen, L.A., He, H. and Pham-Huy, C. 2006. Chiral drugs: an overview. Int. J. Biomed. Sci. 2: 85-100.

Xu, T., Li, J.H., Momen, R., Huang, W.H., Kirk, S.R, Shigeta, Y. and Jenkins, S. 2019. Chirality-helicity equivalence in the S and R stereoisomers: a theoretical insight. J. Am. Chem. Soc. 141:5497-5503.

Chapter Fourteen

Alkanes (nomenclature, properties, and reactions)

HYDROCARBONS

Hydrocarbons are a class of organic compounds containing only carbon and hydrogen atoms. There are four main types of hydrocarbons: alkanes, alkenes, alkynes (aliphatic compounds), and aromatic compounds, besides alkadienes and polyenes. Aliphatic compounds (non-aromatic hydrocarbons) can be open-chain compounds (acyclic) or closed-chain compounds (cyclic). Alkanes just have single C-C bonds. Alkenes just have one double C=C bond (and all other single C-C bonds for alkenes higher than ethane). Alkynes just have one triple C≡C bond (and all other single C-C bonds for alkynes higher than ethyne). Alkadienes have two double C=C bonds. Polyenes have more than two double C=C bonds. The simplest alkane is methane. The simplest alkene is ethene. The simplest alkyne is ethyne. The iconic aromatic compound is benzene. The simplest alkadiene is butadiene. The simplest polyene is hexatriene. See Fig. 14.1 for their representations.

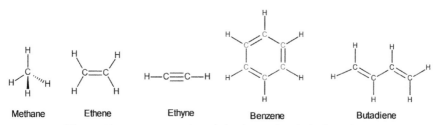

Methane   Ethene   Ethyne   Benzene   Butadiene

Figure 14.1   Representation of simplest or iconic hydrocarbons.

ALKANES

The general formula of alkanes is C_nH_{2n+2}. Besides being the simplest alkane, methane is also the most abundant. It is present in the atmosphere, in the oceans (e.g., in methane clathrate – methane trapped in solid, crystal water) and in the Earth's crust. Ethane and propane are the second and third simplest alkanes and they made up the natural gas along with methane (in largest amount) and butane. All of these compounds are important fossil fuels, along with gasoline, diesel, and kerosene. Alkanes can also be used as solvents, lubricating oil, paraffin wax, and asphalt. Natural gas is obtained from Earth's crust reservoir and higher alkanes are obtained from petroleum crude oil after its refining process. In the first and main unit of the petroleum refining process, there is a huge distillation tower where the separation of main alkanes takes place. See Table 14.1 below where 1C, 2C, and so on, represents alkane with one carbon atom, alkane with two carbon atoms, and so on.

Table 14.1   Alkanes from petroleum crude oil after first refining unit in atmospheric and vacuum distillation tower.

Boiling point range/°C	Range of obtained alkanes according to their carbon size	Corresponding commercial usage
Bellow 20	1C-4C	Natural gas
20-60	5C-6C	Solvents
60-100	6C-7C	Solvents
40-200	5C-10C	Gasoline
175-325	12C-18C	Kerosene
250-400	12C and higher	Diesel
Non-volatile liquids	20C and higher	Wax and lubricating oil
Non-volatile solids	20C and higher	Asphalt

All carbon atoms in alkanes have sp^3 hybridization and tetrahedral geometry where two bonds are in the plane and other two outside the plane. Alkanes can be linear (without branching) or branched. The most stable geometry of linear alkanes is represented by zig-zag conformation in bond line formula (or dashed-wedged line notation), although other conformations of the main chain exist in dynamic equilibrium (Fig. 14.2).

Branched alkanes have two important structural parts: the main chain or backbone and the side chain (secondary structure attached to the main chain). The side chain can be a simple (or linear) branching or a complex branching. Simple/linear branching is a linear side chain, for example, methyl, ethyl, propyl, butyl... so on attached to a main chain. On the other hand, complex branching (or branched side chain or branched alkyl group) has "branching in branching", for example, methylethyl, methylpropyl, ethylbutyl, attached to a main chain. Figure 14.3 depicts two examples of branched alkanes and some examples of branching (not attached to the main chain), where the non-bonded line represents the place where the branching attaches to the main chain.

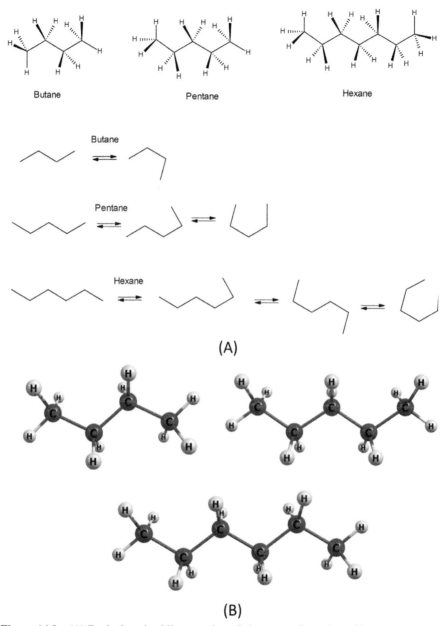

Figure 14.2 (A) Dashed-wedged line notation of zig-zag conformation of butane, pentane, and hexane, along with their bond line formula representing dynamical equilibrium among all conformers of their corresponding main chain; (B) stick-and-ball model of optimized geometries of butane, pentane, and hexane.

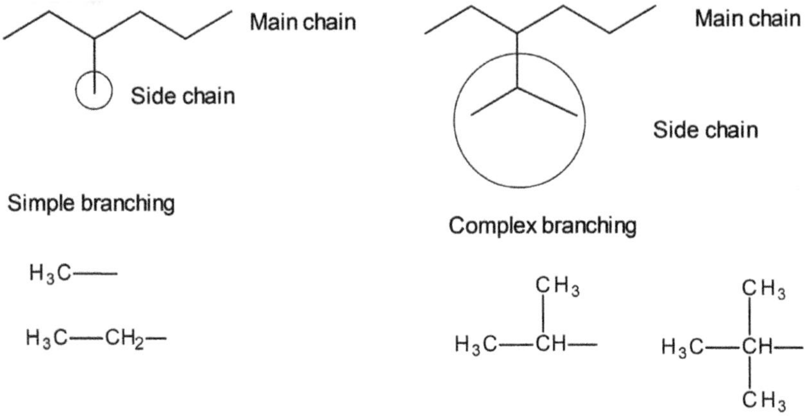

Figure 14.3 Bond line formula of branched alkanes and condensed formula of some simple branching and complex branching, where non-bonded line represents the place where the branching is attached to the main chain.

NOMENCLATURE OF LINEAR ALKANES

The nomenclature of akanes have two parts: prefix and suffix. The suffix is the same for every alkane: -ane. The prefix varies according to the size of the linear alkane. See the Table 14.2 for more information.

Table 14.2 Nomenclature of the linear alkanes.

Carbon size	Prefix	Nomenclature
1C	Meth	methane
2C	Eth	ethane
3C	Prop	propane
4C	But	butane
5C	Pent	pentane
6C	Hex	hexane
7C	Hept	heptane
8C	Oct	octane
9C	Non	nonane
10C	Dec	decane
11C	Undec	undecane
12C	Dodec	dodecane
13C	Tridec	tridecane
14C	Tetradec	tetradecane
15C	Pentadec	pentadecane
16C	Hexadec	hexadecane
17C	Heptadec	heptadecane
18C	Octadec	octadecane
19C	Nonadec	nonadecane
20C	Eicos	eicosane

ALKYL GROUP/SUBSTITUENT

Alkyl group or alkyl substituent is a type of side chain (or branching) derived from alkane by removing one hydrogen atom, giving the following molecular formula: C_nH_{2n+1}. The missing hydrogen is replaced by a dash indicating the position where the alkyl group is (or will be) attached to the main chain. Usually the R symbol is used as a general representation of an alkyl group.

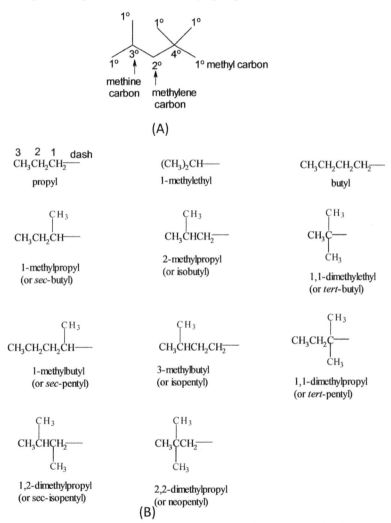

(A)

(B)

Figure 14.4 (A) Types of carbon atoms in one alkane in bond line formula; (B) some alkyl groups (simple/linear and complex/branched) in condensed formula along with their nomenclatures.

The nomenclature of alkyl group derives from its corresponding alkane nomenclature changing "ane" suffix into "yl" suffix. The numbering scheme in alkyl groups is the following: the carbon atom attached to the main chain (where the dash is placed) has the first number (one) and the subsequent carbon atoms have 2,3... and so on. This numbering scheme is important for nomenclature of branched (or complex) alkyl groups. In some cases it is important to know the type of carbon atom as well for the nomenclature of alkyl group.

There are four types of carbon atoms in any organic molecule: primary carbon (where only one carbon atom is bonded to it), secondary carbon (where two carbon atoms are attached to it), tertiary carbon (where three carbon atoms are linked to it), and quaternary carbon (where four carbon atoms are bonded to it). The primary carbon is also called methyl carbon; the secondary carbon is also called methylene carbon; and the tertiary carbon is called methine carbon. See the example in Fig. 14.4(A).

As mentioned above, the nomenclature of some alkyl groups (some branched alkyl groups) are related to the type of carbon atoms in their moieties. For example, *sec*-butyl or *sec*-pentyl is an alkyl group with four or five atoms where the C1 (the carbon atom bond to the main chain) is a secondary carbon). Please do not confuse 1C (alkane with one carbon atom with C1 position of the first carbon). In the same sense, *tert*-butyl or *tert*-pentyl is an alkyl group where the C1 is a tertiary carbon. Less intuitive prefixes are iso and neo. As for iso prefix, in isobutyl and isopentyl, for example, the penultimate carbon atom in the "main chain" of the branched alkyl group is tertiary (i.e., it has a methyl branching attached to it). As to neo prefix, in neopentyl, for example, the penultimate carbon atom in the "main chain" of the branched alkyl group is quaternary (i.e., it has two methyl bonded to it). See the examples in Fig. 14.4(B). Alternatively, these branched alkyl groups have another nomenclature from IUPAC. For example, *sec*-butyl is also 1-methylpropyl; isopentyl is 3-methylbutyl; neopentyl is 2,2-dimethylpropyl (see Fig. 14.4(B)).

IUPAC RULES FOR NOMENCLATURE OF BRANCHED ALKANES

First rule: The order of appearance of side chains in the nomenclature is given by alphabetic order no matter whether prefixes di-, tri-, tetra-, and so on, appear in the side chains (when two, three, or four, and so on, respectively, equivalent side chains are attached in the main chain).

Second rule: Identify the main chain which has the highest number of carbon atoms.

Third rule: The numbering scheme (from right to left or from left to right along the main chain) is based on the following criteria in the order:
1. The one that it will give the smallest numbers (lowest sum) to side chains. See the example in Fig. 14.5(A).
2. When both numbering schemes (from right to left and from left to right in a horizontal main chain) give the same positions to the side chains, then choose the numbering scheme that gives the smallest number to the side chain that appears firstly in the nomenclature. See the example in Fig. 14.5(B).

Note: There are cases in which both criteria (1 and 2) fail, then any option for the numbering scheme is valid (ex.: 2,4-dimethylpentane).

Fourth rule: The complex or branched side chain will appear between parentheses in the nomenclature of the alkane with its own numbering scheme. See the example in Fig. 14.5(C).

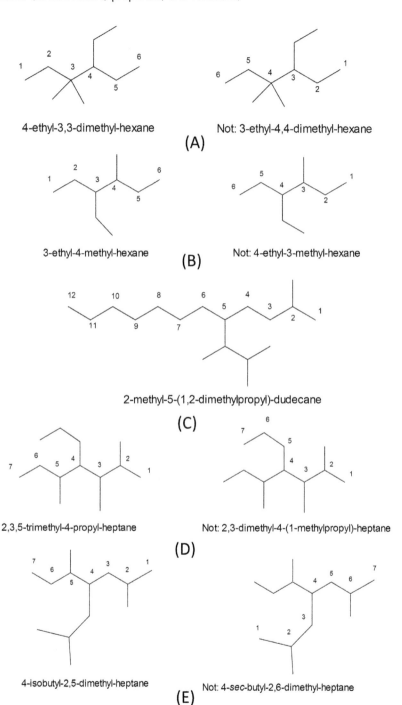

4-ethyl-3,3-dimethyl-hexane

Not: 3-ethyl-4,4-dimethyl-hexane

(A)

3-ethyl-4-methyl-hexane

Not: 4-ethyl-3-methyl-hexane

(B)

2-methyl-5-(1,2-dimethylpropyl)-dudecane

(C)

2,3,5-trimethyl-4-propyl-heptane

Not: 2,3-dimethyl-4-(1-methylpropyl)-heptane

(D)

4-isobutyl-2,5-dimethyl-heptane

Not: 4-sec-butyl-2,6-dimethyl-heptane

(E)

Figure 14.5 Correct and incorrect nomenclature for branched alkanes as example for (A) second rule, (B) third rule, (C) fourth rule, (D) fifth rule first option, (E) fifth rule second option.

Fifth rule: When two (or more) candidates for the main chain exist, that is, when they have the same size, one chooses the main chain based on the following criteria in the order:

1. One chooses the main chain which has the highest amount of side chains. See the example in Fig. 14.5(D).
2. If two or more candidates for the main chain also have the same number of side chains, then one chooses the main chain which gives the smallest numbers for the side chains. See the example in Fig. 14.5(E).

Note: There are cases in which both criteria fail, then any option for choosing the main chain is valid, e.g. Fig. 14.5(A) and Fig. 14.5(B).

CONFORMATIONAL ANALYSIS OF ETHANE

As previously discussed in the last chapter, any molecule with at least two carbon atoms (or carbon-oxygen or carbon-nitrogen atoms) bonded by a single bond has, at least, two conformers and at most six conformers for each single bond ($360/60 = 6$). In each pair of conformers, one conformer is minimum in PES and the other is maximum in PES.

In ethane, there is only one single C-C bond and there is only one pair of conformers (Fig. 14.6(A): synperiplanar (a transition state whose vectors associated with its imaginary frequency are represented in Fig. 14.6(B)) and antiperiplanar (a minimum at PES). Ethane synperiplanar conformer is also called eclipsed because all C-H/C-H bonds are in fact eclipsed and its antiperiplanar conformer is called staggered. *Then, ethane oscillates between two antiperiplanar rotamers passing through an energy barrier of its eclipsed conformer with 2.7 kcal mol^{-1} or 11.29 kJ mol^{-1}.* In synperiplanar (or eclipsed) conformer there is an electronic repulsion between eclipsed C-H/C-H bonds of vicinal carbons (C1 and C2) as it is indicated in Fig. 14.6(A). Each C-H bond at C1 is eclipsed with each C-H bond at C2 in ethane, yielding three C-H/C-H bonds repulsion. The electrostatic force for each eclipsed C-H/C-H repulsion is given below:

$$F_{elect} = 4\left(K \frac{ee'}{r^2} \right)$$

Where e is one electron of C-H bond at C1, e' is one electron of eclipsed C-H bond at C2, and r is the distance between the eclipsed C-H bonds.

According to the Boltzmann distribution equation below and using $T = 298$ K, we have:

$$\frac{N_i}{N_j} = \exp\left[-\left(E_i - E_j\right)/RT\right]$$

$$R = 8.314 \, JK^{-1}mol^{-1}$$

$$T = 298 \, K$$

$$RT = 2477.5 \, Jmol^{-1}$$

$$\frac{N_i}{N_j} = \exp\left[11286/2477\right] = e^{4.56} = 96$$

The ratio of antiperiplanar/synperiplanar, N_i/N_j, is 96. That is, for each synperiplanar conformer there are 96 antiperiplanar conformers at 298 K.

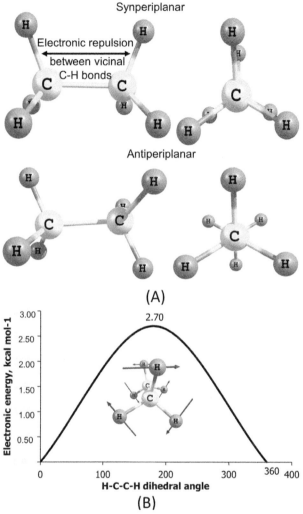

Figure 14.6 (A) Optimized geometries of synperiplanar and antiperiplanar conformers of ethane; (B) PES of the conformational analysis of ethane at ωB97XD/6-31G(d,p) level of theory.

CONFORMATIONAL ANALYSIS OF BUTANE

From C1-C2 and C3-C4, butane has two conformers (synperiplanar and antiperiplanar) for each pair, similar to ethane. Please try to check this by yourself. However, from C2-C3, there are four conformers: synperiplanar and antiperiplanar (like all other) plus synclinal and anticlinal. Synperiplanar and anticlinal are the eclipsed conformers (with eclipsed C-Me/C-Me and C-H/C-H bonds in synperiplanar and eclipsed C-H/C-H and C-H/C-Me bonds in anticlinal where Me means methyl group) as depicted in Fig. 14.7(A). Consequently, both of them are transition states whose vectors of their imaginary frequencies are shown in Fig. 14.7(B). The optimization of synperiplanar yields syclinal and optimization of anticlinal yields the antiperiplanar (Fig. 14.7(A)).

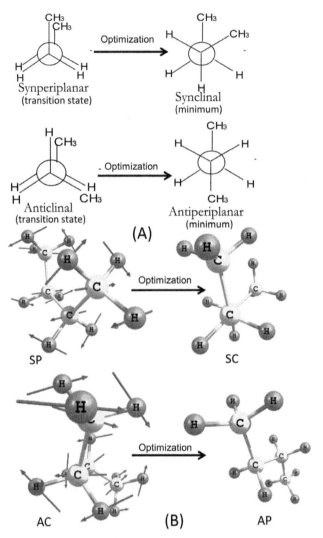

Figure 14.7   (A) Newman projection of conformers of butane at C2-C3 pair; (B) stick-and-ball model of the same conformers.

As shown in Fig. 14.8, synclinal and antiperiplanar are minimum in PES. Antiperiplanar is the absolute minimum and synclinal is the relative minimum with 0.44 kcal mol^{-1} more energy due to the electronic repulsion between the methyl groups. Antiperiplanar always passes through anticlinal transition state in either direction of the PES, whose energy barrier is 3.26 kcal mol^{-1}. On the other hand, depending on the direction of the PES, synclinal passes through anticlinal barrier (with 2.82 kcal mol^{-1}) or synperiplanar barrier (with 4.77 kcal mol^{-1}). From all vicinal carbon pairs (C1-C2, C2-C3, and C3-C4) butane has eight conformers. Nonetheless, butane has only six different conformers (in energy) because the conformers from C3-C4 are similar in energy than the conformers from C1-C2.

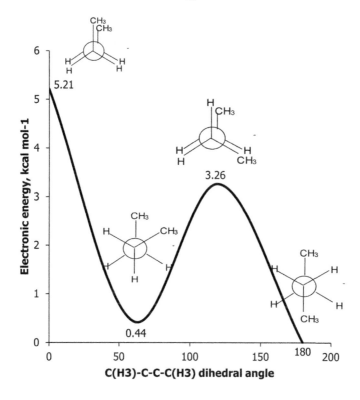

Figure 14.8 Potential energy surface from the conformational analysis of butane from C2-C3 pair at ωB97XD/6-31G(d,p) level of theory.

PERFORMING CONFORMATIONAL ANALYSIS IN HIGHER ALKANES

The steps for performing a conformational analysis of an alkane are: (1) draw the bond-line formula of the corresponding alkane; (2) draw the dashed-wedged notation from the pair of carbon atoms where it is intended to analyze the conformers and put in condensed formula the alkyl group if it is higher than a methyl or use names such as "Me" for methyl, "Et" for ethyl, "Prop" for propyl, and so on (as convention, draw the antiperiplanar form in dashed-wedged notation as a starting point); (3) Put an eye from where you will see the first carbon of the carbon pair; (4) from this perspective draw the corresponding Newman projection (this is the first conformer, an antiperiplanar conformer); (5) rotate 60° clockwise (as a convention) the carbon at the front or the carbon at the back (as convention, we usually rotate the carbon at the back) of the Newman projection of the first conformer, then you generate a second conformer; (6) continue this rotation procedure until "the sixth conformer and check whether there are two or four repeated conformers, i.e., conformers with the same energy according to the vicinal match. If only the fifth conformer is repeated (that is, all other before the fifth are distinguished), then you can stop the conformational analysis in the fifth conformer (Fig.14.9(B) and (C)). If the substituents of both carbons or either carbon (from the conformational analysis) are the same, then you can stop the conformational analysis in the second conformer (for example, in ethane and 2,2,3,3-tetramethylbutane from C2-C3).

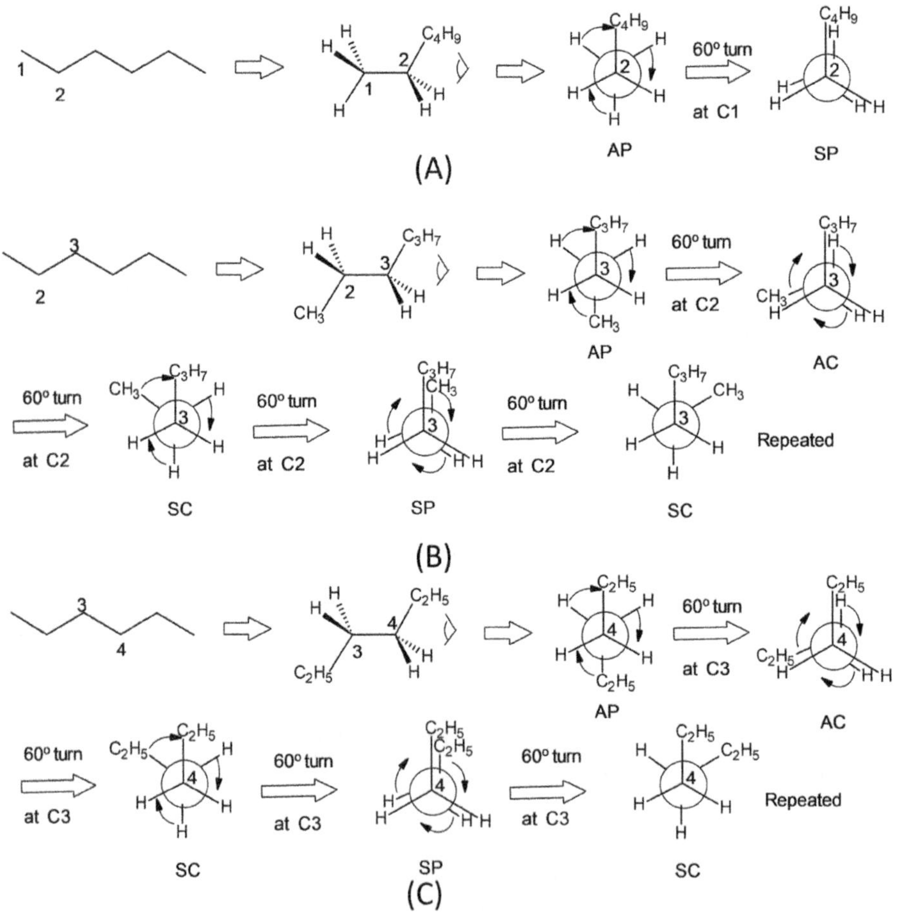

Figure 14.9   Conformational analysis of hexane from (A) C1-C2,
(B) C2-C3, and (C) C3-C4 pairs.

For example, perform the conformational analysis of hexane. Firstly, from C1-C2 pair. One can see in Fig. 14.9(A) the procedure previously described where one finds two conformers AP (short for antiperiplanar) and SP (short for synperiplanar). In this case, it is not necessary to proceed to the fifth conformer (second part of the step 6) because all substituents in C1 are the same (hydrogen atoms). Afterwards, analyze the C2-C3 pair where one gets four different conformers, AP, SP, SC (short for synclinal), and AC (short for anticlinal). The fifth conformer is the repeated form of the SC conformer since no energy difference exists between these two SC forms (see Fig. 14.9(B)). Similarly, the conformational analysis from C3-C4 pair gives four conformers (see Fig. 14.9(C)), but with energies different from the AP, SP, SC, and AC conformers from C2-C3 analysis. Then, from C1 to C4 there are 10 conformers and from C4 to C6 there are six more conformers which resemble the conformers from C1-C2 and C2-C3 pairs. Then, hexane has ten different conformers, each with a distinguished energy.

When analyzing a branched hexane, for example, 3-methylhexane (as in Fig. 14.10(A)), we observe from C2-C3 pair six distinguished conformers: AP, AC1, SC1, SP, SC2, AC2, where one can note that the energies of AC1 and AC2 are different. In AC1 the eclipsed bonds are C-H/C-Prop and two C-H/C-Me while in AC2 the eclipsed bonds are C-H/C-Prop, C-Me/C-Me, and C-H/C-H. Likewise, the energies of SC1 and SC2 are different. In SC1, the angle between C-Prop and C-Me is 60° (for the leftmost methyl) and the angle between both C-Me is 180° (for the rightmost methyl), while in SC2 the angle between both C-Me is 60°. A similar result arises from the conformational analysis of 2,3-dimethylhexane, giving six distinguished conformers from the C2-C3 pair as well (see Fig. 14.10(B)).

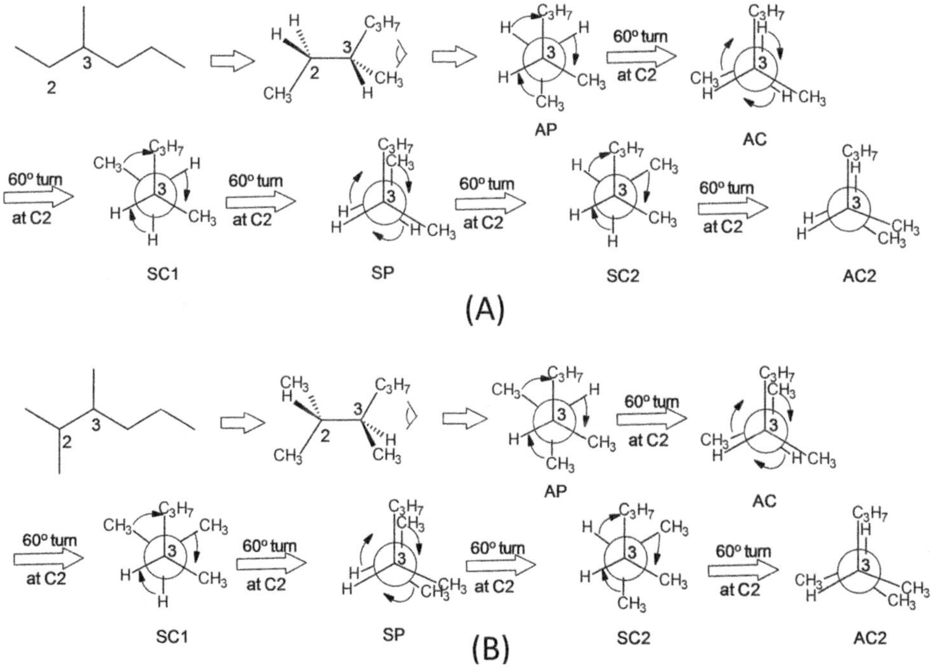

Figure 14.10 Steps for the conformational analysis of (A) 2-methylhexane and (B) 2,3-dimethylhexane.

ISOMERS OF ALKANES

Alkanes with one carbon to three carbon atoms do not have isomers. Then they exist as methane, ethane, and propane. The first alkane with isomer has four carbon atoms: butane and methylpropane. The alkane with five carbon atoms has three isomers: pentane, 2-methylbutane, and 2,2-dimethylpropane. The alkane with six carbon atoms has five isomers: hexane, 2-methylpentane, 3-methylpentane, 2,3-dimethylbutane, 2,2-dimethylbutane. The alkane with seven carbon atoms has nine isomers: heptane, 2-methylhexane, 3-methylhexane, 2,4-dimethylpentane, 2,3-dimethylpentane, 2,2-dimethylpentane, 3,3-dimethylpentane, 3-ethylpentane, and 2,2,3-trimethylbutane. By writing down the bond-line formula for all these alkanes, one will realize that the starting point for finding isomers is the linear alkane; then one can start to put methyl branching in the one-minus-carbon main chain in all possible positions; thereafter by putting two methyl branches in the two-minus-carbon main chain in all possible positions, and so on.

STABILITY OF BRANCHED ALKANES

In a range of isomers of an alkane, the most stable alkane is the most branched alkane. This statement can be experimentally observed by the combustion reaction of the series isomers. For example, the combustion reaction for alkane isomers with eight carbon atoms is:

$$C_8H_{18} + \frac{25}{2}O_2 \rightarrow 8CO_2 + 9H_2O$$

As the energy of the products is the same for all isomers in combustion reaction, the difference of the heat of combustion is given by the energy of the alkane isomer. The more stable the isomer is, the less negative the heat of combustion. This experimental data is in good agreement with the corresponding theoretical data. In Table 14.3, we show the theoretical heat of combustion (ΔH_{comb}) of some isomers with eight carbon atoms where there is a linear increase of the methyl branching (from 0 to 4). From zero methyl branching to four methyl branching, there is an increase in the heat of combustion (becomes less negative), which means that the most branched isomer is the most stable and the least branched is the least stable. *Therefore, branching stabilizes the alkane.* This branching stabilization might be attributed to 1,3-alkyl-alkyl interaction (also called protobranching or higher electron delocalization. However, data in Table 14.3 indicates that intramolecular hydrogen-hydrogen bond plays a secondary role in the stabilization of higher branched alkanes (Monteiro and Firme 2014). Figure 14.11 shows that 2,2,3-trimethylpentane and 2,2,3,3-tetramethylbutane have two and three hydrogen-hydrogen bonds, respectively, while other less branched isomers have no intramolecular interaction. To observe the importance of the hydrogen-hydrogen bond in the stabilization of the branched alkanes, one can see in Table 14.3 that stabilization from protobranching or electron delocalization corresponds to

0.3-0.4 kcal mol^{-1}. These values come from the $\Delta\Delta H_{comb}$ difference when going from octane to 3-methylheptane (0.3 kcal mol^{-1} stabilization) and from 3-methylheptane to 3,3-dimethylhexane (0.4 kcal mol^{-1}), all of them with no hydrogen-hydrogen bond. However, when going from 3,3-dimethylhexane to 2,2,3-trimethylpentane (with two hydrogen-hydrogen bonds), the stabilization increases to 1.1 kcal mol^{-1} (more than twice the stabilization in alkane with no hydrogen-hydrogen bond). The same trend exists when going from 2,2,3-trimethylpentane to 2,2,3,3-tetramethylbutane, where the stabilization is two-fold higher than that when there is no intramolecular interaction. Then, hydrogen-hydrogen bond doubles the stabilization of the branched alkanes which have intramolecular interaction in comparison with branched alkanes with only protobranching or electron delocalization as the stability factor.

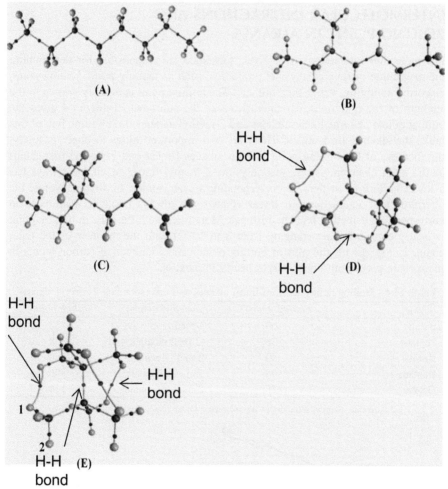

Figure 14.11   Molecular graph from QTAIM of (A) octane, (B) 3-methylheptane, (C) 3,3-dimethylhexane, (D) 2,2,3-trimethylpentane, and (E) 2,2,3,3-tetramethylbutane. See the Acknowledgment section.

Table 14.3 Amount of branching, amount of H-H bonds, and heats of combustion (ΔH_{comb}) in branched alkanes.[a]

Octane isomers[b]	Methyl branching	H-H bonds	ΔH_{comb} (kcal mol^{-1})	$\Delta\Delta H_{comb}$ (kcal mol^{-1})	$\Delta\Delta H_{comb}$ difference
Octane	0	0	−1647.6	2.6	(2.6-2.3) 0.3
3-methyl-heptane	1	0	−1647.3	2.3	(2.3-1.9) 0.4
3,3-dimethyl-hexane	2	0	−1646.9	1.9	(1.9-0.8) 1.1
2,2,3-trimethyl-pentane	3	2	−1645.8	0.8	(0.8-0) 0.8
2,2,3,3-tetramethyl-butane	4	3	−1645.0	0	-

(a) Data from ref. (Monteiro and Firme 2014).
(b) Optimized at ωB97XD/6-311G++(d,p) level of theory.

INTERMOLECULAR INTERACTIONS AND BOILING POINT IN ALKANES

Intermolecular interactions of a given substance are responsible for determining the magnitude of several physical properties, such as melting point, boiling point, viscosity, solubility, vapor pressure, etc. The boiling point is directly related to the strength of the intermolecular interactions of the substance. Table 14.4 gives the boiling points of some linear alkanes and 2-methyl alkanes. In the same line of this table, the alkanes are isomers. There are two important things to observe: firstly, the increase of the boiling point in both alkanes (linear and 2-methyl branching) as the main chain increases; secondly, the 2-methyl branched alkane always has a lower boiling point than its corresponding linear isomer. In another words, the intermolecular interactions in linear alkanes are always stronger than those in corresponding 2-methyl branched isomer. Moreover, the difference in boiling point is nearly constant chain (ranging from 8 to 12°C) with the increase of the main chain. In Fig. 14.12, the plot of boiling point versus number of carbon atoms in main chain gives both linear curves nearly in parallel.

Table 14.4 Boiling point, in °C, of linear alkanes and corresponding 2-methyl alkanes.

Alkane	Boiling point/°C	2-methyl alkane	Boiling point/°C
Butane	0	2-methylpropane	−12
Pentane	36	2-methylbutane	28
Hexane	69	2-methylpentane	60
Heptane	98	2-methylhexane	90
Octane	126	2-methylheptane	116

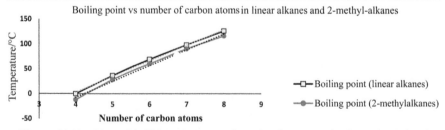

Figure 14.12 Plots of boiling point vs number of carbon atoms in the main chain of linear and 2-methyl branched alkanes.

By using the bimolecular model in our theoretical calculations, i.e., optimized alkane complexes followed by QTAIM analysis, we have observed that there is a direct relation between the number of hydrogen-hydrogen bonds in alkanes and their corresponding boiling points (Monteiro and Firme 2014). Please see Fig. 11.10(A) in chapter eleven. However, the coefficient of determination (R^2) is 0.905, which implies that another factor (besides hydrogen-hydrogen bond) plays a secondary role in determining the boiling points of alkanes. In Fig. 14.13, the intermolecular interactions as hydrogen-hydrogen bonds are shown in butane, pentane, and hexane optimized complexes from the QTAIM molecular graph.

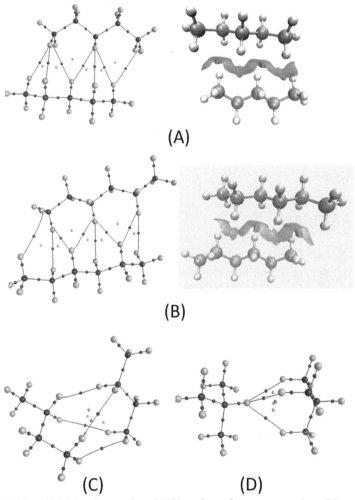

Figue 14.13 (A) Molecular graph and NCI surface of pentane complex; (B) molecular graph and NCI surface of hexane complex; (C) molecular graph of butane complex; and (D) molecular graph of 2-methylpropane.

It is important to add that a broader view of the intermolecular interactions is given by the NCI theory. While QTAIM shows only bond paths, NCI shows the whole surface of intermolecular interactions. When considering NCI analysis of the same complexes (as shown in Figs. 14.13(A) and (B)), we observe the spanning of the intermolecular interactions in a surface along (and between) their main chains. According to the definition of NCI, this surface encompasses the hydrogen-hydrogen bonds, but also shows secondary interactions (weaker than hydrogen-hydrogen bonds according to the coloring scheme of the surface).

As shown in chapter eleven, the NCI surface between the main chains in alkane complexes in Fig. 14.13(A) is called S function (with sign(λ_2)ρ coloring scheme) or reduced density gradient, RDG, which includes the gradient of the charge density, the same used for finding bond critical points of molecular graph (Contreras-García et al. 2011). This surface informs intra/intermolecular repulsion (in red), weaker van der Waals interactions (in light green), stronger van der Waals interactions (in dark green), and H-bonds (in light blue). More details are given in the end of chapter two.

Figure 14.13(D) depicts the molecular graph of the optimized 2-methylpropane complex in the same level of theory used in our previous paper (Monteiro and Firme, 2014) in order to compare to its isomer butane which has a higher boiling point. From the molecular graphs we can see that 2-methylpropane complex has three hydrogen-hydrogen paths, while butane complex has four hydrogen-hydrogen paths. This QTAIM result is in accordance with the expected lower boiling point in 2-methylpropane.

ALKYL RADICAL

Radical is a neutral (not charged) molecule or atom with one or more unpaired (or single) electron. Due to its single electron(s) radical is an open-shell molecule. The term open-shell is related to the shells of an atom (isolated or in a molecule) given by the main quantum number, n, where the highest shell has one atomic orbital with an unpaired electron. When a molecule has one atom with one ore more single electrons, it is called an open-shell molecule. On the other hand, when all electrons are spin-paired this molecule is called closed-shell, for example, in alkanes.

Alkyl radicals can be rationalized as alkane derivatives missing one hydrogen atom (with one proton and one eletron), yielding an unpaired electron in one carbon atom. Depending on the alkyl radical, the hybridization varies from sp^2 to sp^n (where $2 < n < 3$). In methyl radical the geometry of carbon is trigonal plane with sp^2 hybridization, where the unpaired electron occupies a p orbital as it is depicted in its GVB orbital in Fig. 14.14(A). In this case, GVB(7)/6-31G++(d,p) level of theory is used in methyl radical optimized geometry from DFT. Figures 14.14 (B), (C) and (D) also depict the optimized geometries of ethyl radical (a primary alkyl radical), 2-propyl radical (a secondary alkyl radical), and tert-butyl radical (a tertiary alkyl radical) respectively. As we move from primary to tertiary radicals, one can see that the geometry of the carbon atom with one unpaired electron becomes less trigonal plane and more pseudo tetrahedral. In all of these cases (except for methyl radical), the carbon hybridization is sp^n (where $2 < n < 3$) where n(ethyl radical) < n(2-propyl radical) < n(tert-butyl radical).

Radical stability follows the same trend of carbocation stability, that is, primary radical < secondary radical < tertiary radical, based on the same reasoning: the more alkyl groups bonded to the carbon with an unpaired electron the more delocalized (and more stable) is this delocalized single electron. This is true except for some radicals with a very highly delocalized single electron which gives these radicals an usual stability (Carey and Sundberg 2007), then, for the vast majority of radicals are very unstable and very reactive.

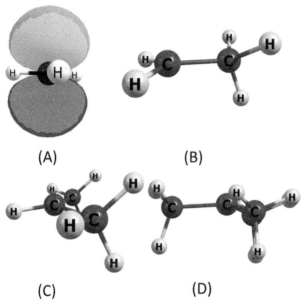

Figure 14.14 (A) GVB *p* orbital in optimized geometry of methyl radical; (B) optimized geometry of ethyl radical; (C) optimized geometry of 2-propyl radical; (D) optimized geometry of *tert*-butyl radical from ωB97XD/6-31G++(*d,p*) level of theory.

FREE RADICAL VERSUS POLAR MECHANISM

In chemistry, there are two distinguished reaction mechanisms: polar reaction and (free) radical reaction. In polar reaction (as previously discussed in chapter ten), there is a movement of two electrons (represented by a full arrow) from the nucleophile towards the electrophile, represented as A^- and B^+, respectively in Fig. 14.15(A). Usually, radicals including alkyl radicals are not regarded as nucleophiles nor electrophiles, then the mechanism involving these species cannot be called "polar". In radical mechanism, a radical might react with closed shell molecule or with another radical (open shell molecule) where there is a movement of a single electron from both molecules to form a new covalent bond. In Fig. 14.15(A), we show the reaction between two radicals (*A•* and *B•*).

Figure 14.15(B) shows a chart about the two general types of mechanisms: polar (two electron movement from nucleophile towards electrophile) and (free) radical (single electron movement from one radical to other radical or a closed-shell molecule). Polar mechanism occurs through a step-wise mechanism (or mechanism in steps) or concerted mechanism. In chapter eight, we have shown an example of a polar concerted mechanism:

$$NaBr + CH_3Cl \rightarrow NaCl + CH_3Br$$

And in chapter ten, we have shown an example of a polar mechanism in steps:

$$(CH_3)_3 \, C\text{-OH} + HCl \rightarrow (CH_3)_3 \, C\text{-Cl} + H_2O$$

Concerted mechanism just has a single step where bond cleavage and bond formation occurs at the same time. Step-wise mechanism has two or more steps where bond cleavage and bond formation occurs in different moments of the reaction.

Free radical reaction has several types of mechanisms, but those of great importance for hydrocarbon chemistry are the substitution (which will be discussed in the next section) and addition (which will be discussed in chapter seventeen).

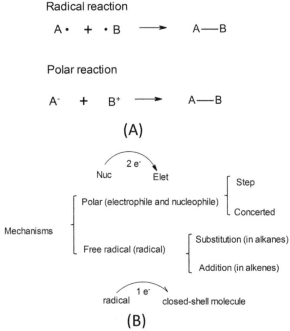

Figure 14.15 (A) Schematic representation of radical and polar reactions; (B) chart of distinguished cases of polar and radical mechanisms.

HALOGENATION OF ALKANES VIA RADICAL SUBSTITUTION

Owing to the facts that the electronegativity difference between carbon and hydrogen atoms is very small and that alkane only hassigma bonds (which are stronger and closer to the bonding atoms than π bonds), alkane is neither a nucleophile nor an electrophile. Then, alkanes cannot react by polar mechanism, but only through free radical reaction. Moreover, the only type of radical reaction alkanes undergo is the substitution radical reaction, besides combustion reaction (which is also a radical reaction and leads to complete cleavage of C-C and C-H bonds).

Most radical reactions have three main steps: initiation (where the first radical is formed), propagation (a cyclic step which repeats until there is radical present in the media), and the termination step (where the radicals are extinguished).

Substitution and addition radical reactions undergo these three steps. The substitution reaction in alkanes is a halogenation reaction where an alkyl halide is formed (Fig. 14.16). Firstly, light (represented as $h\nu$) breaks the weak bond in halogen molecule (for example, in chlorine molecule) to form two halide radicals (e.g., two chlorine radicals) in the initiation step. In the next step, the halide radical abstracts one hydrogen atom of the alkane (e.g., in ethane) to form a new bond and molecule (hydrogen halide, e.g., HCl) and a new radical (the alkyl radical, e.g., ethyl radical). As soon as an alkyl radical is formed, it starts to react with unreacted halogen molecule to form alkyl halide and a new halide radical (for example, ethyl radical forming a new bond with a chlorine atom after cleavage of the chlorine-chlorine bond). As the concentration of alkyl halides and halide radical increases, the termination step increases as well until the end of the reaction.

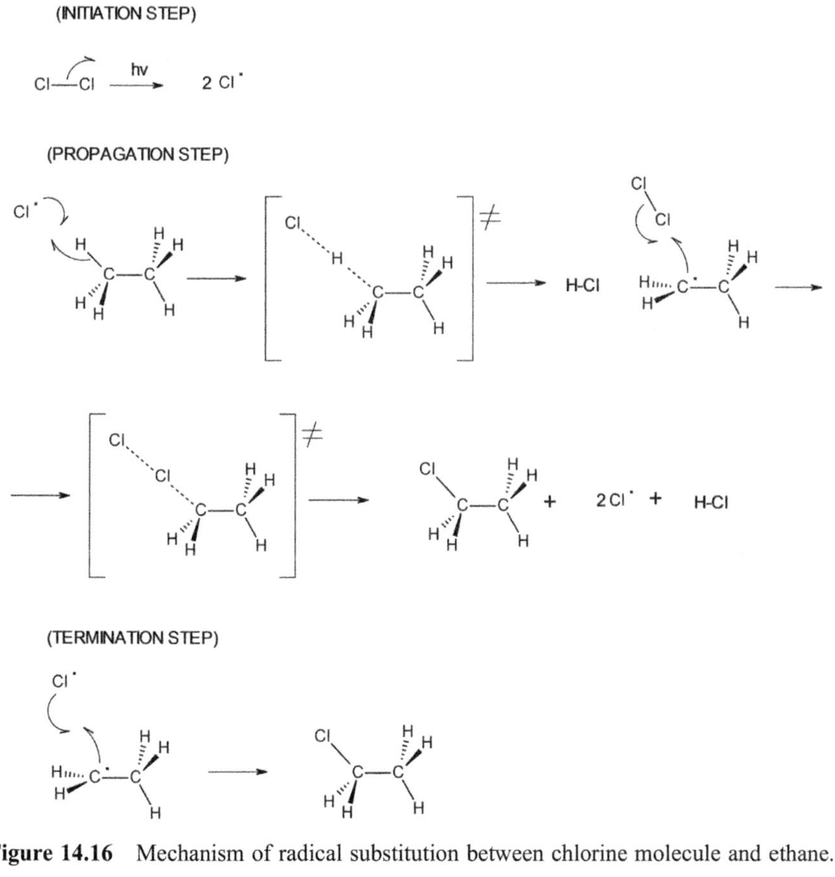

Figure 14.16 Mechanism of radical substitution between chlorine molecule and ethane.

From ωB97XD/6-31G++($2d,p$) level of theory, we have found the first transition state depicted in Fig. 14.16. In Fig. 14.17, the intrinsic reaction coordinate (IRC) from this transition state (abstraction of hydrogen atom of ethane from chlorine radical) is depicted, where reactants (chlorine radical and ethane) appear on the right and products/intermediates (HCl and ethyl radical) appear on the left of the IRC curve. The second transition state (according to the scheme in Fig. 14.16) was not found from this level of theory, even by using more robust techniques of calculation. Then, probably the second transition state might not exist.

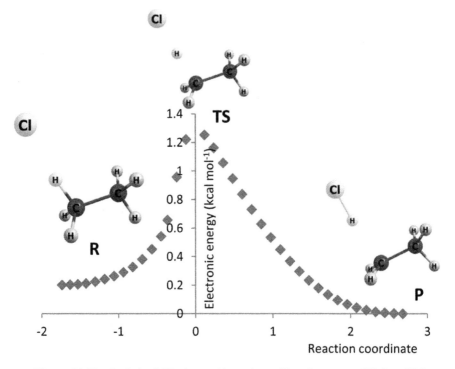

Figure 14.17   Optimized TS along with products (P) and reactants (R) from IRC calculation from Cl---H---CH_2CH_3 transition state.

EXERCISES

1. Represent the following alkanes in bond-line formula:
   a) 2-methyl-3-*sec*-propyl-hexane
   b) 5-(2-methylpropyl)-nonane
   c) 4-*tert*-butyl-2,6-dimethyl-heptane
   d) 7-ethyl-2,4,6-trimethyl-undecane
   e) 3,4-diethyl-2,7-dimethyl-nonane

   Hint   Use zig-zag conformation in horizontal direction to build the main chain.

2. Give the nomenclature for the branched alkanes according to the Fig.14.18.

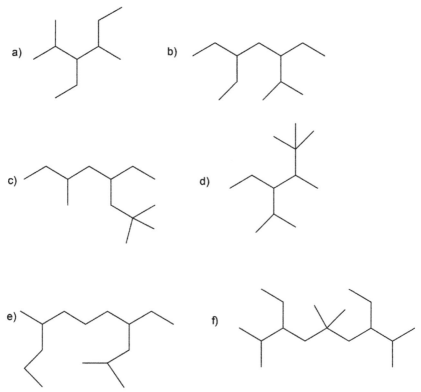

Figure 14.18   Bond-line formula of branched alkanes.

Hint: In that exercise, the main chain is not necessarily in the horizontal direction.

3. Find the conformers of C1-C2 pair in butane.

4. How many distinguished conformers does propane have?

5. By using the Boltzmann distribution equation find the ratio of (a) antiperiplanar/ synperiplanar and (b) antiperiplanar/anticlinal populations at 298 K according to the PES in Fig. 14.8 for conformation analysis at C2-C3 carbon pair in butane.

6. Perform conformational analysis of 2,3-dimethylpentane from dashed-wedged line notation and Newman projection for each conformer. Afterwards, draw its corresponding potential energy surface.

7. Perform conformational analysis of 2,3,3-trimethylhexane and 2,2,3,3-tetra-methylhexane from C2-C3 pair in both cases.

8. 2-Methylpropane reacts with chlorine molecule under light to form 48% of isobutyl chloride and 29% of tert-butyl chloride. Give the mechanism for both products, including transition states.

9. Predict the stability order and boiling point order for:
   (a) Pentane; 2-methylbutane; 2,2-dimethylpropane
   (b) Hexane; 2-methylpentane; 3-methylpentane; 2,3-dimethylbutane

Acknowledgment

Figure 14.11 was reprinted with permission from Monteiro and Firme (2014). Hydrogen-hydrogen bonds in highly branched alkanes and in alkane complexes: a DFT, ab initio, QTAIM and ELF study. J. Phys. Chem. A 118: 1730-1740. Copyright 2014 American Chemical Society.

REFERENCES CITED

Monteiro, N.K.V. and Firme, C.L. 2014. Hydrogen-hydrogen bonds in highly branched alkanes and in alkane complexes: a DFT, ab initio, QTAIM and ELF study. J. Phys. Chem. A 118: 1730-1740.

Carey, F.A. and Sundberg, R.J. 2007. Advanced Organic Chemistry, Part A: Structure and Mechanisms. 5th ed., Springer, New York.

Contreras-Garcia, J., Johnson, E.R., Keinan, S., Chaudret, R., Piquemal, J.-P., Beratan, D.N. and Yang, W. 2011. NCIPLOT: A program for plotting non-covalent interactions. J. Chem. Theory Comput. 7: 625-632

Chapter Fifteen

Cycloalkanes, Bicyclic, and Caged Hydrocarbons

NOMENCLATURE AND PROPERTIES OF CYCLOALKANES

Cycloalkanes have a closed circuit of the saturated main chain (i.e., a closed chain or cyclic chain), while in alkanes there is an open-chain or acyclic main chain. There are some similarities between alkenes and cycloalkanes: (1) the general formula is similar to alkenes and cycloalkanes: C_nH_{2n}; (2) the CC bonds in cycloalkanes (as in alkenes) are not free to rotate and then; (3) substituted cycloalkanes might have spatial estereoisomerism (cis/trans) as alkenes.

Except for cyclopropane, all other cycloalkanes have a puckered ring (or non-planar structure). Cyclopropane has a planar structure by a simple geometric imposition (since three atoms or three points cannot occupy more than one plane!). The most common cycloalkanes are depicted in Fig. 15.1, where there appear two different views for each of them: the upper view (which resembles regular geometric objects) and side view or lateral view or perspective view (which shows their real structure). In this lateral or perspective view, except for cyclopropane, all cycloalkanes have a non-planar structure which can be made up of two or three planes. This perspective view is related to the most stable conformer in each cycloalkane.

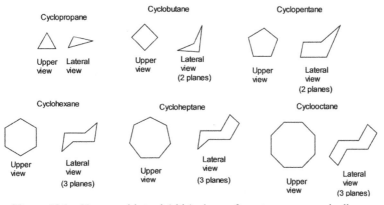

Figure 15.1   Upper and lateral (side) views of most common cycloalkanes.

A more realistic structure of cycloalkanes is shown in Fig. 15.2 – an optimized geometry of cyclobutane (A), cyclopentane (B), cyclohexane (C), and cyclooctane (D) in side view and in upper view. Except for the lateral views of cyclobutane and cyclohexane, Fig. 15.2 clearly indicates that their optimized structures in both views are not as regular as depicted in their corresponding bond-line formulas.

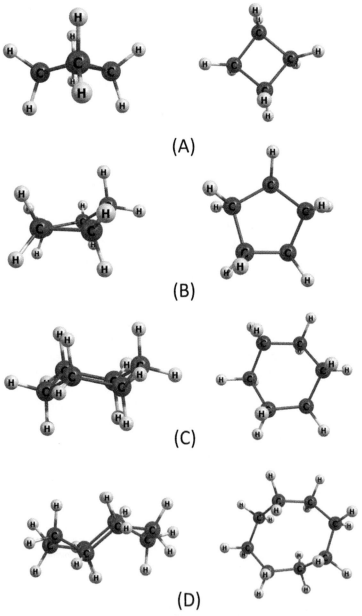

(A)

(B)

(C)

(D)

Figure 15.2 Optimized geometries from side and upper views of (A) cyclobutane, (B) cyclopentane, (C) cyclohexane, (D) cyclooctane.

Cycloalkanes might have one or more susbtituents groups (alkyl groups or halogen atoms) attached to their rings. Single substituted alkanes are named with the substituent group followed by the name of the cycloalkane, for example, methylcyclopropane. Two substituents in the same carbon appear in the nomenclature according to alphabetical order, for example, 1-ethyl-1-methylcyclobutane (Fig. 15.3).

The numbering scheme in cycloalkanes is clockwise or counter-clockwise, while in alkanes is from left to right or right to left. When two or more substituents are attached to different carbon atoms in the ring, the number 1 is given to the substituent that gives the smaller sum of numbers for each substituent. For example, 3-bromo-1,1-dimethylcyclohexane (sum=3+1+1=5) and not 1-bromo-3,3-dimethylcyclohexane (sum=1+3+3=7). Another example is, 2-bromo-1-chloro-3-methylcyclopentane (sum=2+1+3=6) and not 5-bromo-1-chloro-4-methylcyclopentane (sum=5+1+4=10), nor 1-bromo-2-chloro-5-methylcyclopentane (sum=1+2+5=8). Another example is 1-chloro-4-ethyl-2-methylcyclopentane (sum=1+4+2=7) and not 1-chloro-3-ethyl-5-methylcyclopentane (sum=1+3+5=9). These examples are depicted in Fig. 15.3.

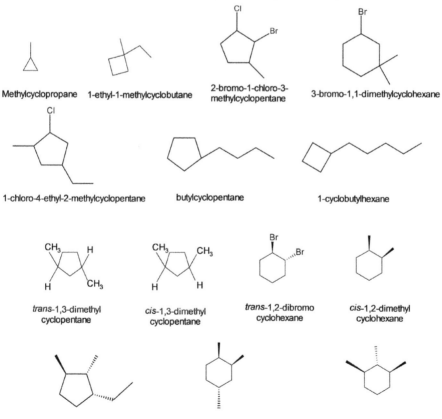

Figure 15.3 Representation and nomenclature of some substituted cycloalkanes.

Cycloalkane can have a longer alkyl group than ethyl and methyl as long as this alkyl chain is smaller than the ring itself. For example, butylcyclohexane. On the other hand, when the alkyl chain is longer than the ring, the alkyl chain is the main chain and the cycloalkyl group is the side group. For example, 1-cyclobutylhexane in Fig. 15.3.

As previously mentioned, non-geminal disubstituted cycloalkanes might have spatial isomerism *cis* and *trans*. For example, *trans*-1,3-dimethylcyclopentane, *cis*-1,3-dimethylcyclopentane, *trans*-1,2-dibromocyclohexane, and *cis*-1,2-dimethyl-cyclohexane (Fig. 15.3).

When there are three substituents in the cycloalkane, one possibility is to use *cis trans* or *cis cis* or *trans trans* to name them. For example, *cis trans*-1-ethyl-2,3-dimethylcyclopentane, *cis trans*-1,2,4-trimethylcyclohexane, and *trans trans* 1,2,3-trimethylcyclohexane, as depicted in Fig. 15.3.

From the data in Table 15.1, one can see that every cycloalkane has a higher melting point and boiling point than its corresponding alkane. Similar to the alkanes, as the size of the ring in the cycloalkane increases, the melting and boiling points increase as well. Like alkanes, the main intermolecular interaction in cycloalkanes is the hydrogen-hydrogen bond. Figure 15.4 shows the optimized geometry of the cyclobutane complex and the corresponding molecular graph from QTAIM where there are eight hydrogen-hydrogen bonds, while in butane complex (in Fig. 14.13(C)), there are four hydrogen-hydrogen bonds. Then, a cycloalkane has more hydrogen-hydrogen bonds than its corresponding alkane, which accounts for the results in Table 15.1.

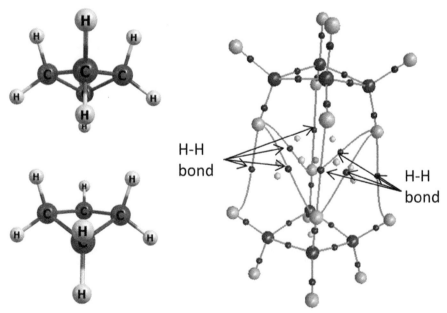

Figure 15.4 Optimized geometry and molecular graph according to QTAIM of cyclobutane complex using ωB97XD/6-311G++(d,p) level of theory.

Table 15.1   Melting point and boiling point, in degree Celsius, of some cycloalkanes and corresponding alkanes.

Cycloalkane	Melting point/C	Boiling point/°C	Alkane	Melting point/°C	Boiling point/°C
Cyclopropane	−127.6	−32.7	Propane	−187.7	−42.1
Cyclobutane	−50	12.5	Butane	−138.3	−1.0
Cyclopentane	−93.9	49.3	Pentane	−129.8	36.1
Cyclohexane	6.6	80.7	Hexane	−95.3	68.7
Cycloheptane	−12	118.5	Heptane	−90.6	98.4
Cyclooctane	14.3	148.5	Octane	−56.8	125.7

ANGLE STRAIN, RING STRAIN, AND TORSIONAL STRAIN

Angle strain is the increase of potential energy of a molecule due to the deviation from the ideal (hybridized) angle of a determined atom. The angle strain is measured in energy units, but can be estimated by degree unit. For example, the angle strain in the methylene carbons in cyclophanes (aromatic rings bonded to alkyl bridges in a closed circuit – see chapter nineteen) varied from 0.5 to 9.7° (Firme and Araújo 2018). In this case, the angle strain in the alkyl bridges is the deviation from the expected bond angle of a sp^3 hybridized carbon.

Torsional strain is the increase of the potential energy (measured in energy units) of a molecule due to the deviation of an ideal dihedral angle (see chapter eight) associated with its corresponding most stable conformer or geometry (for a rigid aliphatic or cyclic structure). Torsional strain can also be estimated from degree units. For example, in alkyl bridges of cyclophanes, the torsional strain is measured as a deviation from 60° in H-C-C-H dihedral angle, where 0° is associated with the eclipsed and most unstable geometry of the alkyl bridge (in the case of cyclophanes, there are no conformers in the alkyl bridges because these bridges are rigid). In cyclophanes, the torsional strain ranged from 1 to 57° (Firme and Araújo 2018). Torsional strain in planar cycloalkanes is responsible for the instability of this conformer as it will be shown in the next sections.

Ring strain can be estimated as the sum of individual angle strains of each atom in the ring. Ring strain is also given in energy unit, but can also be estimated as degree units. For example, in cyclopropane, the geometric (and virtual) C-C-C bond angle is 60°. Then, each carbon atom has (109.5−60) 49.5° angle strain and the ring strain of cyclopropane is (49.5×3) 148.5°. The ring strain equation (given in degree units) of a cycloalkane is:

$$RS = \sum_{i}^{n} AS_i \therefore AS_i = \left| iA_i - rA_i \right|$$

Where RS is ring strain; AS_i is the angle strain of the ith atom; n is the total number of atoms in the ring; iA_i is the ideal bond angle (of the expected hybridization) of ith atom; and rA_i is the real bond angle of the ith atom.

The ring strain in methylenecyclopropane (where one carbon atom in the ring has sp^2 hybridization) is:

$$RS = 2 \times (109.5 - 60) + (120 - 60) = 159°$$

The ring strain in cyclopropene (where two carbon atoms in the ring have sp^2 hybridization) is:

$$RS = (109.5 - 60) + 2 \times (120 - 60) = 169.5°$$

Then, the increasing order of above-mentioned three-membered molecules is: cyclopropane < methylenecyclopropane < cyclopropene.

The ring strain as a function of angle strain having degree unit is only an estimate. The more appropriate value of ring strain is given in energy units per mole from strain energy equations that take into account torsional strain and (all) angle strain(s). All angle strains in cycloalkane correspond to every C-C-C, C-C-H, and H-C-H angle strain in each methylene group in the ring. As we have already noted, there are several equations to obtain the strain energy of cyclophanes. Although their equations and values are different, they agree qualitatively (Firme and Araújo 2018). This is also the same for the ring strain of cycloalkanes, i.e., there are several distinguished ring strain equations for cycloalkanes, for example, the heat of combustion, the standard strain energy, and heat of cyclization.

The heat of combustion of cycloalkanes can be used to obtain the ring strain according to the equation below:

$$RS = \left| \left(\frac{\Delta H_{comb}}{n(C)} \right) + 139.86 \right|$$

Where ΔH_{comb} is the heat of combustion of a cycloalkane; $n(C)$ is the number of carbon atoms in the ring of the cycloalkane; and 139.86 kcal mol^{-1} is obtained from a methylene group in an unstrained model (in this case, cyclohexane using ωB97XD/6-311G++(d,p) level of theory).

The standard ring strain energy (SE) is obtained from the heat of formation of the cycloalkane from its elements (ΔH_f) and subtracting from the hypothetical strainless model, e.g., cyclohexane with $\Delta H_f = -29.5$ kcal mol^{-1} and -4.92 kcal mol^{-1} per methylene group (Wiberg 1986).

Table 15.2 Theoretical heat of combustion (ΔH_{comb}), in kcal mol^{-1}, theoretical ring strain (RS), in kcal mol^{-1}, from ωB97XD/6-311G++(d,p) level of theory, standard strain energy (SE), in kcal mol^{-1}, and theoretical heat of cyclization of alkanes (ΔH_{cycliz}), in kcal mol^{-1} from ωB97XD/6-311G++(d,p) level of theory.

	ΔH_{comb}/ kcal mol^{-1}	RS/ kcal mol^{-1}	SE/ kcal mol^{-1}	ΔH_{cycliz}/ kcal mol^{-1}
Cyclopropane	−447.48	9.30	27.5	39.78
Cyclobutane	−584.78	6.33	26.3	37.40
Cyclopentane	−706.07	1.35	6.2	19.08
Cyclohexane	−839.17	0.00	0.1	12.55
Cycloheptane	−984.63	0.80	6.2	18.39
Cyclooctane	−1136.17	2.16	9.7	30.87

In Table 15.2, there is the data of the theoretical heat of combustion (ΔH_{comb}) along with the corresponding theoretical ring strain (RS), using the last equation and the corresponding standard strain energy (SE) of some cycloalkanes. There is a linear relation between theoretical RS and standard SE of cycloalkanes with coefficient of determination (R^2) 0.9400. Table 15.2 also shows the values of the theoretical heat of cyclization of alkanes to form the corresponding cycloalkane and hydrogen (ΔH_{cycliz}). Excluding cyclooctane, there is a coefficient of determination close to unit (0.9983) for the plot of ΔH_{cycliz} vs SE.

CYCLOPROPANE

Strain energy of cyclopropane (27.5 kcal mol^{-1}; 9.2 kcal mol^{-1} per CH$_2$) is close to that from cyclobutane (26.5 kcal mol^{-1}; 6.6 kcal mol^{-1} per CH$_2$), despite the fact that their C-C-C bond angles (60° for cyclopropane and 88.5° for cyclobutane) and C-C-C angle strains (49.5° in cyclopropane and 20° in cyclobutane) are rather different. According to the C-C-C angle strain values in cyclopropane and cyclobutane, the strain energy per methylene group in cyclopropane was supposed to be more than two-fold the strain energy per methylene in cyclobutane. But, strain energy per methylene group in cyclopropane is only 40% higher than that in cyclobutane. The question is: what makes the cyclopropane much less unstable than it was supposed to be? The answer is: σ C-C bonds in cyclopropane bend outside the ring in order to decrease each C-C-C angle strain and the cyclopropane strain energy (or ring strain). From QTAIM calculation, the C-C-C bond path angle is 78°. Then, σ C-C-C bond bends the ring outwards from 60° to 78°. As a consequence, the actual C-C-C angle strain in cyclopropane is 31.5° (and not 49.5°), which is only 11.5° (or 57.5%) higher than that in cyclobutane (see exercise 4). This bent bond in C-C-C bonds in cyclopropane is called "banana bond" because of the supposed resemblance between the shape of the fruit and the C-C bonding orbital.

The banana bonds in cyclopropane can be viewed in GVB σ C-C singly-occupied orbitals. Figure 15.5(A) shows one σ C-C with its pair of singly-occupied GVB orbitals of cyclopropane which are clearly bent outwards the ring (higher volume of each GVB orbital outside than inside the ring), mainly when compared to the two singly occupied GVB orbitals of one σ C-C bond in cyclopentane (Fig. 15.5(B)). Another way to observe the bent bonds in cyclopropane is through the deformation electron density. The deformation electron density map is the difference between the electron density of a determined molecule and the electron density of all isolated atoms from this molecule. Electron density deformation map shows the electron density in chemical bonds. In the case of cyclopropane, it clearly shows the C-C σ bonds bent outwards the ring (Fig. 15.5(C)), while in cyclopentane the electron density deformation map of its C-C σ bonds are symmetrically distributed outwards and inwards the ring (Fig. 15.5(D)).

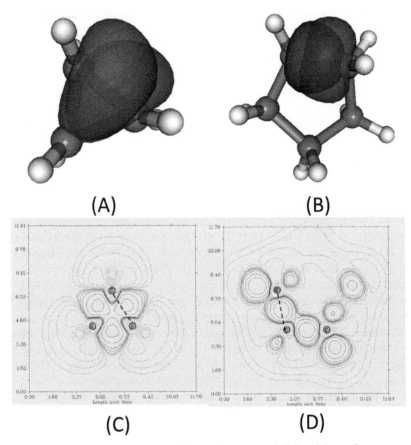

Figure 15.5 GVB singly occupied orbitals of (A) one C-C bond in cyclopropane; and (B) of one C-C bond in cyclopentane; deformation electron density map of (C) cyclopropane; and (D) cyclopentane obtained from MultiWFN software (Lu and Chen 2012). The GVB positive lobe (in non-bonding region) was omitted for simplicity.

The banana bond in cyclopropane resembles the π-bond in alkenes because both of them are bent outwards the C-C axis. As a consequence, cyclopropane and substituted cyclopropanes react similarly to alkene molecules in eletrophilic addition (more details in chapter seventeen). Moreover, cyclopropane is the unique cycloalkane that reacts through polar reaction. For example, the hydrobromination of cyclopropane and propylene yield 99.7% and 98.5% *n*-propyl bromide, respectively (Lacher et al. 1950).

CYCLOBUTANE AND CYCLOPENTANE

Except for cyclopropane, all other cycloalkanes have, at least, two conformers. For example, cyclobutane and cyclopentane have two distinguished conformers, while cyclohexane have three distinguished conformers.

The high-lying conformer of cyclobutane is planar ring which is also a transition state (i.e., it has one imaginary frequency associated with the torsional vibration to remove its planarity). In planar, transition state conformer of cyclobutane, both C1-H/C2-H and C3-H/C4-H are eclipsed (Fig. 15.6(A)), that is, they have a torsional stress associated with an eclipsed conformer, which accounts for its higher energy. In the puckered geometry of the cyclobutane conformer (Fig. 15.6(B)), both these eclipsed C1-H/C2-H and C3-H/C4-H are removed, which decreases the torsional strain in cyclobutane. Nonetheless, both planar and puckered geometries of cyclobutane have eclipsed C1-H/C3-H and C2-H/C4-H conformations (Fig. 15.6(C) and (D)). Then, puckered geometry only partly removes the eclipsing from planar geometry (from vicinal carbon atoms) because it maintains the C-H/C-H eclipsing in its opposite carbons.

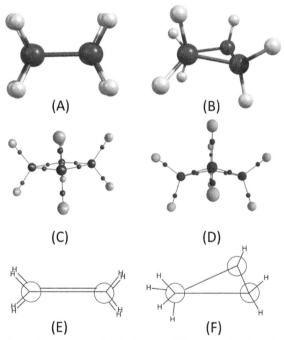

Figure 15.6 Optimized geometries of (A) planar cyclobutane and (B) puckered cyclobutane, along with their corresponding molecular graphs in (C) and (D), respectively, and their corresponding Newman projections in (E) and (F), respectively.

As discussed in the previous chapter, the electrostatic force for each eclipsed C-H/C-H repulsion is given by the equation:

$$F_{elect} = 4\left(K \frac{ee'}{r^2} \right)$$

When the vicinal C-H/C-H eclipsing is removed from planar to puckered geometry in cyclobutane, there is a energy decrease of 0.88 kcal mol^{-1} according to B3LYP/6-31G(d,p) level of theory. Figure 15.7 shows the potential energy surface involving the conformers of cyclobutane.

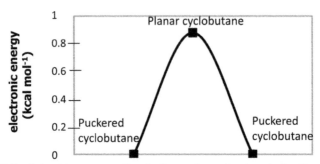

Figure 15.7 Potential energy surface in the conformational analysis of cyclobutane according to B3LYP/6-31G(d,p) level of theory.

In the case of cyclopentane, there are also two conformers (one transition state and one minimum). However, its transition state conformer is not a planar structure as the TS conformer in cyclobutane. Noticeably, the minimum (at PES) conformer of cyclopentane partly decreases the torsional strain from its TS conformer (see Fig. 15.8). No TS conformer was found at B3LYP/6-31G(d,p) level of theory. At ωB97XD/6-31G(d,p) level of theory, the energy barrier from minimum to TS conformer in cyclopentane is 0.6 kcal mol^{-1}. There is a slight relief of torsional strain from TS to minimum conformer of cyclopentane.

It is important to emphasize that no comparison with energy barrier can be appropriately done because the levels of theory used are different, but one can infer that both barriers are relatively low and much smaller than the energy barrier in ethane (from its eclipsed conformer to its staggered conformer).

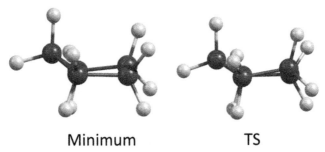

Minimum   TS

Figure 15.8 Optimized geometries of (A) cyclopentane TS conformer and (B) cyclopentane minimum conformer at ωB97XD/6-31G(d,p) level of theory.

Each C-C-C angle in cyclopentane is 103.2°, 105.2°, 106°, 105.2°, and 103.2°. The averaged C-C-C angle in cyclopentane is 104.6°. The corresponding ring strain of cyclopentane according to its angle strain is 24.5° (see below).

$$RS = \sum_{i}^{5} |iA_i - rA_i| = 5 \times |109.5 - 104.6| = 24.5°$$

The ring strain (from sum of angle strain equation) in cyclopentane is more than three times lower than that in cyclobutane (see exercise 4).

CYCLOHEXANE

There are three distinguished conformers of cyclohexane: chair and boat conformers (which are minimum at the PES) and half-chair conformer (transition state at the PES). Figure 15.9 shows the optimized geometries of these three conformers. Except for the chair conformer, boat and half-chair conformers are not symmetric. See their side views where the four carbons in the middle (excluding the carbons at rightmost and leftmost in the side view) are not planar as they are in chair conformer. Moreover, from upper view, half-chair conformer seems to have the most distorted structure.

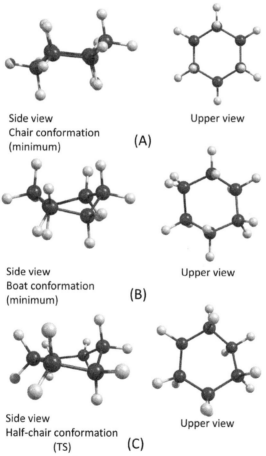

Side view
Chair conformation   (A)
(minimum)

Upper view

Side view
Boat conformation   (B)
(minimum)

Upper view

Side view
Half-chair conformation
(TS)   (C)

Upper view

Figure 15.9   Optimized geometries of (A) chair conformer, (B) boat conformer, and (C) half-chair conformer of cyclohexane from B3LYP/6-31G(d,p) level of theory.

Bond-line formulas in Fig. 15.10(B) give a simplification of the actual structures of half-chair and boat structures (where the four carbons in the middle, according to side view, are planar). The closest representation to their optimized structures is given in Fig. 15.10(C).

The C-C-C angle in cyclohexane conformers are: 111.5° (chair conformer), 112.2°, and 111.2° (boat conformer), 114.7°, 109.9°, and 118.1° (half-chair conformer). Then, ring strain is the highest in the half-chair and the smallest in the chair conformations due to their corresponding angle strain values (see exercise 6). As a consequence, we have the following potential energy surface where the difference between chair and half-chair conformations is about 12 kcal mol^{-1}, which is four-fold the energy difference between eclipsed and staggered conformers of ethane (Fig. 15.10(A). As a consequence, one chair conformation goes to the first TS (first half-chair conformation) and then to the boat conformation, which subsequently goes to the second TS (second half-chair conformation) and from this TS towards the second chair conformation (Figs. 15.10(B) and (C)).

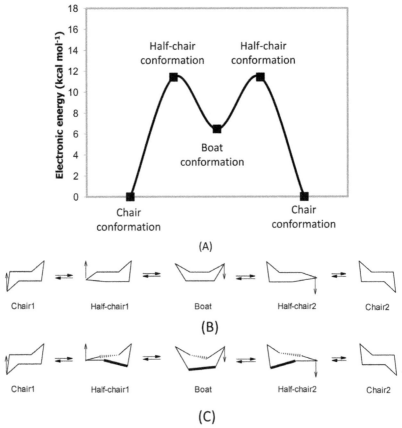

Figure 15.10   (A) Potential energy surface of cyclohexane conformers from B3LYP/6-31G(d,p) level of theory. Bond-line formula of all cyclohexane conformers in (B) simplified form and, in (C) a representation closer to the actual structures.

The chair conformation of cyclohexane is the most stable structure of a cycloalkane. It has the lowest ring strain. However, cyclohexane does have a ring strain. By using the ring strain equation as a sum of all angle strain in the chair conformation:

$$RS = \sum_{i}^{6} |iA_i - rA_i| = 6 \times |109.5 - 111.5| = 12°$$

The ring strain in cyclohexane (12°) is twice lower than that from cyclopentane.

When using cyclohexane as a reference model of zero strain energy (as it is shown in *RS* and *SE* equations in Table 15.2), then obviously cyclohexane has a zero ring strain. But, as we have seen before, the ring strain in cyclohexane (from the sum of all angle strain) is not zero! Similarly, the theoretical heat of cyclization of alkanes (Table 15.2) also indicates that ring strain in cyclohexane is not zero, although it is the smallest of all cycloalkanes.

From the view indicated by the eye in Fig. 15.11 of the chair conformer, one can see stick-and-ball and the corresponding Newman projection of cyclohexane chair conformation. From this Newman projection one can see that each carbon pair at the right and at the left (which are four carbons in the middle of the side view) are staggered with torsional angles close to 60° which also (along with low angle strain) accounts for the very low ring strain of cyclohexane in comparison with other cycloalkanes.

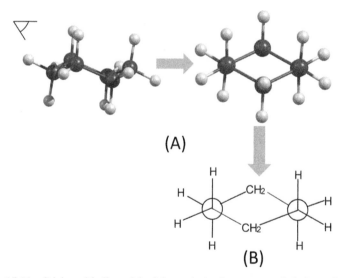

(A)

(B)

Figure 15.11   Stick-and-ball model of the optimized geometry of chair conformation of cyclohexane from side view and from the view of the eye, and its corresponding Newman projection.

Cyclohexane chair conformation has a very important particularity. It has axial and equatorial C-H bonds (as indicated in Fig. 15.12). The axial C-H bonds are the upwards and downwards vertical bonds. The equatorial C-H bonds are the upwards and downwards diagonal bonds. The upward equatorial bonds are those from the leftmost carbon and the two carbons at the right and the downwards equatorial bonds are those from the leftmost carbon and the two carbons at the left side of the Fig. 15.12.

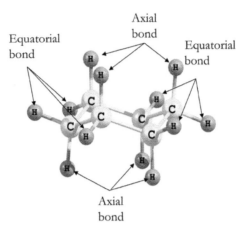

Figure 15.12   The C-H equatorial and axial bonds of cyclohexane chair conformation.

For the beginners in organic chemistry, it might be a difficult task to correctly draw cyclohexane with its axial and equatorial bonds. We show one way to draw it in Fig. 15.13 by following the steps from (A) to (G). Firstly, draw two parallel lines in horizontal direction (one more to the right and the other more to the left). Secondly, put two dots, one above (at the left of the two lines) and the other below (at the right of the two lines) the parallel lines (or vice versa). Then, link the parallel lines with their closest dot to form the chair conformation in (C). Afterwards, draw the upwards axial bonds and the downwards axial bonds (D and E). Next, by knowing that each carbon atom has sp^3 hybridization and tetrahedral geometry, the fourth bond (the equatorial bond) is drawn by following the tetrahedral geometry (F). In the carbon with downwards axial bond, draw an upwards diagonal bond while in the carbon with upwards axial bond draw a downwards diagonal bond (G). The parallel lines in (A) can also be drawn in diagonal direction which gives a diagonal chair conformation (H). One can use either horizontal chair conformation (G) or diagonal chair conformation (H). Note that the angle between axial and equatorial bonds at the same carbon should not be 90° or less since it is a sp^3 carbon.

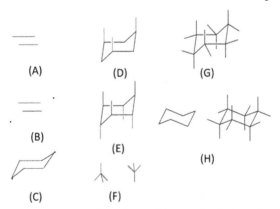

Figure 15.13   Steps to draw the cyclohexane in chair conformation with its axial and equatorial bonds.

It is very important to observe that when interchanging both chair conformations (as depicted in Fig. 15.14), four changes occur: (1) all downwards axial bonds change into downwards equatorial bonds (e.g., C-H$_2$ in left chair to right chair) and (2) all upwards axial bonds change into upwards equatorial bonds (e.g., C-H$_4$ in left chair to right chair), as well as, (3) all downwards equatorial bonds change into downwards axial bonds (e.g., C-H$_3$ from left chair to right chair), and (4) all upwards equatorial bonds change into upwards axial bonds (e.g., C-H$_1$ from left chair to right chair). Although the equatorial and axial bonds change when going from one chair to the other chair conformer in cyclohexane, no energy difference exists between them (see Fig. 15.10). This is not the case for many substituted cyclohexanes, as shown in the next section.

Figure 15.14   Bond-line formula of both cyclohexane (diagonal) chair conformations with their axial and equatorial bonds.

DERIVATIVES OF CYCLOHEXANE

Monosubstituted cyclohexanes (such as methylcyclohexane, isopropylcyclohexane, *tert*-butylcyclohexane, bromocyclohexane, etc.) have two distinguished chair conformers with slight to very different energies. There is an equilibrium constant between them, which is higher than a unity when going from the least stable conformer to the highest stable conformer. The least stable conformer has the alkyl or halogen group at axial position while the highest stable conformer has this group at the equatorial position. Mostly, the larger the substituent group at axial position the higher the energy difference between both chairs and the equilibrium constant (towards the highest stable conformer). One exception is isopropyl group where both methyl groups are pointing outwards and does not influence the steric effect of isopropyl group in comparison to methyl group in destabilizing the corresponding chair conformation with this group in axial position.

Figure 15.15 depicts the optimized geometries of methylcyclohexane, isopropylcyclohexane, and *tert*-butylcyclohexane in both chair conformations along with their corresponding Gibbs free energy difference (ΔG) between both chair conformers. The corresponding equilibrium constant can be calculated from the equation below at 298.15 K (see chapter four).

$$\Delta G = -RT \ln K \therefore K = e^{-\frac{\Delta G}{RT}}$$

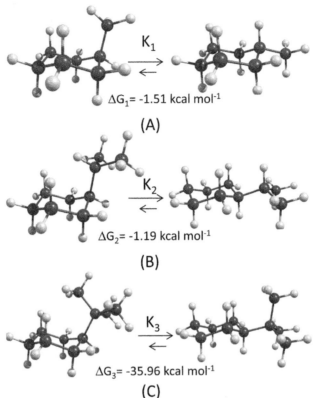

Figure 15.15 Optimized geometries of both chair conformations and their corresponding Gibbs free energy difference of (A) methylcyclohexane, (B) isopropylcyclohexane, and (C) *tert*-butylcyclohexane calculated at ωB97XD/6-311G++(d,p) level of theory.

According to the above equation and the calculated values of Gibbs free energy difference depicted in Fig. 15.15, the axial-equatorial equilibrium constants for methylcyclohexane, isopropylcyclohexane, and *tert*-butylcyclohexane are $K_1 = 1.84$, $K_2 = 1.62$, and $K_3 = 2.02 \times 10^3$, respectively. As the volume of the substituent group increases, the trend to form equatorial conformer increases sharply.

By using the Boltzmann distribution equation, one might also obtain the population ratio between the chair conformations of substituted cyclohexanes. In case of single molecule reaction or transformation, the population ratio equals the equilibrium constant (chapter four).

$$\frac{N_i}{N_j} = K = \exp\left[-\left(E_i - E_j\right)/RT\right] = e^{-\frac{\Delta G}{RT}}$$

Then, the population ratios (amount of chair conformer with larger substituent at equatorial position over the amount of chair conformer with larger substituent at axial position) for methylcyclohexane, isopropylcyclohexane, and *tert*-butylcyclohexane are 1.84, 1.62, and 2.02×10^3, respectively.

Figure 15.16 shows the NCI surface (0.5 isosurfaces of S function) in colored scheme (which actually appears in black-and-white here) along with the corresponding

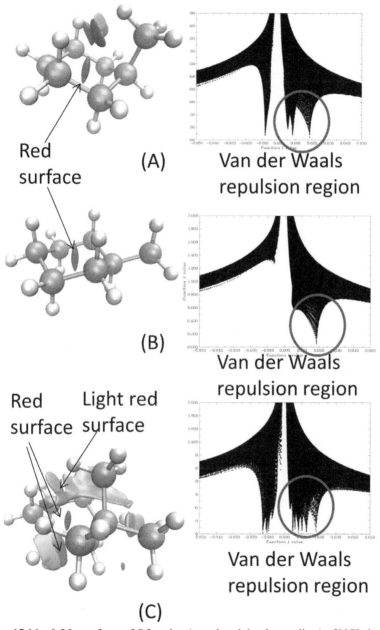

Figure 15.16 0.5 Isosurfaces of S function (or reduced density gradient) of NCI along with corresponding plot of S function versus $\text{sign}(\lambda_2)\rho$ in chair conformation of (A) axial C-Me bond, and (B) equatorial C-Me bond of methylcyclohexane, and (C) axial C-t-Bu bond of *tert*-butylcyclohexane.

S function versus sign(λ_2)ρ plot of both chair conformation (axial and equatorial positions of the methyl group) of methylcyclohexane (Figs. 15.16(A) and (B), respectively) and axial position of *tert*-butyl group in chair conformation of *tert*-butylcyclohexane (Fig. 15.16(C)). See more details about NCI analysis at the end of chapter two.

From Fig. 15.16(B), one can see that the repulsion region of methylcyclohexane with methyl group at equatorial position is only related to the ring. As for the other chair conformation (where methyl group is in axial position), in Fig. 15.16(A), the repulsion region is higher (left part of sign(λ_2)ρ axis of the plot of S function versus sign(λ_2)ρ), which is in accordance with its higher free Gibbs energy. When comparing to the chair conformation of tert-butyl group in axial position of *tert*-butylcyclohexane (Fig. 15.16(C)), this repulsion region is much greater, which is also in accordance with the higher energy difference between chair conformers of *tert*-butylcyclohexane.

In the literature, this repulsion in axial position of a substituent group in chair conformation of a monosubstituted cyclohexane (as evidenced by the NCI analysis in Fig. 15.16) is called 1,3 diaxial repulsion because it involves van der Waals repulsion between C-hydrogen (or other higher substituent) and C-substituent group both in axial position.

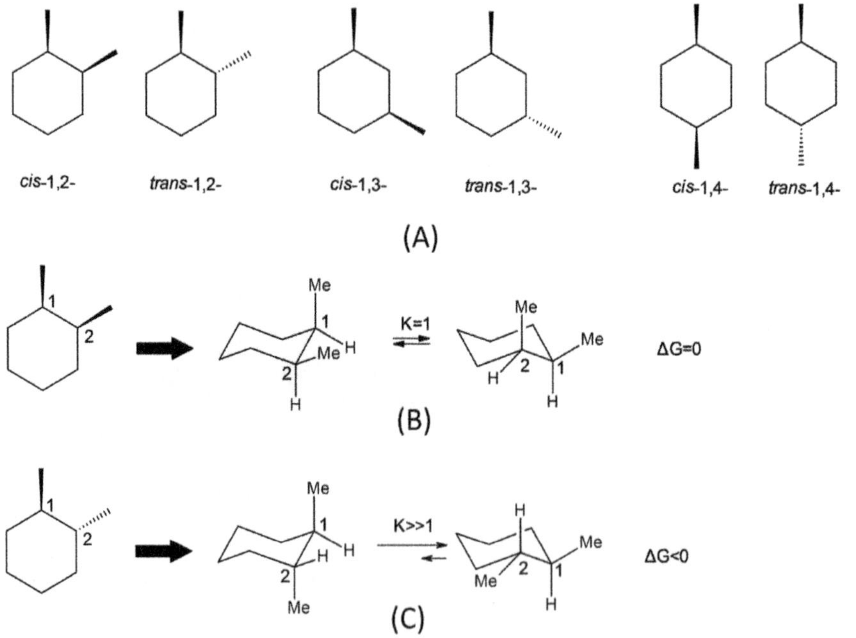

cis-1,2-   trans-1,2-   cis-1,3-   trans-1,3-   cis-1,4-   trans-1,4-

(A)

(B)

(C)

Figure 15.17   (A) Upper view of dimethylcyclohexanes; upper view and side view of both chair conformations of (B) 1,2-*cis*-dimethylcyclohexane and (C) 1,2-*trans*-dimethylcyclohexane.

Disubstituted cyclohexanes have two spatial stereoisomers: *cis* and *trans*. Substituent groups in disubstituted cyclohexane might be in 1,2 or 1,3 or 1,4 positions (see examples in Fig. 15.17(A) where the substituent group is methyl or -Me). Each of them might have two distinguished chair conformers or not. If in each chair conformation, there is one substituent group at axial position and other at equatorial position and these groups are equal, then both chair conformations are the same, which leads to $\Delta G=0$ and $K=1$. This is the case of the chair conformations of *cis*-1,2-dimethylcyclohexane (Fig. 15.17(B)). For the *trans*-1,2-dimethylcyclohexane (Fig. 15.17(C)), there is one chair conformer with both methyl groups at axial positions (with much higher repulsion and energy) and other chair conformation with both methyl groups at equatorial positions (with much smaller energy), which yields $\Delta G<0$ and $K\gg1$ in the direction of a more stable conformer.

All *trans*-1,2-, 1,3-, and 1,4-disubstituted cyclohexanes have two stereoisomers (*R,R*) and (*S,S*), while *cis*-1,2-, 1,3-, and 1,4-disubstituted cycloalkanes are achiral. In fact, *any trans-disubstituted cycloalkane (not having a plane of symmetry) has two enantiomers* (Smith and March 2007).

NOMENCLATURE OF BICYCLIC HYDROCARBONS

There are three types of biclyclic hydrocarbons: bridged, fused, and spiro. Bridged and fused bicyclic hydrocarbons have two bridgedhead carbon atoms as the extremes of a chain connecting the two rings in the bicyclic compound. The bridgedhead carbon atoms are the beginning and the end of the chain which belongs to both rings. If in this chain (in common between the two rings) there are one or more carbon atoms (so-called bridge atoms) between the two bridgedhead carbon atoms, then the bicyclic hydrocarbon is called bridged bicyclic hydrocarbon. If there is no bridge atom between the two bridgehead atoms, the bicyclic hydrocarbon is called fused bicyclic hydrocarbon. Another type of bicyclic hydrocarbon is called spiro bicyclic hydrocarbon when there is only one carbon atom linking the two rings.

Like cycloalkanes, bicyclic hydrocarbons can be viewed from upper and side views, as depicted in Fig. 15.5. In upper view they are apparently flat, but actually they have puckered geometry, as indicated by the side view.

As for the bridged and fused bicyclic hydrocarbons, the nomenclature begins with "bicyclo" followed by squared brackets containing three *x.y.z* numbers [*x.y.z*] and the name of the corresponding alkane (as a sum of all atoms in the compound). No hyphen is used in this nomenclature. The *x* number is the amount of carbon atoms in the biggest ring (disregarding the bridge and bridgedhead atoms); the *y* number is the amount of carbon atoms in the smallest ring (not taking for granted the bridge and bridgedhead atoms); and *z* is the amount of bridge atoms (not considering the bridgedhead atoms). The numbering scheme of bridged and fused bicyclic hydrocarbons starts from one of the bridgedhead carbon atom and follows the biggest ring towards the other bridgedhead carbon atom and goes on to the smallest ring until reaching the first bridgedhead carbon atom. Figure 15.18 depicts the upper and side views, along with corresponding optimized geometry, of bicyclo[1.1.0]butane, bicyclo[3.2.1]octane, bicyclo[2.2.1]heptane, bicyclo[2.2.2] octane, and bicyclo[4.2.0]octane. The fused bicyclic hydrocarbons are bicyclo[1.1.0] butane and biclyclo[4.2.0]octane and the others are bridged bicyclic hydrocarbons.

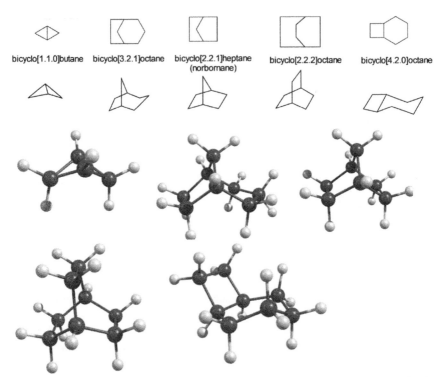

Figure 15.18 Bond-line formula in upper and side views of some fused and bridged bicyclic hydrocarbons, along with their corresponding optimized geometry at PM6 semi-empirical method.

Regarding the spiro bicyclic hydrocarbons, the nomenclature begins with "spiro" followed by two $x.y$ numbers in the squared brackets $[x.y]$ and it ends with the name of the corresponding alkane. The x number is the amount of carbon atoms in the smallest ring (not considering the carbon atom as a junction of both rings) and the y number is the amount of carbon atoms in the biggest ring (also not taking into account the carbon atom in common in both rings). Figure 15.19 shows spiro[3.3]heptane and spiro[3.4]octane in bond-line formula along with their corresponding optimized geometries.

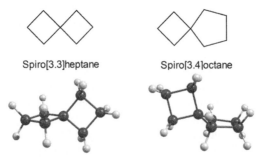

Figure 15.19 Bond-line formula and optimized geometries of spiro compounds.

DECALIN

Decalin or bicyclo[4.4.0]decane is a type of fused bicyclic hydrocarbon used as a solvent in several resins and used as fuel additives. Decalin has two stereoisomers: *cis* and *trans*. Figure 15.20 shows the optimized geometries of both isomers (Figs. 15.20(A) and (B)). According to ωB97XD/6-311G++(*d,p*) level of theory, *trans*-decalin is 2.51 kcal mol^{-1} more stable than *cis*-decalin.

The stereoisomer *trans*-decalin has a TS conformer where the four carbon atoms in the middle of each cyclohexane moiety from side view are not planar as they are in chair conformation of cyclohexane (Fig. 15.20(C)). Then, both cyclohexane moieties of TS conformer of *trans*-decalin are much more puckered than those from its minimum conformer where the four carbon atoms in the middle of each cyclohexane moiety are planar (as in cyclohexane).

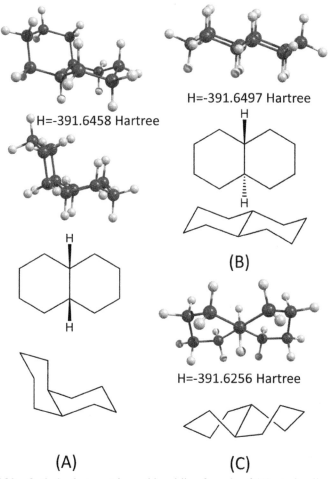

Figure 15.20 Optimized geometries and bond-line formula of (A) *cis*-decalin, (B) *trans*-decalin, and (C) TS conformer of *trans*-decalin in side and upper views along with their corresponding absolute enthalpies in Hartree from ωB97XD/6-311G++(*d,p*) level of theory.

CAGED HYDROCARBON

From the perspective of topology, a linear molecule has only two charge density critical points: nuclear attractor (charge density accumulation in all three directions) and bond critical point (charge density accumulation in two out of three directions); a cyclic molecule has a third critical point: the ring critical point; and a caged molecule has a fourth critical point: the cage critical point (see the QTAIM section in chapter two). A ring critical point has dispersion of charge density in two (out of three) directions, while the cage critical point has a charge density dispersion in all three directions. Then, a caged hydrocarbon is a set of more than two fused cyclic moieties in order to form a cage.

Examples of saturated caged hydrocarbons are adamantane (Fig. 15.21(A)), tetrahedrane (Fig. 15.21(B)), cubane (Fig. 15.21(C)), and prismane (Fig. 15.21(D)). Prismane (or Ladenburg benzene) is a much less stable isomer of benzene which was first synthesized in early 1970s (Katz and Acton 1973). Cubane is a "cubic alkane" which was first synthesized in the early 1960's (Eaton and Cole 1964). All of them are optimized geometries and minimum at PES (no imaginary frequency) at ωB97XD/6-31G++(d,p) level of theory.

In a simple way, cubane can be thought as a "fusion" of two cyclobutanes releasing four equivalents of hydrogen molecule, and prismane as a "fusion" of two cyclopropanes releasing three equivalents of hydrogen molecule.

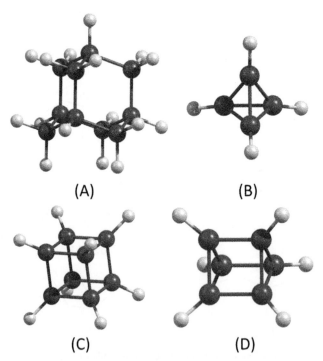

(A)   (B)

(C)   (D)

Figure 15.21   Optimized geometries of (A) adamantane, (B) tetrahedrane, (C) cubane, and (D) prismane at ωB97XD/6-31G++(d,p) level of theory.

Although tetrahedrane is obtained theoretically as a minimum at PES, it was never synthesized till date and will probably never be obtained at the same conditions as the other saturated caged hydrocarbons were. As we have noted in a previous work: "Tetrahedrane is one of the most strained organic molecule, with highly symmetrical structure and unusual bonding being known as a Platonic solid. It has a strain energy of 586 kJ.mol^{-1}" (Monteiro et al. 2014). Nonetheless, the tetra-*tert*-butyltetrahedrane (whose optimized geometry is shown in Fig. 15.22(A)) can be formed from tetra-*tert*-butylcyclobutadiene by photolysis using *n*-octane as a solvent (Maier 1991). As we have shown in this same work: "There are 23 intramolecular hydrogen-hydrogen (H-H) bonds between *tert*-butyl groups according to the virial graph" (see Fig. 15.22(B)) which accounts for its stabilization which affords its synthesis (Monteiro et al. 2014).

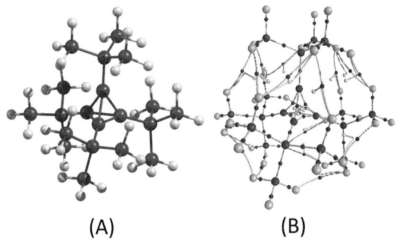

(A)   **(B)**

Figure 15.22   (A) Optimized geometry and (B) virial graph of tetra-*tert*-butyltetrahedrane. Data obtained from our previous work (Monteiro et al. 2014).

In our previous work (Monteiro et al. 2014), we have used GVB to rationalize the difference of stability of tetrahedrane and cubane. Figure 15.23 shows selected GVB singly-occupied orbitals from a specific C-C bond (named VB(C-C)1 and VB(C-C)') and one GVB orbital of a carbon from a vicinal bond (named VB(C-C)2) of tetrahedrane (Fig. 15.23(A)) and cubane (Fig. 15.23(B)). The overlap between these singly occupied orbitals shows an expected highest value for the overlap between singly occupied orbitals from a C-C bond (0.828 and 0.810). In addition, one can see a higher overlap between singly occupied orbitals of vicinal C-C bonds from tetrahedrane (0.279) than that in cubane (0.224). One can assume a direct relation between electronic repulsion from vicinal bonds and the overlap of GVB singly orbitals from vicinal bonds (here represented by VB(C-C)$_1$/VB(C-C)$_2$). Therefore, GVB overlap indicates a higher electronic repulsion between vicinal C-C bonds, which partly accounts for the higher instability of tetrahedrane. Another reason comes from the highest strain energy of tetrahedrane in comparison to cubane (see exercise 11).

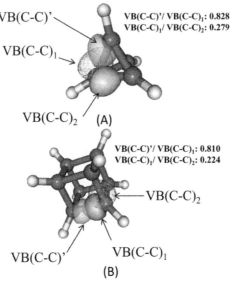

Figure 15.23 Selected GVB singly occupied orbitals of (A) tetrahedrane and (B) cubane along with their corresponding overlap values. Figure adapted from (Monteiro et al. 2014). See the Acknowledgment section. The GVB positive lobe (in non-bonding region) was omitted for simplicity.

EXERCISES

1. By knowing that each C-C-C bond angle in cyclobutane is 88.5°, give the ring strain (in degrees) of cyclobutane.
2. Given the ring strain (in kcal mol^{-1}) of the following four-membered ring molecules (Fig. 15.24), explain this result using the ring strain equation in degrees.

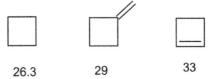

| 26.3 | 29 | 33 |

Figure 15.24 Bond-line formula of some four-membered rings along with their corresponding strain energies in kcal mol^{-1}.

3. Explain why C-C bonds in cyclopropane bend outwards the ring.
4. a) Calculate angle strain of virtual and real cyclopropane and angle strain of cyclobutane. b) Calculate total ring strain (in degrees) and ring strain by methylene group of virtual and real cyclopropane and cyclobutane by knowing the C-C-C angles of virtual and real cyclopropane are 60° and 78° and C-C-C angle of cyclobutane is 88.5°.
5. Explain why planar conformer in cyclobutane has a higher energy than its puckered conformer.

6. By assuming averaged C-C-C angle for boat and half-chair conformers (111.7° and 114.2°, respectively), calculate the ring strain (in degree unit) of all cyclohexane conformers where C-C-C angle of chair conformer is 111.5°. The expected sp^3 carbon angle is 109.5°.

7. Give both chair conformers in side view of cis-1,3- and trans-1,3-dimethylcyclohexane.

8. Predict K and ΔG for the chair conformations of cis-1,4- and trans-1,4-dimethylcyclohexane.

9. Give the chair isomers in side view of cis-cis-1,2,3-tribromocyclohexane and predict K and ΔG.

10. Use free online softwares(*) to calculate the population ratio of chair conformers of fluorocyclohexane and compare to methylcyclohexane using PM6 semi-empirical method for both of them.
(*) For example, ChemCompute (Perri and Weber 2014).

11. By considering the strain energy of saturated caged hydrocarbon as a sum of ring strain energy of its faces and by taking account ring strain in degrees, give the corresponding value of strain energy (in this case in degrees) of cubane and tetrahedrane, which can explain partly the difference of stability between them.

12. Do the conformational analysis of methylcyclohexane and compare to that of cyclohexane.

13. Give the chair conformers of cis-1,2-diethylcyclohexane and trans-1,4-dibromocyclohexane and indicate the most stable for each case.

14. Give both chair conformers of α- and β-pyranose depicted in Fig. 15.25(A). In Fig. 15.25(B) the chair conformer of another stereoisomer is shown in order to prove that in planar structure, dashed and wedged bonds in pyranose (or in any cyclohexane derivative) represent downwards axial/equatorial bond and upwards axial/equatorial bond, respectively.

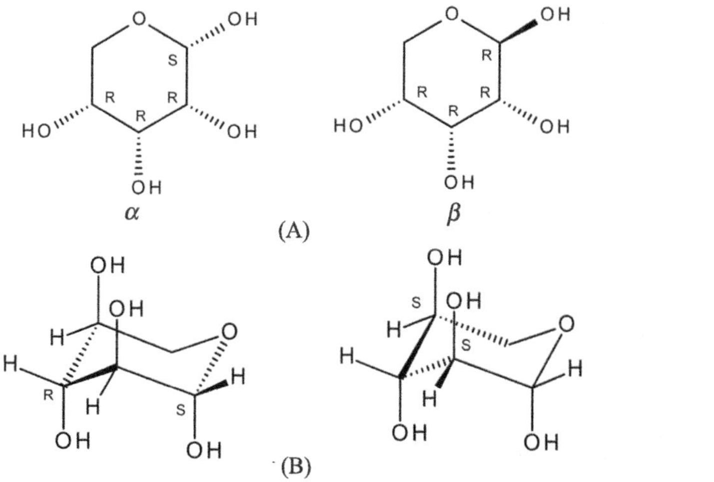

Figure 15.25   (A) Structures of an α- and β-pyranose; (B) chair conformation of a stereoisomer of above-mentioned pyranose.

Acknowledgment

Figure 15.23 was adapted from Figure 6 of our own previous work (http://dx.doi. org/10.1039/C4NJ01271B). We would like to thank Royal Society of Chemistry for permitting the publishing of part of our work in this book.

REFERENCES CITED

Eaton, P.E. and Cole, T.W. 1964. Cubane. J. Am. Chem. Soc. 86: 3157-3158.

Firme, C.L. and Araújo, D.M. 2018. Revisiting electronic nature and geometric parameters of cyclophanes and their relation with stability – DFT, QTAIM and NCI study. Comp. Theor. Chem. 1135: 18-27.

Katz, T.J. and Acton, N. 1973. Synthesis of prismane. J. Am. Chem. Soc. 95: 2738-2739.

Lacher, J.R., Walden, C.H. and Lea, K.R. 1950. Vapor phase heats of hydrobromination of cyclopropane and propylene. J. Am. Chem. Soc. 72: 331-333.

Lu, T. and Chen, F. 2012. Quantitative analysis of molecular surface based on improved Marching Tetrahedra algorithm. J. Mol. Graph. Model. 38: 314-323.

Maier, G. 1991. Unusual molecules – unusual reactions from tetra-tert-butyl-tetrahedrane to the dimer of carbon sulfide. Pure Appl. Chem. 63: 275–282.

Monteiro, N.K.V., de Oliveira, J.F. and Firme, C.L. 2014. Stability and electronic structures of substituted tetrahedranes, silicon and germanium parents – a DFT, ADMP, QTAIM and GVB study. New J. Chem. 38: 5892-5904.

Perri, M.J. and Weber, S.H. 2014. Web-based job submission interface for the GAMESS computational chemistry program. J. Chem. Educ. 91: 2206-2208.

Smith, M.B. and March, J. 2007. March's Advanced Organic Chemistry. Reactions, Mechanisms and Structure. John Wiley and Sons Inc. Hoboken, New Jersey.

Wiberg, K.B. 1986. The concept of strain in organic chemistry. Angew. Chem. Int. Ed. Engl. 25: 312-322.

Chapter Sixteen

Alkenes (nomenclature and properties)

INTRODUCTION AND NOMENCLATURE

Alkenes are hydrocarbons with one and only one double CC bond having a condensed formula (C_nH_{2n}) similar to that from cycloalkanes. Both carbon atoms of the double CC bond are called vinylic carbons and $-CH=CH_2$ is called the vinyl group. Both vinylic carbons are sp^2 hybridized and they have planar trigonal geometry so that the vinyl group is planar with π orbital above and below this plane (Fig. 16.1(B)). As it was discussed in chapter six, the π bond comes from the overlap between one p atomic orbital (from one vinylic carbon) and the other p atomic orbital (from the other vinylic carbon). Besides the π bond, each vinylic carbon has three σ bonds: two σ bonds with non-vinylic substituent (see Fig. 16.1(C)) and one σ bond with the other vinylic carbon as it was shown in Fig. 6.8 in chapter six. It is important to remember (as it was mentioned in chapter two) that according to delocalization index of QTAIM, there are only two electrons in the bonding region (not four electrons!) and the other electrons are located in the vinylic carbon atoms.

Figures 16.1(A) and (B) show ethene along with its four GVB σ and π orbitals from CC bond (remember that GVB orbitals are singly-occupied so that each bond, π or σ, is made up of two GVB orbitals). The σ GVB orbitals are slightly distorted to one of the vinylic carbon (Fig. 16.1(A), while each π GVB orbital has more pronounced concentration in one of the two sp^2 carbon atoms (Fig. 16.1(B)). As expected from the above-mentioned observation, the overlap of both σ GVB orbitals (0.867) is higher than that from π GVB orbitals (0.673).

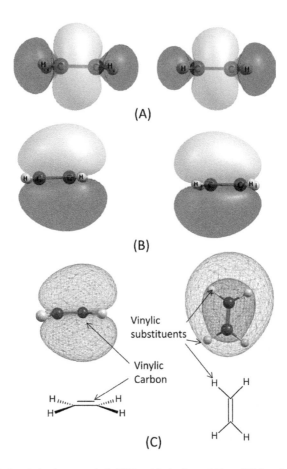

Figure 16.1 Pair of singly-occupied GVB orbitals from (A) σ CC bond; and (B) from π bond in ethene calculated at 6-31G++(d,p) basis set; and (C) side view (at left) and front view (at right) of one singly-occupied GVB π bond from ethene along with their corresponding bond-line formula representations.

As already shown in chapter six, a CC double bond is stronger than a CC single bond accounted for by their corresponding electrostatic force expressions below. See more details in chapter six. As a consequence of the inverse relation between bond strength and bond length, the bond length of a CC single bond (1.54 Å in average) is longer than that from a CC double bond (1.35 Å in average).

$$F_{\text{single CC}} = 4\left(K \frac{eZ_{\text{eff}(C)}}{r_{C_s1\sigma}^2} \right)$$

$$F_{\text{double CC}} = 4\left(K \frac{eZ_{\text{eff}(C)}}{r_{C_d1_\sigma}^2} \right) + 4\left(K \frac{eZ_{\text{eff}(C)}}{r_{C_d1_\pi}^2} \right)$$

Then: $F_{\text{double CC}} > F_{\text{single CC}}$

The vinyl group can be visualized from two distinguished views. In the side view (at the left of Fig. 16.1(C)), only the vinylic carbons are in the plane of the paper and all vinylic substituents are outside the paper's plane. Two vinylic substituents are at the front of the paper's plane and the other two vinylic substituents are at the back of paper's plane. This view is the most suitable for mechanism representation, as we will show in the next chapter. In the upper view (at the right of Fig. 16.1(C)), both vinylic carbon atoms plus their vinylic substituents are in the plane of the paper. While in the side view of the vinyl group, one can see that the π orbital lies above and below the plane of the vinyl group, in the upper view one can see only one lobe of the π orbital (where the other lobe is behind the vinyl group).

The IUPAC nomenclature of alkenes uses -ene suffix and the numbering scheme of the main chain (from the left to the right or vice versa) is given so that the first vinyl carbon atom in the chain has the lowest number. Remember that for alkanes the numbering scheme is in accordance with the lowest sum of numbers to its side chains, but in alkenes, side chains do not determine the numbering scheme. The number of the first vinyl carbon in the main chain appears right before the -ene suffix in alkenes nomenclature for a chain higher than three carbon atoms, for instance, but-1-ene. See more examples in the next section. When both numbering schemes for a branched alkene give the same lowest number to the first vinyl carbon, then the chosen numbering scheme will be the one that gives the lowest number to the side chain, e.g., 2-methylbut-2-ene and not 3-methylbut-2-ene.

ISOMERISM

Some alkenes have *cis/trans* stereoisomerism, while others do not. For example, ethene, propene, methylpropene, 2-methylbut-2-ene and 2,3-dimethyl-but-2-ene do not have stereoisomerism (Fig. 16.2(A)). The requisite for stereoisomerism in alkenes implies that each vinylic carbon has one (and only one) most voluminous non-vinylic substituent. If one or both vinylic carbon(s) has/have two equal substituents there is no stereoisomerism, i.e., $XHC=CHX$ or $X_2C=CX_2$. Similarly, if three equal substituents are found in the vinilyc carbon atoms, i.e., $X_2C=CHX$, there is no stereoisomerism.

Examples of stereoisomerism in alkenes are shown in Fig. 16.2(B), where you see only one most voluminous non-vinylic substituent at each sp^2 carbon. For example, in *cis/trans*-3-methylpent-2-ene, its C3 has one ethyl and one methyl substituent groups and its C2 has one methyl and one hydrogen substituent groups.

Ethene has no isomer while propene has cyclopropane as an isomer. Remember that cycloalkanes and alkenes have the same condensed formula. The isomers of but-1-ene are *cis*-butene, *trans*-butene, 2-methylpropene, and cyclobutane. Here we give one procedure to find all stereoisomers of an alkene: firstly, change the position of the double bond in the highest main chain giving stereoisomers when it is possible (Fig. 16.3(A)); secondly, start to put branches in the main chain in one alkene with double bond in one specific position, methyl by methyl group, noting the possibility of stereoisomerism as well, and then do the same for the other alkene with double bond in another specific position, and so on (Fig. 16.3(B)).

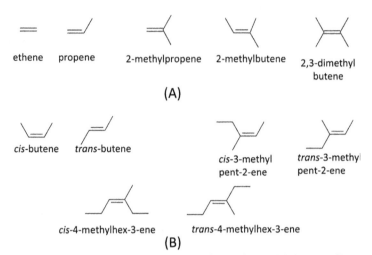

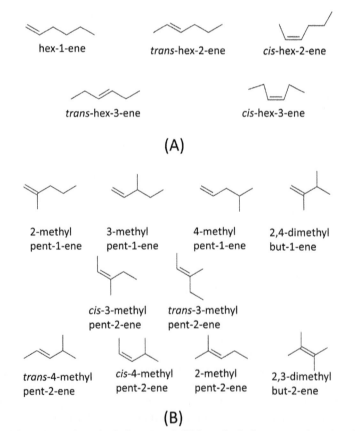

Figure 16.2 (A) Some alkenes with no stereoisomerism and (B) some alkenes with stereoisomerism.

Figure 16.3 (A) non-branched alkenes and (B) branched alkene stereoisomers of C_6H_{12} in bond-line formula.

STABILITY OF *cis/trans* STEREOISOMERS

The stability of stereoisomers of alkenes is very important for the stereoselectivity of elimination reactions (see the next volume). According to ωB97XD/6-31G(d,p) level of theory, *trans*-butene is 1 kcal mol^{-1} more stable than *cis*-butene. The stereoisomer *cis*-butene has three conformers with respect to the rotation of C(vinylic)-C(methyl) bond (Fig. 16.4): one minimum at PES (where two hydrogen atoms from the methyl groups are as close as possible, making an attractive hydrogen-hydrogen bond, as depicted in Fig. 16.5(A)) and two transition states at PES (one first order TS, with one imaginary frequency (Fig. 16.4(B)), and one second order TS, with two imaginary frequencies (Fig. 16.4(C)). The increasing order of energy is: minimum at PES < first order TS < second order TS. Then, the most stable conformer of *cis*-butene has two hydrogen atoms (one from each of its methyl groups), making an attractive interaction. When this H-H bond is broken (with one or two rotations of Csp^2-Csp^3 rotation), the conformer increases its energy.

Figure 16.4 shows the NCI surface (0.5 isosurfaces of S function) in colored scheme (which actually appears in black-and-white here) along with the corresponding S function versus sign(λ_2)ρ plot of three conformers of *cis*-butene. See chapter two for more details about NCI. Figure 16.5(A) also presents the molecular graph of *cis*-butene as minimum at PES (potential energy surface).

The minimum at PES of *cis*-butene has the highest intensity of attractive interaction and one bond critical point from H-H bond and the lowest intensity of repulsive interaction relative to C-H/C-H repulsive interaction from methyl groups (Fig. 16.4(A)). while its second order TS has the lowest intensity of attractive interaction and highest intensity of repulsive interaction (Fig. 16.4(C)). The first order TS of *cis*-butene has intermediate attractive and repulsive interactions (Fig. 16.4(B)). One reason for the higher energy of *cis*-butene (as minimum at PES) with respect to its *trans* stereoisomer is the repulsive C-H/C-H interaction (from methyl groups), although it has concomitantly attractive H-H bonds. In addition, C-C-C angle in *trans* stereoisomer (125.1°) is closer to 120° than that in *cis* stereoisomer (127.8°) of *cis*-butene. Then, *cis* stereoisomer has a higher angle strain than *trans* stereoisomer.

CONFORMERS OF 1-ALKENES IN GAS PHASE

In gas phase, molecules have high translational and rotational motions, which prevent intermolecular interactions in most cases. Hence, the microscopic analysis of conformers in gas phase can be done by single molecules or monoalkenes.

The 1-butene or but-1-ene has two conformers as minimum at the PES and one maximum at the PES (Fig. 16.5). The most stable conformer (skew conformer) has the methyl group in C4 in an inward position (Fig. 16.5(A)), while the least stable conformer (syn conformer) in this methyl group is outwards (Fig. 16.5(B)).

Propene itself has two conformers as well: one minimum at PES and the other TS at PES. The enthalpy difference is about 1.4 kcal mol^{-1} between them. Both

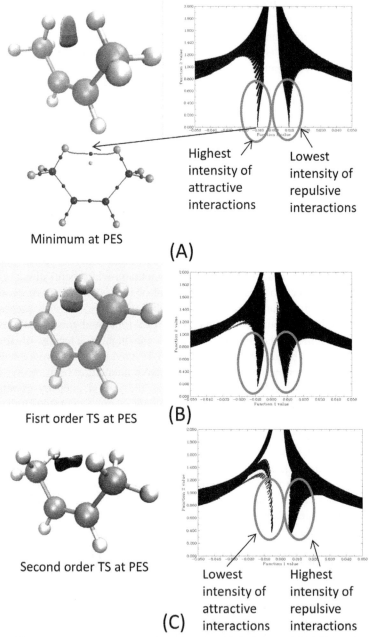

Figure 16.4 Optimized geometries of conformers of *cis*-butene: (A) minimum at PES; (B) first order TS; (C) second order TS. 0.5 Isosurfaces of S function (or reduced density gradient) of NCI along with corresponding plot of S function versus $\mathrm{sign}(\lambda_2)\rho$ of (A) minimum at PES; (B) first order TS; and (C) second order TS of *cis*-butene along with molecular graph of its minimum at PES conformer in (A). Courtesy of Hindawi Publisher (Firme 2019).

conformers have one methyl hydrogen in the same plane as the vinylic carbon and hydrogen atoms. In the low-lying conformer (the minimum at PES), this methyl hydrogen is in U-shape (coplanar with the (C1)-vinylic hydrogen), and in the TS conformer this methyl hydrogen is placed outwards (see Fig. 8.10 from chapter eight).

The pent-1-ene has three conformers as minimum at the PES (Fig. 16.5(C)), but none of them is in a linear, straight conformation. All of the three conformers are bent inwards, turning to the double bond. Even when the starting geometry is a linear, straight main chain, the final, optimized conformer of pent-1-ene has a bent-inward conformation (Fig. 16.5(D)).

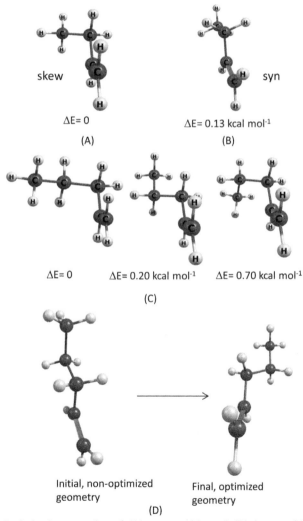

Figure 16.5 Optimized geometries of (A) most stable and (B) least stable conformers of but-1-ene; (C) optimized geometries of conformers of pent-1-ene; and (D) initial, non-optimized linear, straight conformer and final, optimized conformer of pent-1-ene from ωB97XD/6-311G++(d,p) level of theory. Courtesy of Hindawi Publisher (Firme 2019).

Figure 16.6 shows the plots of electronic energy versus C-C-C-C dihedral angle for but-1-ene and pent-1-ene. One can see that the rotational barrier ranges from 2.2 to 4.9 kcal mol^{-1} according to the electronic energy as energetic parameter. The rotational barriers in the conformational analysis of the C-C-C-C dihedral angle of other studied 1-alkenes are in the same range.

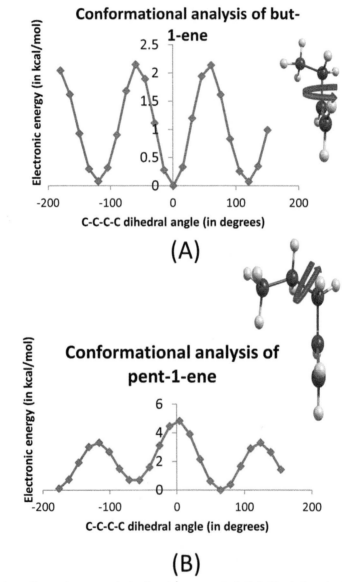

Figure 16.6 Electronic energy, in kcal mol^{-1}, versus C-C-C-C dihedral angle, in degrees, for (A) but-1-ene and (B) pent-1-ene as indicated by the curved arrow of the corresponding molecule. The calculated values are in marked points in each plot. Courtesy of Hindawi Publisher (Firme 2019).

The same trend occurs for hex-1-ene which has five optimized conformers as minimum at the PES $\Delta E=0$, where the second most stable conformer ($\Delta E=0.01$ kcal mol^{-1}) is very close in energy to the most stable conformer (Fig. 16.7). The most unstable conformers of hex-1-ene have less bent-inward conformations (Fig. 16.7). Then, we can state that *the most stable conformation of any 1-alkene has a bent-inward conformation.*

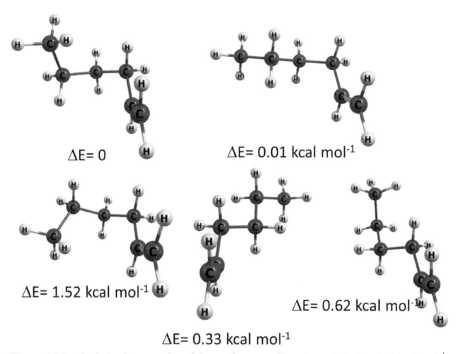

$\Delta E= 0$ $\Delta E= 0.01$ kcal mol^{-1}

$\Delta E= 1.52$ kcal mol^{-1}

$\Delta E= 0.62$ kcal mol^{-1}

$\Delta E= 0.33$ kcal mol^{-1}

Figure 16.7 Optimized geometries of the conformers of hex-1-ene $\Delta E=0$ to 0.19 kcal mol^{-1} from ωB97XD/6-311G++(*d,p*) level of theory. Courtesy of Hindawi Publisher (Firme 2019).

Figure 16.8 shows the most stable conformers of hept-1-ene and its least stable conformers ($\Delta E=2.07$ kcal mol^{-1}). Likewise, the most stable conformers of hept-1-ene have a bent-inward conformation, while the least stable conformers of hept-1-ene have a lesser bent-inward conformation. In addition, no linear, straight conformation of hex-1-ene and hept-1-ene exists. These results show a distinguished behavior between linear alkanes and 1-alkenes: *the most stable conformation of linear alkanes is a straight zig-zag conformation, while in 1-alkenes is a bent-inward conformation.*

In all cases of monomers of 1-alkenes, the bent-inward geometry favors π bond interaction with methyl/methylene hydrogen/carbon atoms according to NCI (see Firme 2019). All monomers of 1-alkenes as minima at the PES have the propene moiety as minimum at the PES, where the (C3)-methylene hydrogen is coplanar with (C1)-vinylic hydrogen.

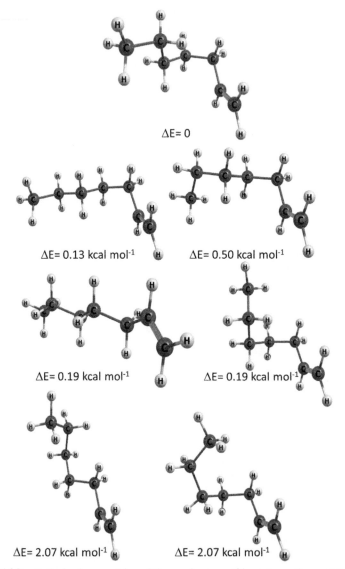

Figure 16.8 Optimized geometries of the conformers of hept-1-ene from ωB97XD/6-311G++(d,p) level of theory. Courtesy of Hindawi Publisher (Firme 2019).

STABILITY OF ALKENES

The stability of alkenes is important for the regioselectivity and stereoselectivity (as previously mentioned) of elimination reactions to yield two or more alkenes as main products (see the next volume).

There are two ways of measuring the stability of alkenes experimentally: (1) heat of hydrogenation which is limited to the stability of π bond and

(2) the heat of combustion which evaluates the overall stability of alkenes but it cannot be used for substituted alkenes with non-hydrocarbon substituent groups. Moreover, this data can also be obtained theoretically. It is important to emphasize that hydrogenation reaction of alkenes yields corresponding alkanes as products, and total combustion reaction gives carbon dioxide and water as products.

We have used Gaussian G4 method and B3LYP density functional to obtain the thermodynamic and topological data of a series of alkenes and fluoro-substituted alkenes (De Freitas and Firme 2013). Table 16.1 shows the theoretical heat of hydrogenation and the parametrized free Gibbs energy of combustion of the studied alkenes, along with their CC delocalization index and double CC bond length. The heat of combustion for distinguished carbon chain size of alkenes has to be parametrized in order to be used for comparison. The ethene is used as reference and the parametrized equation is:

$$\Delta G_{comb(p)} = \frac{2\Delta G_{comb}}{N_C}$$

Where N_C is the number of carbon atoms in the alkene.

Both thermodynamic data indicate that these reactions are exergonic and exothermic, as expected. Then, as the stability of the alkene increases, both $\Delta G_{comb(p)}$ and ΔH_{hyd} decrease in modulus (or they become less negative). Considering hydrogen substituent as neutral (since it does not donate charge density nor remove charge density), we can establish that ethene has zero substituent; propene, fluoroethene and styrene have one substituent; butene, difluoroethene, and methylpropene have two substituents; 2-methylbut-2-ene and trifluoroethene have three substituents and; 2,3-dimethylbut-2-ene and tetrafluoroethene have four substituents.

Table 16.1 Parametrized free Gibbs energy of combustion ($\Delta G_{comb(p)}$) from B3LYP/6-311++G(d,p) level of theory, heat of hydrogenation (ΔH_{hyd}) from G4 method, both in kcal mol^{-1}, delocalization index of double CC bond, DI (C=C), from B3LYP method, and CC bond length from G4 method.[a]

Molecule (number of substituents at the vinylic carbon)	$\Delta G_{comb(p)}$ (kcal/mol) B3LYP	ΔH_{hyd}/ kcal. mol^{-1} G4	DI (C=C)	CC Bond length/Å
Ethene (0)	−298.27	−32.48	1.90	1.327
Propene (1)	−293.28	−29.31	1.84	1.329
trans-butene (2)	−291.54	−27.04	1.79	1.331
cis-butene (2)	−292.04	−28.35	1.79	1.333
2-methylpropene (2)	−291.17	−27.53	1.79	1.332
2-methyl-2-butene (3)	−290.77	−26.42	1.75	1.337
2,3-dimethyl-2-butene (4)	−291.68	−26.81	1.72	1.344
Fluoroethene (1)	-	−30.71	1.80	1.321
1,2-difluoroethene (2)	-	−33.00	1.69	1.324
1,1,2-trifluoroethene (3)	-	−39.20	1.60	1.324
Tetrafluoroethene (4)	-	−49.65	1.55	1.323
Styrene (1)	−246.42	−28.22	1.77	1.334

(a) Data from our work (De Freitas and Firme 2013).

The theoretical data between $\Delta G_{comb(p)}$ and ΔH_{hyd} is in good agreement. As a consequence, the order of stability of the studied alkenes is given in Fig. 16.9(A). When considering only alkyl/aryl substituents, *as the number of alkyl substituents at the vinylic carbon atoms increases, the stability of the corresponding alkene also increases*. On the other hand, when the substituent is one phenyl group in styrene, it becomes more stable than *cis*-butene, probably because of the instability factors of *cis* isomer previously discussed.

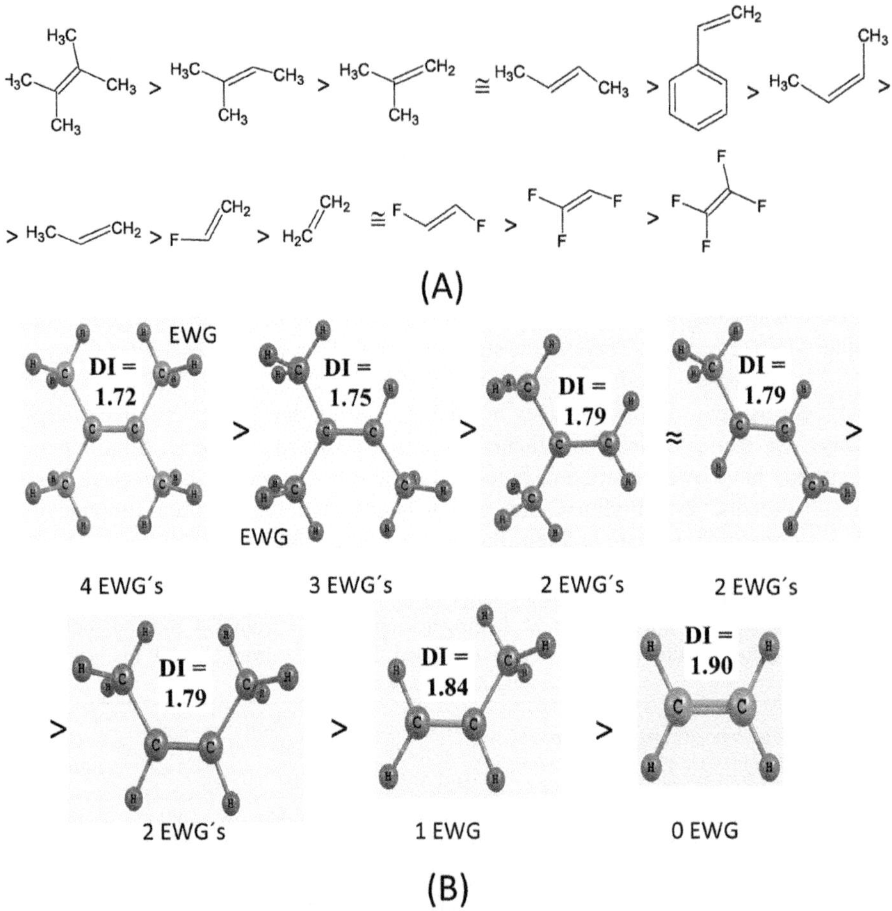

(A)

(B)

Figure 16.9 (A) Relative stability of alkenes from free Gibbs energy calculated by G4 method. (B) G4 optimized geometries of studied alkenes along with their DI (C=C).

An important parameter in Table 16.1 is the QTAIM delocalization index of double CC bond, DI (C=C), which is the amount of shared electrons between two atoms. As we already mentioned in QTAIM section from chapter two, the amount of shared electrons in single, double, and triple bonds is, in average, one, two, and three, respectively, and not two, four, and six as expected from Lewis theory, which is not based in quantum chemistry calculations (Firme et al. 2009).

Both DI (C=C) and the CC double bond length indicate that as the number of substituents at the vinylic carbon increases (from 0 to 4), the CC bond length increases and the DI (C=C) decreases (see Fig. 16.9(B)). Then, *each alkyl substituent removes the charge density of the CC double bond partly, i.e., for alkenes, the alkyl groups are (moderate) electron withdrawing group (EWG)*. The inverse relation between DI (C=C) and stability of alkenes is clear in Fig. 16.9(B). In addition, the QTAIM atomic partial charge of the vinylic carbon also decreases in modulus (it becomes less negative) when the vinylic carbon is bonded to the methyl group (see De Freitas and Firme 2013).

The remaining question is how an alkyl group acting as moderate EWG in alkene leads to its stabilization? Probably, as the charge density in the vinylic carbon atoms decreases in modulus (when bonded to alkyl groups), their Z_{eff} increases more than the decrease of DI (C=C), which leads to stronger double CC bonds, as one can see in the electrostatic force equation of the double CC bond below.

$$F_{elect(C=C)} = 2\left(\frac{Z_{eff(\text{vinylic C})} \cdot DI(C=C)}{r^2} \right)$$

Another possible reason is the repulsive interaction between π electrons and σ electrons from σ C-C and/or C-H bonds out of the plane containing the vinylic carbons. As the number of alkyl groups increase and each group withdraws moderately and partially the π density of the double bond, this repulsive π bond-σ bond deacreases and the stability of the alkene increases. In Fig. 16.10(C) the overlapping values between CC π bond and σ CH bond in the plane of vinylic carbons (overlap nearly zero) and between CC π bond and σ CH bond out of the plane of vinylic carbons is depicted, which indicates the repulsive interaction between double bond and vicinal CH bonds out of the plane of vinylic carbons. Finally, the alkyl groups promote electron delocalization of the π electrons, which also lead to stabilization.

As for fluoro groups attached to alkenes, fluoroethene is more stable than ethene and 1,2-difluoroethene is as stable as ethene. These facts reveal an interesting characteristic of fluorine atom as substituent: it is an electron withdrawing group (EWG) by inductive effect and electron donating group (EDG) by resonance effect. The DI of CC double bond of fluoroethene is similar to that from propene and DI(C=C) of 1,2-difluoroethene is close to that from 2-methylpropene. In these cases, the electron donating (ED) capacity of fluoro substituents is higher than the removal of electron density. On the other hand, in trifluoroethene and tetrafluoroethene, having the lowest DI (C=C) of the studied series, the electron withdrawing capacity of fluoro substituents is higher than the electron donation. This information is in accordance with GVB overlap matrix between fluorine lone pairs (LP1 and LP2) and C-F singly occupied orbitals (VB(CF)$_1$ and VB(CF)$_2$). These overlaps are much higher in fluoroethene (with higher influence of electron donating capacity) than in trifluoroethene (with higher influence of electron withdrawing capacity). See GVB orbitals and overlap data in Fig. 16.10(A) and (B) for fluoroethene and trifluoroethene, respectively. Then, when two fluorine atoms are attached to the same vinylic carbon atom, the EW capacity is higher than its ED capacity.

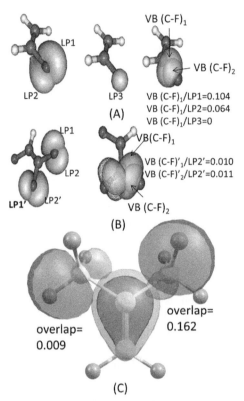

Figure 16.10 GVB orbitals of fluorine lone pairs (LP) and GVB singly occupied C-F bonds in (A) fluoroethene and (B) trifluoroethene, along with their overlap values. Courtesy of Springer-Verlag (see the Acknowledgment). Ref.: (De Freitas and Firme 2013); (C) GVB orbitals of CC π bond and two σ CH bonds of methylpropene along with their corresponding overlap values.

INTERMOLECULAR INTERACTIONS IN ALKENES

Unlike alkanes, the alkenes might have three different types of intermolecular interactions: (1) the so-called hydrogen-hydrogen bond; (2) hydrogen-carbon interaction and; (3) carbon-carbon interaction. In the search for the most stable complex for *cis*-butene and *trans*-butene, we have found three conformers for both of them (Figs. 16.11(A) and (B)). From the QTAIM analysis of their corresponding molecular graphs (Figs. 16.11(C) and (D)), we see these three types of intermolecular interactions in these alkenes. Some of these interactions exist in ethene and propene complexes (Figs. 16.11(E) and (F)). The conformation analyses of ethene and propene complexes have been previously done (Jalkanen et al. 2005) and their low-lying complexes are shown in Fig. 16.11 (E). As a consequence, the analysis of intermolecular interactions in alkenes is more complex than that in alkanes, where all three types of intermolecular interaction might be present or some of them, while in alkanes only hydrogen-hydrogen bonds are present. Then, the use of bimolecular model to analyze some of alkene's properties (e.g., boiling point) has to be limited to certain types of alkenes so that there is only one type of intermolecular interaction, which enable some linear relation as is depicted in 1-alkenes, for instance (see ahead).

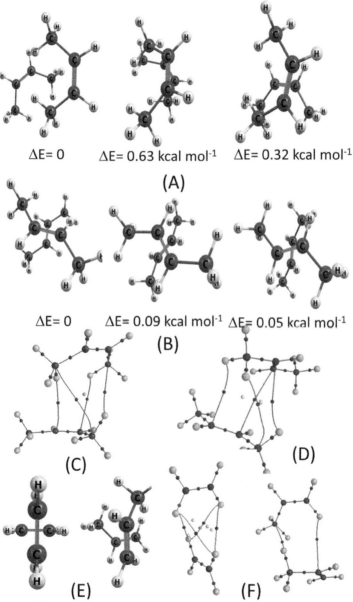

$\Delta E= 0$ $\Delta E= 0.63$ kcal mol^{-1} $\Delta E= 0.32$ kcal mol^{-1}

(A)

$\Delta E= 0$ $\Delta E= 0.09$ kcal mol^{-1} $\Delta E= 0.05$ kcal mol^{-1}

(B)

(C)   (D)

(E)   (F)

Figure 16.11 Optimized complex conformers of (A) *cis*-butene and (B) *trans*-butene along with their molecular graphs in (C) and (D), respectively. Optimized complexes of (E) ethene and propene and (F) their corresponding QTAIM molecular graphs from ωB97XD/6-311G++(*d,p*) level of theory. Courtesy of Hindawi Publisher (Firme 2019).

BOILING POINT OF 1-ALKENES AND THEIR CONFORMATIONAL ANALYSIS IN LIQUID PHASE

Unlike gas phase, in liquid phase the molecules have slower translational and rotational motions so that the intermolecular interactions might occur. In that case, from microscopic perspective, the simplest method to interpret the interactions and conformational analysis in liquid phase is the bimolecular model. It is important to add that the bimolecular model (or higher order model for some cases) is the simplest and most effective way to analyze a macroscopic property (which depends on "zillions" of molecules interacting with each other) through a microscopic standpoint (which depends on two or more molecules interacting with each other), as we have done successfully to rationalize the boiling points of alkanes (Monteiro and Firme 2014).

Table 16.2 shows the boiling point of some 1-alkenes along with the number of hydrogen-hydrogen bonding in the corresponding complexes (number of H-H bonds). As the carbon chain increases, the boiling point of the corresponding alkene also increases, because the number of H-H bonds increases proportionally. Table 16.2 also presents the electronic energy of complex formation of the corresponding 1-alkenes based on the bimolecular model.

Table 16.2 Boiling point, in °C, and electronic energy of complex formation (ΔE(complex)), in Hartree, from ωB97XD/6-311G++(d,p) level of theory of some 1-alkenes.

Alkene	Boiling point/°C	ΔE(complex)/ Hartree	Number of H-H bonds
Propene	−47.6	−0.0050	2[a]
But-1-ene	−6.5	−0.0089	6[a]
Pent-1-ene	30	−0.0102	8
Hex-1-ene	64	−0.0115	10
Hept-1-ene	94	−0.0113	13
Oct-1-ene	121	−0.0141	15

(a)Two hydrogen-carbon intermolecular interactions.

Like 1-alkenes in gas phase (single molecules) which have lots of conformers for the main chain, there are also lots of conformations in their corresponding bimolecular complexes. Surprisingly, unlike the 1-alkenes in gas phase where the most stable conformation is associated with bent-inward conformation, the conformation of each 1-alkene in the most stable conformer of their complexes is nearly straight zig-zag conformation (similar to alkanes) when the double bond at one terminal of their chain is not taken for granted. Take for example, the conformers of pent-1-ene, where in the most stable conformer each single pent-1-ene has nearly straight zig-zag conformation (Fig. 16.12). Further examples of conformational analysis of bimolecular complexes of 1-alkenes are shown in the Appendix section 3. Similarly to pent-1-ene, in their most stable complexes, each alkene molecule adopts a straight, zig-zag conformation.

Another similarity with alkane complexes is that in the most stable conformation of the bimolecular complexes of 1-alkenes, there are nearly only hydrogen-hydrogen

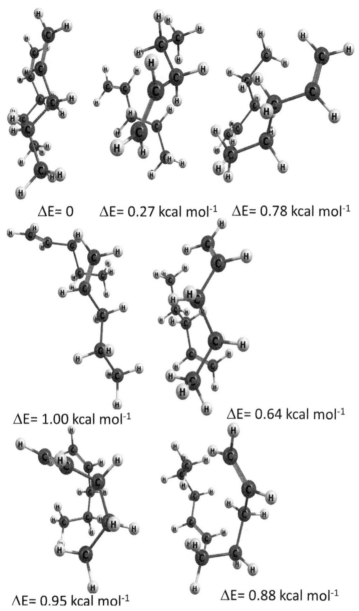

ΔE= 0   ΔE= 0.27 kcal mol^{-1} ΔE= 0.78 kcal mol^{-1}

ΔE= 1.00 kcal mol^{-1} ΔE= 0.64 kcal mol^{-1}

ΛE= 0.95 kcal mol^{-1} ΔE= 0.88 kcal mol^{-1}

Figure 16.12 Optimized complex conformers of pent-1-ene from ωB97XD/6-311G++(d,p) level of theory. Courtesy of Hindawi Publisher (Firme 2019).

bonds (Fig. 16.13(A) to (E)), for but-1-ene, pent-1-ene, hex-1-ene, hept-1-ene and oct-1-ene complexes, respectively, which enables linear relations with boiling point, for example. Like alkanes, the number of intermolecular interactions- according to the bond paths of the QTAIM molecular graphs- of alkene complexes increases as the number of their carbon size increases. Propene complex has two intermolecular interactions; *trans*/*cis*-butene and but-1-ene complexes have from 4 to 7 intermolecular interactions; pent-1-ene and *trans*-pent-2-ene complexes have 7 to 8 intermolecular interactions; hex-1-ene complex has 10 intermolecular interactions, and so on. *One important difference relative to the type of intermolecular interactions according to QTAIM between alkanes and alkenes is that in the former there is only hydrogen-hydrogen bond and in the latter besides H-H bond, there are also carbon-hydrogen and carbon-carbon intermolecular interactions.*

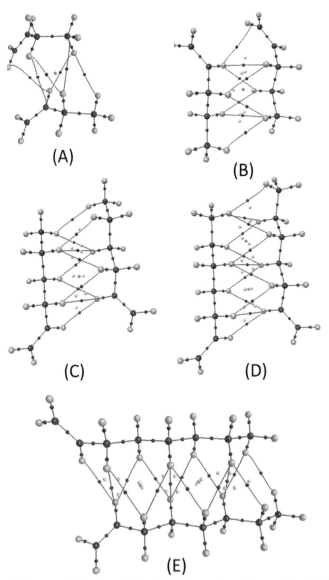

Figure 16.13 QTAIM molecular graphs of the most stable complex of (A) but-1-ene, (B) pent-1-ene, (C) hex-1-ene, (D) hept-1-ene, and (E) oct-1-ene from ωB97XD/6-311G++(d,p) level of theory. Courtesy of Hindawi Publisher (Firme 2019).

The linear relations between the electronic energy of complex formation (in Hartree) of the 1-alkenes with their corresponding boiling points (Fig. 16.14(A)) and with their corresponding number of hydrogen-hydrogen bonds (Fig. 16.14(B)) have coefficients of determination higher than 0.91. When comparing number of hydrogen-hydrogen bonds to corresponding boiling points of 1-alkenes, the linear relation is outstanding (Fig. 16.14(C)), showing that the bimolecular model can be applied to predict physical properties of 1-alkenes.

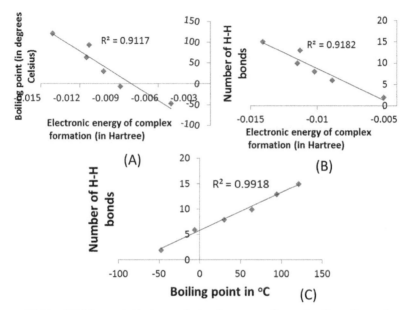

Figure 16.14 (A) Linear plot between electronic energy of complex formation and corresponding boiling points of 1-alkenes; (B) linear plot between electronic energy of complex formation and number of hydrogen-hydrogen bonds in 1-alkenes; (C) number of hydrogen-hydrogen bonds versus boiling points. Courtesy of Hindawi Publisher (Firme 2019).

EXERCISES

1. Draw in bond-line formula all isomers with C_5H_{10} condensed formula.
2. By knowing the C-C-C angles of *trans* and *cis* stereoisomers of butene (*trans*: 125.1° and *cis*: 127.8°), give their angle strains in degrees.
3. Give the chemical equations for hydrogenation and combustion reactions of ethene, propene, and but-1-ene. In addition, explain why thermodynamic data of combustion reaction has to be parametrized (for comparison), while thermodynamic data of the hydrogenation reaction does not.

Acknowledgment

Figure 16.10 was published in our own article whose copyrights belong to Springer-Verlag, and we appreciate the publisher's permission to print this figure in this book. Figures 16.4 to 16.8 and 16.11 to 16.14 were published in our own article, whose copyrights belong to Hindawi, and we appreciate the publisher's permission to print these figures in this book.

REFERENCES CITED

De Freitas, G.R.S. and Firme, C.L. 2013. New insights into the stability of alkenes and alkynes, fluoro-substituted or not: a DFT, G4, QTAIM and GVB study. J. Mol. Model. 19: 5267-5276.

Firme, C.L., Antunes, O.A.C. and Esteves, P.M. 2009. Relation between bond order and delocalization index of QTAIM. Chem. Phys. Lett. 468: 129-133.

Firme, C.L. 2019. Deeper insights in conformational analysis of cis-butene and 1-alkenes as monomers and dimers: QTAIM, NCI and DFT approach. J. Chem. Article ID 2365915.

Jalkanen, J-P., Pulkkinen, S. and Pakkanen, T.A. 2005. Quantum chemical interaction energy surfaces of ethylene and propene dimers. J. Phys. Chem. A. 109: 2866-2874.

Monteiro, N.K.V. and Firme, C.L. 2014. Hydrogen-hydrogen bonds in highly branched alkanes and in alkane complexes: a DFT, *ab initio*, QTAIM and ELF study. J. Phys. Chem. A 118: 1730-1740.

Chapter Seventeen

Alkenes (reactions)

INTRODUCTION

Unlike alkanes, alkenes undergo polar reactions of basically three types: pericyclic reactions, polymerization reactions, and electrophilic addition reactions. Most polar reactions in alkenes are stepwise, but alkenes also undergo concerted reactions. Similarly to alkanes, alkenes also undergo radical reactions, following radical addition instead of radical substitution as in alkanes.

An important point in representing the mechanism of these reactions is the use of the most appropriate, correct view of the alkene. As shown before (Fig. 16.1 in chapter sixteen), there are two views for representing alkenes: upper view where the π-bond electrons/orbital is shown above the plane containing vinylic carbon and their substituents (i.e., π-bond electrons are outside the plane of the paper); and side view where π-bond electrons/orbital is within the plane of the paper and the vinylic substituents are behind and at the front of the plane of the paper. In all alkene reactions, the π-bond electrons play the main role and then any appropriate mechanistic representation should take into account the correct approach of π-bond electrons to the reactant within the plane of the paper (since we are limited to a bi-dimensional representation in the paper) yielding the effective collision. As it was discussed in chapter ten, the effective collision requires reactants to approach each other at the correct orientation and with the minimum energy.

The only alkene's view that can correctly represent the approach between alkene's π-bond electrons and its reactant within the plane of the paper is the side view. On the other hand, from the upper view, the approach between π-bond electrons and its reactant might result in a non-effective collision and does not lead to the expected transition state. In other words, *the alkene's side view is the correct orientation for alkene's reaction yielding an effective collision, while the alkene's upper view is the incorrect orientation for a bi-dimensional representation of a reaction since it yields a non-effective collision.*

As one can see from Fig. 17.1(A), *alkene's reactant has to approach above or below the π-bond electrons for effective collision and only the side view can represent this correctly.* Otherwise, from the upper view, the alkene's reactant (represented by *R*) collides laterally with the alkene and cannot react with its π-bond electrons.

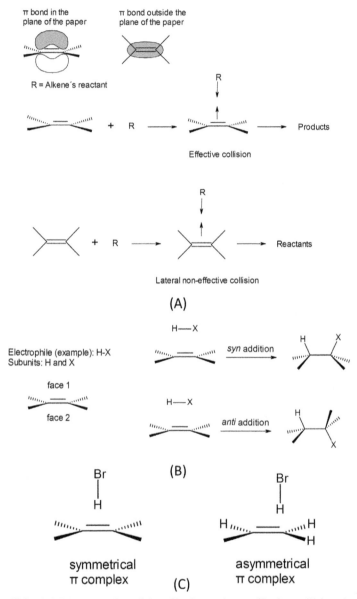

Figure 17.1 (A) Representation of the effective and non-effective collisions involving alkene; (B) representation of *syn* and *anti* addition; (C) representation of π complexes.

The addition of a reactant to alkene occurs by the cleavage of two bonds in reactants (the alkene's π bond and electrophile's σ bond) followed by the formation of two σ bonds (addition of two subunits of the electrophile in both vinylic carbon atoms, one subunit to each of them). By supposing that the electrophile has two subunits (for example, proton and halide ion from hydrogen halide electrophile, so-called HX), the addition of the two subunits of the electrophile in alkene might occur in two different ways: *anti* and *syn* addition (Fig. 17.1(B)). Since alkene has two sp^2 (planar trigonal) carbon atoms, the alkene has two faces. Then, *the two subunits of the electrophile might be added to the same face of the alkene (syn addition) or to opposite faces of the alkene (anti addition).*

All polar (radical-free) addition of electrophiles in alkenes have a π complex as an intermediate prior to the transition state of the rate determining step. When the alkene is symmetrical (ethene, 2,3-dimethylbutene, for example) the electrophile (e.g., HX) is placed in the middle of the π bond. It forms a symmetrical π complex where the interatomic distance between electrophile atom and vinylic carbons are (nearly) equivalent. When the alkene is asymmetrical (propene, methylpropene, for example), the electrophile is directed to the least substituted vinylic carbon (see Fig. 17.1(C)). It forms an asymmetrical π complex where the interatomic distance between electrophile atom and vinylic carbons are rather different. Then, the π complex gives the first orientation for Markovnikov's regioselectivity.

Most electrophilic addition reactions in alkenes supposedly undergo carbocation or derivative of carbocation (e.g., β-chlorocarbocation) intermediate. However, our theoretical results indicate that carbocation is not a critical point in the PES, but only an "instantaneous" entity in the PES, i.e., its lifetime is shorter than that from transition state (which is indeed rather short).

The rate determining step is the transition state precedent to this intermediate or instantaneous entity. *These electrophilic addition reactions in alkenes are an example of late transition state, i.e., the transition state and its subsequent intermediate (or product) have close energy and geometry similarities.* As a consequence, any factor (substituent, for instance) (de)stabilizing the transition state also (de)stabilizes the subsequent intermediate and vice versa because they have close similarities in their geometry. This is the main point in Hammond's postulate (see chapter ten).

An important, comprehensive description of a polar addition mechanism is given in the "Polar addition of hydrogen halide to alkenes: apolar solvent" section in the "second order" subsection.

Due to the fact that in electrophilic addition to alkene one π bond is interchanged into one σ bond in the product, all of these reactions are exothermic (heat is released at the end of the reaction). The entropy of these reactions decrease since we have 2:1 reactant:product ratio, and then the sign of Gibbs free change will be dependent on the reaction temperature (see chapter four).

SYMMETRY/ASYMMETRY IN ALKENES AND MARKOVNIKOV REGIOSELECTIVITY

The Markovnikov regioselectivity in alkene's addition reaction can be reasoned on the intermediate prior to TS_{RDS} or on the TS_{RDS} itself. As for the former (intermediate prior to TS_{RDS}), it is based on the Hammond's postulate for late *transition state (which is the rate determining step in these reactions) that precedes the corresponding carbocation or carbocation derivative (supposing this intermediate could exist).* Nonetheless, the Markovnikov regioselectivity can only be applied to "relatively asymmetric" alkenes.

The terms "relatively asymmetric" alkene and "relatively symmetric" alkene are applied to *the number of alkyl groups attached to the vinylic carbon* in the alkene regardless of its *cis/trans* stereoisomery (if it exists). If one alkyl group is attached to each vinylic carbon (1:1 alkyl group), the alkene is relatively symmetric, in spite of whether both alkyl groups are similar or not. Take for example, *cis*-butene, *trans*-butene, *cis*-pent-2-ene, *trans*-pent-2-ene, *cis*-hex-3-ene, and *trans*-hex-3-ene. If two alkyl groups are attached to each vinylic carbon (2:2 alkyl groups), the alkene is also a relatively symmetric alkene, in spite of whether all alkyl groups are similar or not. Take for example, 2,3-dimethylbut-2-ene, *trans*-3,4-dimethylhex-3-ene, and *cis*-3,4-dimethylhex-3-ene. Another type of relatively symmetric alkene is ethene without any alkyl group (0:0 alkyl group). *Any other alkene (which is not a relatively symmetric alkene with 1:0, 2:0, or 2:1 alkyl groups) is regarded as a relatively asymmetric alkene, and the Markonikov rule is only applied to these alkenes.*

In other words, *relatively symmetric alkenes have zero, one, or two alkyl groups in each vinylic carbon, no matter its stereoisomery or similarity or not in the alkyl groups, while relatively asymmetric alkenes exist in one of the following possibilities: a) one alkyl group in one vinylic carbon and zero alkyl group in another vinylic carbon; or b) two alkyl groups in one vinylic carbon and zero alkyl group in another vinylic carbon; or c) two alkyl groups in one vinylic carbon and one alkyl group in another vinylic carbon.* See examples in Fig. 17.2.

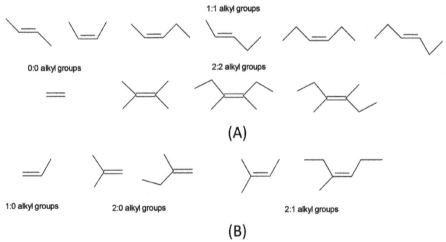

Figure 17.2 Examples of (A) relatively symmetric alkenes and (B) relatively asymmetric alkenes.

When considering the non-existence of carbocation as an intermediate, Markovnikov rule can be understood on the population of two transition states in the rate determining step. The most stable transition state has the larger population which will give a greater amount of corresponding product. The most stable transition state is the one where the (formal) positive partial atomic charge is located in the most substituted vinylic carbon during the rate determining step. This is a more appropriate alternative understanding of Markovnikov rule.

Figure 17.3(A) shows one example of a Markovnikov reaction, that is, a reaction that is ruled by Markovnikov regioselectivity. The 2-methylbut-2-ene reacting with hydrogen bromide gives two transition states because this alkene is relatively asymmetric. The most stable TS (partial charge in tertiary carbon) is formed in larger amount while the less stable TS (partial charge in secondary carbon) is formed in minor amount. As a consequence, the main product is formed from the subsequent step of bromide ion addition to the most stable carbocation. In this reaction, a racemic mixture is the minor product from the addition of bromide ion to the less stable carbocation (in minor amount). On the other hand, for a relatively symmetric alkene (e.g., *trans*-butene) reacting with hydrogen bromide yields only one carbocation and no regioseletivity exists as a consequence (Fig. 17.3(B)).

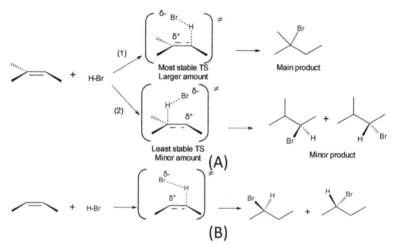

Figure 17.3 (A) Example of Markovnikov reaction and (B) example of non-Markovnikov reaction.

Therefore, the Markovnikov rule is determined kinetically by the transition state of the rate determining step of electrophile addition to a relatively asymmetric alkene which gives two reaction paths. The reaction path with lower barrier is the one which forms the most transition state (where positive partial atomic charge is in the most substituted vinylic carbon). The reaction path with higher barrier gives the less stable transition state (with smaller population). In addition, as indicated in previous section, the Markovnikov regioselectivity starts at the asymmetrical π complex (Fig.17.1(C)), although it is determined at the following transition state.

Figure 17.4 shows the pictorial representation of SEP of the hydrobromination of propene. It forms an asymmetrical π complex followed by two transition states. The low-lying TS forms partial positive charge in a secondary carbon where two alkyl groups delocalize this charge. The high-lying TS forms partial positive charge in a primary carbon where only one alkyl group delocalize this charge which gives a higher energy to this transition state.

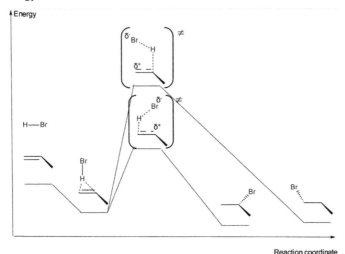

Figure 17.4 Pictorial representation of two reaction paths of a Markovnikov reaction based on the assumption of the existence of carbocation as an intermediate in the PES.

POLAR ADDITION OF HYDROGEN HALIDE TO ALKENES: INTRODUCTION

Hydrogen halides, HX, are gases at room temperature and when dissolved in water, form corresponding acids (hydrohalic acids). Polar addition of hydrogen halide to alkenes can be done by dissolution of gaseous hydrogen halide in solvents such as acetic acid (polar, protic solvent) or dichloromethane/pentane/heptane (apolar solvent) or bubbling HX directly to the alkene. *It is very important to note that the type of solvent used will influence the kinetics of this reaction.* The kinetics of this reaction is also influenced by the reactivity of HX which is: HI > HBr > HCl > HF.

The polar addition of HX is the one that excludes radical addition after removing solvent impurities and reactants are radical-free. The polar addition of hydrogen halide to alkene in apolar solvent might follow several distinguished rate laws under second order, third order, and fourth order kinetics (Mayo and Katz 1947). In fact there is a competition between three mechanisms: second order kinetics (first order in HX), third order kinetics, i.e., one HX catalyzed reaction (second order in HX), and fourth order kinetics, i.e., two HX's catalyzed reaction (third order in HX).

$$r = k_2[alkene][HX]$$

$$r = k_3[alkene][HX]^2$$

$$r = k_4[alkene][HX]^3$$

$$r = k_4[alkene]^2[HX]^2$$

Nonetheless, Mayo and collaborators found that the addition of HX in alkenes in apolar solvents (pentane or heptane) majorly follow fourth order kinetics, being first order in alkene, and third order in hydrogen halide (Mayo and Katz 1947; Mayo and Savoy 1947).

$$r = k_4[alkene][HX]^3$$

Thirty years later, using more advanced techniques and different reaction conditions, Haugh and Dalton also found fourth order kinetics (third order in HCl and first order in propene) for the addition of hydrogen chloride to propene (Haugh and Dalton 1975).

Mayo and Katz found that when traces of water or catalyst are present in the reaction media, the preferred reaction kinetics is second order (Mayo and Katz 1947).

$$r = k_2[alkene][HX]$$

Although very few experimental results were presented, Sergeev and coworkers found that hydrogen bromide addition to liquid 1-alkenes followed a third order kinetics, first order in alkene, and second order in hydrogen bromide (Sergeev et al. 1982).

$$r = k_3[alkene][HX]^2$$

Then, when using apolar solvent or no solvent in polar addition of HX to alkenes, three competing mechanisms exist: (1) second order kinetics (first order in both alkene and HX);(2) third order kinetics (first order in alkene and second order in HX); and (3) fourth order kinetics (first order in alkene and third order in HX).

According to our calculations (Firme 2019), *fourth order kinetics has the lowest barrier and it is preferred in the absence of a catalyst (impurities) and water at low temperatures, while second order kinetics has the highest barrier* (Firme 2019). The enthalpy of activation of hydrogen bromide addition to propene for second order, third order, and fourth order kinetics is 30.06, 23.34, and 19.89 kcal mol^{-1}, respectively. The stabilization of π complex is 3, 5, and 8 kcal mol^{-1} for second, third, and fourth order kinetics, respectively. In the third order and fourth order kinetics, one might say that the reaction is catalyzed by one or two hydrogen halides, respectively. *These hydrogen halides as catalysts form one or two hydrogen bonds in the TS which is/are responsible in decreasing the energy barrier of third and fourth order kinetics, respectively, with respect to that from second order kinetics* (Firme 2019).

For all three competing mechanisms *the Markovnikov regioselectivity is determined by the subsequent TS$_{RDS}$.* As it is observed in IRC calculations (see ahead), *syn addition of hydrogen halide occurs for all rate laws in apolar medium* (Firme 2019).

After TS$_{RDS}$, there is no minimum critical point other than the product. From TS$_{RDS}$ to the product, one might infer that an "instantaneous" carbocation might exist, but it has a lifetime shorter than that of a transition state (which is one vibration lifetime). Then, the halide ion diffusion (which is formed in TS$_{RDS}$) to the opposite face of the "instantaneous" carbocation is slower than its *syn* addition.

We have also done ONIOM calculation using the explicit solvent model (carbon tetrachloride as apolar solvent). The ONIOM is a hybrid method having two or three layers, wherein each layer is a different level of theory which is used in order to enable the calculation of large systems (one layer including solvent molecules and the other layer including carbocation-chloride anion-hydrogen chloride system). After geometry optimization, no carbocation and counter ion was formed but the final product (Firme 2019). Therefore, we can say that *the polar addition of hydrogen halide in alkenes in apolar media occurs through asynchronous concerted mechanism.*

When using polar protic solvent in polar addition of HX to alkenes there are also two competing mechanisms: (1) anti addition (third order kinetics) and (2) syn addition (second order kinetics). Likewise, no carbocation exists as ion pair in simulated polar, protic medium. ONIOM method with explicit solvent model was also used where water was the solvent in the low layer and carbocation of styrene was used with two chloride anions in the high layer. After geometry optimization, one water molecule was added to the carbocation followed by deprotonation of the added water, which yielded an alcohol molecule as the product (Firme 2019). Then, *the mechanism of hydrohalogenation of alkenes in polar, protic medium is also asynchronous concerted.*

POLAR ADDITION OF HYDROGEN HALIDE TO ALKENES: APOLAR SOLVENT

Second Order Kinetics

As mentioned above, there are three competing mechanisms for polar addition of HX to alkenes using apolar solvent. *The second order kinetics has a much higher barrier than third and fourth order kinetics for polar HX addition in apolar solvent.* Then, the polar addition of HX to alkenes in apolar solvent occurs mainly at low temperatures by fourth order kinetics and impurity-free media. However, when there are impurity traces, it was observed that second order kinetics is preferred, since the catalyst will be the impurity instead of one or two hydrogen halide besides the one adding to the alkene (Mayo and Katz 1947).

Let us firstly show the mechanism of the second order kinetics of this reaction. We use hydrogen bromide and propene as an example of this mechanism in Fig. 17.5. However, this is applicable to several other alkenes and also to hydrogen chloride replacing hydrogen bromide.

Firstly, the reactants form a π complex, which is a minimum at PES where HBr is placed vertically towards the alkene's π bond. Then, π bond electrons attack the hydrogen from HBr, promoting H-Br bond cleavage and σ electrons movement towards the bromine atom. During the transition state (which is associated with the rate determining step), alkene's π bond and H-Br σ bond are partially broken to partially form a new bond (between one vinylic carbon and the hydrogen from HBr). Moreover, the opposite vinylic carbon holds a partial positive atomic charge (δ^+) keeping its sp^2 hybridization while the other vinylic carbon forming a new

σ bond changes its hybridization to $sp^{2.5}$ (in a pseudo-tetrahedral geometry) and halogen atom holds a partial negative atomic charge (δ^+).

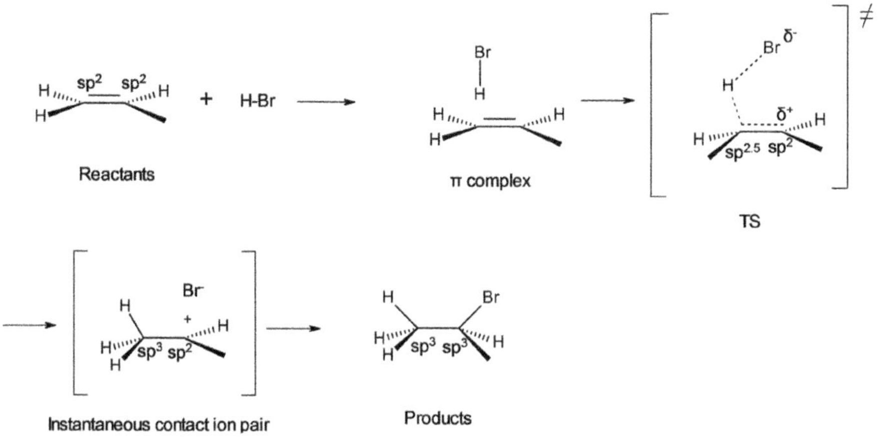

Figure 17.5   Schematic representation of second order mechanism of hydrogen bromide in propene.

It is important to emphasize that a bond being partially broken and/or formed in TS is represented by a dashed line and that for transition states we will use δ instead of q for partial atomic charges (where q is used for atoms in molecules as a minimum at potential energy surface, PES).

After the transition state associated with the rate determining step, TS$_{RDS}$, it forms an "instantaneous" contact ion pair which is a carbocation (with a new sp^3-hybridized carbon and a sp^2-hybridized trivalent carbon or trivalent Csp^2 holding a formal positive atomic charge) and bromide ion as its counter-ion. The term instantaneous is used to represent that this contact ion pair is not a minimum (nor a maximum) at PES. This instantaneous contact ion pair is automatically converted into the product (propyl bromide) during the optimization process (see chapter eight for a better discussion about optimization procedure) without a further transition state. This direct conversion without a transition state is due to the fact that the contact ion pair does not have any sort of physical barrier (such as a solvent molecule) for the attachment between the ions to form a new σ bond. *More straightforwardly, there is no second barrier and no carbocation intermediate in the second order kinetics hydrogen halide addition to alkenes.*

In Fig. 17.6(A), we show the critical points of the PES for the reaction between propene and hydrogen bromide under theoretical second order kinetics from ωB97XD/6-311G++(d,p) level of theory. The π complex is 3 kcal mol^{-1} more stable than the isolated reactants. We can see that the hydrogen atom from hydrogen bromide is pointing towards C1 vinylic carbon (not in the middle of the CC double bond), forming an asymmetrical π complex. Then, π complex is directing the Markovnikov regioselectivity which is determined by the subsequent transition state. Further, QTAIM results prove this interaction between hydrogen (from hydrogen halide) and least substituted vinylic carbon in π complex.

Figure 17.6(B) shows the plot of the corresponding intrinsic reaction coordinate calculation, IRC. *The IRC calculation starts from a previously obtained TS structure to form a critical point before and after this TS.* The IRC method is very important to confirm that the previously obtained TS is the transition sate of the reaction, indeed.

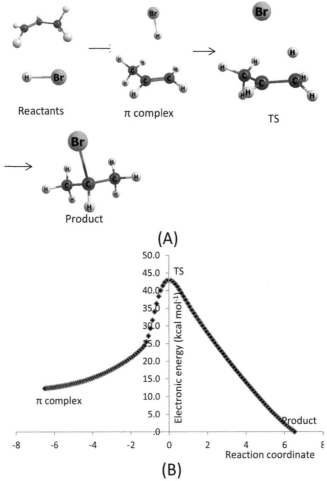

(A)

(B)

Figure 17.6 (A) Initial and final points of IRC calculation of second order kinetics along with optimized TS from the reaction between propene and hydrogen bromide, and its (B) corresponding IRC plot calculated at ωB97XD/6-311G++(d,p) level of theory.

From the TS structure of the second order kinetics towards the final point in the IRC calculation, we observe that there occurs a *syn* addition of hydrogen halide in the alkene. The energy barrier is 30.06 kcal mol^{-1} whose activated complex resembles the schematic structure of TS in Fig. 17.5. After the IRC from calculated TS (in order to confirm the TS), one finds the instantaneous contact ion pair (as depicted in Fig. 17.5), but further optimization calculation of this possible intermediate leads to the product (alkyl bromide). Further attempts to find the second transition state from the contact ion pair failed. Then, the PES of this reaction following the second order kinetics has only one energy barrier. In another words, there is no carbocation intermediate in this reaction. Even tertiary carbocation from methylpropene reacting with hydrogen bromide form instantaneous contact ion pair, which is readily optimized to the corresponding alkyl halide from ωB97XD/6-31G++(d,p) level of theory.

The addition of hydrogen chloride in propene under a theoretical second order kinetics has an energy barrier higher by 3.5 kcal mol^{-1} than that from hydrogen bromide addition, but the mechanism is the same as depicted for the reaction between hydrogen bromide and propene.

Third Order Kinetics

In third order kinetics there are two possible transition states: one forming a six-membered ring and the other nearly similar to that from fourth order kinetics, but missing one hydrogen halide. Nonetheless, only the latter has been confirmed for further IRC calculation (Firme 2019).

The third order kinetics needs further experimental study since there are few presented results that confirm this mechanism (Sergeev et al. 1982). However, *this mechanism (third order kinetics) is quite plausible, since it has an intermediate barrier between second order and fourth order kinetics for hydrogen halide addition to alkene.*

The mechanism of third order kinetics (Fig. 17.7(A)) is quite similar to that from fourth order kinetics (Fig. 17.8), except for the fact that in the former there is only one hydrogen halide as catalyst in both π complex and transition state, while in the latter there are two hydrogen halide catalysts in both π complex and transition state.

Similar to the mechanism of second order kinetics, there is no second barrier and carbocation as an intermediate, since the IRC results indicate that after the TS$_{RDS}$ it forms the products directly (Fig. 17.8(B)). The IRC calculations *also indicate the syn addition of hydrogen bromide to propene, that is, the addition of both proton and bromide ion to the same face of propene.* Similar to second order kinetics, the Markovnikov regioselectivity is initially directed from π complex and determined by the subsequent transition state. The QTAIM results in both critical points show the interaction between hydrogen (from hydrogen halide) and least substituted vinylic carbon in both critical points (Fig. 17.7(C)). Figure 17.7(C) also depicts the interaction between bromine atom from catalyst hydrogen bromide (the one which does not participate in the addition directly) and the vinylic hydrogen bonded to the positively charged carbon in the TS. This intermolecular interaction does not exist in the TS of the second order kinetics and it plays a secondary role in lowering the energy barrier of the third order kinetics with respect to second order kinetics. The major factor for lowering the energy barrier on third order kinetics is the delocalization of the negative charge of the bromide ion to the second molecule of hydrogen bromide by means of the hydrogen bond. A similar reasoning exists in the fourth order kinetics as well.

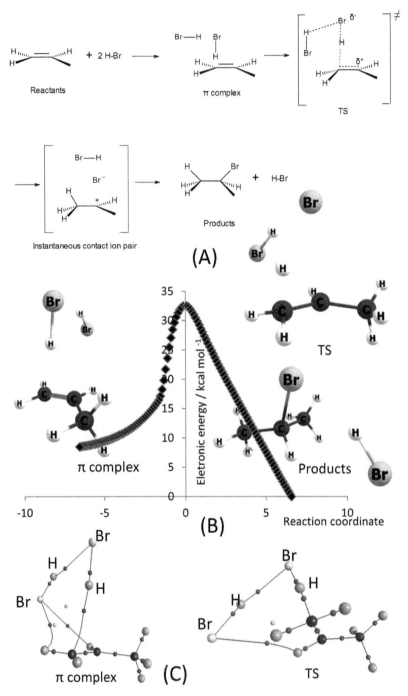

Figure 17.7 (A) Mechanism of third order kinetics of hydrogen bromide addition to propene; (B) corresponding IRC calculation from ωB97XD/6-311G++(d,p) level of theory; and (C) corresponding QTAIM virial graphs from π complex and transition state critical points.

Fourth Order Kinetics

Mayo and Savoy found a fourth order kinetics for the reaction of hydrogen bromide and propene in pentane solution from −78 to 10°C. The kinetics of this reaction is first order in propene and third order in hydrogen bromide (Mayo and Savoy 1947). Similar kinetics were also found for the hydrogen chloride addition to isobutylene or methylpropene in heptane as solvent at 0°C (Mayo and Katz 1947). In both cases all reactants and solvents were distilled to become free of water, oxygen, and other impurities in order to avoid side reactions (abnormal additions). The rate law or the rate expression, r, for hydrogen halide addition to alkene in apolar solvent is shown below.

$$r = k[alkene][HX]^3 \therefore HX = \textbf{HBr or HCl} \therefore \textbf{Apolar solvent}$$

The fourth order kinetics for the hydrogen halide addition to alkenes is the preferred mechanism at low temperatures because of its lower barrier (10 and 7 kcal mol^{-1} lower than that from second order and third order kinetics, respectively). Figure 17.8 shows the mechanism of the fourth order kinetics between propene and hydrogen bromide. Unlike second order and third order kinetics (where a *syn* addition occurs), in fourth order kinetics the *anti* addition happens (see more details in Fig. 17.9(A)), but this is dependent on the level of theory for the IRC calculation (Firme 2019). Similarly to the second and third order mechanisms, it forms a π complex where the hydrogen atom from one hydrogen bromide is directed towards C1 vinylic carbon (asymmetrical π complex) of propene while two hydrogen bromide molecules and one bromine atom interact with each other through two hydrogen bonds. The calculated geometry of this π complex is shown in Fig. 17.9(A). This π complex is 8 kcal mol^{-1} more stable than their isolated molecules while for second and third order kinetics the stabilization of π complex is 3 and 5 kcal mol^{-1}, respectively.

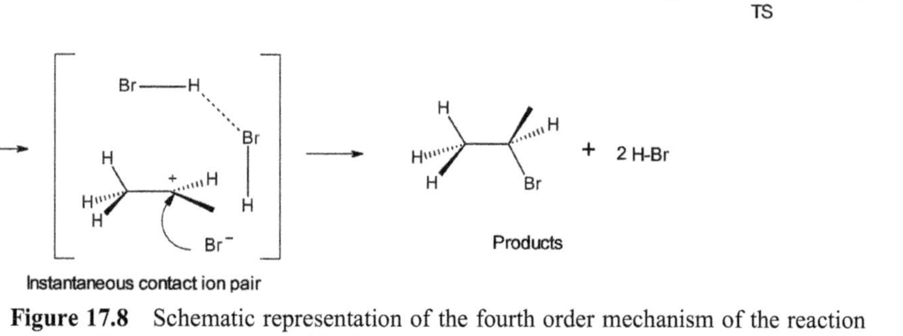

Figure 17.8 Schematic representation of the fourth order mechanism of the reaction between propene and hydrogen bromide showing the *anti* addition.

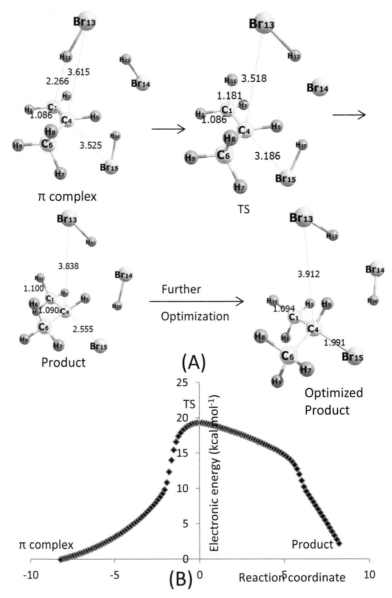

Figure 17.9 (A) Initial and final points of IRC calculation of fourth order kinetics along with optimized TS from the reaction between propene and hydrogen bromide and its (B) corresponding IRC plot calculated at ωB97XD/6-311G++(d,p) level of theory. Bond lengths or interatomic distance, in Angstroms, are shown in dashed lines.

Soon after the π complex, it forms the transition state of the rate determining step where CC π bond and HBr σ bond are partly broken, while σ bond is partly formed between C1 vinylic carbon (with $sp^{2.5}$ hybridization) and hydrogen (from former HBr). In this TS, C2 vinylic carbon has sp^2 hybridization and partial positive charge, while bromine atom has a partial negative charge. After this TS, an instantaneous contact ion pair is formed (which was not observed computationally as a minimum nor a maximum at PES) which leads to the product without any further barrier.

The final point of IRC calculation undergoes further optimization to yield the corresponding alkyl bromide. Then, as already observed for second and third order kinetics, the formation of carbocation is not observed computationally in gas phase (nor in condensed phase from ONIOM calculations), which reinforces experimental observations. However, *in all cases (all three rate laws), the Markovnikov regioselectivity is dictated by the TS_{RDS} since the preferred TS (of lowest energy barrier) has partial positive charge in the most substituted vinylic carbon, yielding 2-chloropropane as product.*

Another important observation during IRC calculation using ωB97XD/6-311G++(d,p) level of theory in fourth order kinetics regards the *anti* addition for this mechanism. While hydrogen (from hydrogen bromide, here named H11) is added to one face of propene, the addition of bromine atom occurs from the other face of propene. However, the bromine atom added to propene is not that bonded to H11, but bromine atom from the other hydrogen bromide (namely Br15). From π complex to the optimized product (Fig. 17.9(A)), one can see the interatomic distance between Br15 and C4 decreasing from 3.525 to 1.1991 Å. However, other levels of theory, MP2/6-311G++(d,p) and ωB97XD/def2-ZVTP, indicate through IRC calculation that the *syn* addition occurs in fourth order kinetics. Accordingly, all levels of theory indicated the *syn* addition of hydrochlorination of alkenes in apolar medium (Firme 2019). *Then, most probably, only syn addition occurs in apolar solvent.*

When using chloride hydrogen reacting with propene in fourth order kinetics, the energy barrier is also 3 kcal mol^{-1} higher than that from bromide hydrogen in same rate law and also 10 kcal mol^{-1} lower than the same reaction between HCl and propene in second order mechanism. Then, *fourth order kinetics has 10 kcal mol^{-1} lower barrier than second order kinetics for both hydrogen halides (HBr and HCl), and reaction with HBr has 3 kcal mol^{-1} lower barrier than that with HCl in both second order and fourth order kinetics.*

POLAR ADDITION OF HYDROGEN HALIDE TO ALKENES: POLAR, PROTIC SOLVENT

Syn and *Anti* Addition

The rate of hydrogen chloride addition to alkenes is greater in apolar solvents than in aprotic, polar solvents, for example, ether as solvent (O'Connor et al. 1939). Later, Fahey and coworkers observed that when using polar, protic solvent (acetic acid or acetic acid and water), the kinetic law is different from that in apolar solvent.

One of the most used polar, protic solvents for this reaction are acetic acid and acetic acid/water. Hydrogen bromide and hydrogen chloride are stronger than acetic acid. Hence, hydrogen bromide or hydrogen chloride are partly deprotonated into bromide ion or chloride ion when reacting with acetic acid or water (Fig. 17.10(A)). Then, there are free halide ions in this reaction medium. Alternatively, one may add halide salt in the reaction medium which gives the same result as using polar, protic solvent.

Fahey and colaborators have done a thorough kinetic experimental study about the hydrogen chloride addition to cyclohexene (Fahey et al. 1970), to deutered cyclohexene (Fahey and Monahan 1970) and to 1,2-dimethylcyclohexene (Fahey and McPherson 1971). They concluded that *two mechanisms occur simultaneously when hydrogen halide addition to alkene occurs in polar, protic solvent: syn addition under second order kinetics and anti addition under third order kinetics.*

$r = k[alkene][\mathrm{HX}] \therefore \mathrm{HX} = \mathbf{HBr} \text{ or } \mathbf{HCl} \therefore \textbf{Polar solvent (syn addition)}$

$r = k[alkene][\mathrm{HX}][X^-] \therefore \textbf{Polar solvent (anti addition)}$

By knowing that every alkene has two faces of its π bond divided by a plane passing through both vinylic carbons and its substituents, then it is possible that both atoms of hydrogen halide attack the same face of the π bond (*syn* addition) or opposite faces of the π bond (*anti* addition).

The *syn* addition means that both the halide ion (bromide ion or chloride ion) and proton are attached to the alkene at the same face of its π bond. Since both atoms attached come from the same hydrogen halide, it has a second order kinetics. On the other hand, in *anti* addition one halide ion (from a previous deprotonated hydrogen halide or added halide salt) is attached to one face of the alkene's π bond, and the proton from the other hydrogen halide molecule is attached to the other face of the alkene's π bond. Hence, *anti* addition has a third order kinetics (first order in alkene, first order in hydrogen halide, and first order in halide ion). The first order in halide ion means that adding a salt of the halide will increase the reaction rate. It is important to emphasize that both kinetics occur simultaneously. A third reaction also occurs: the *anti* addition of acetic acid in alkene (solvolysis). Then, the overall reaction rate expression (rate law) is rather complicated. It is *important to note that the use of cycloalkene rather than open chain alkene is important to verify the anti/ syn addition. Otherwise, it would be impossible.*

Figure 17.10(B) shows the mechanism of the *anti* addition of hydrogen bromide to cyclohexene. Firstly, cyclohexene forms a symmetrical π complex with hydrogen bromide while another bromide ion is interacting with hydrogen atoms of cyclohexene. Next, the $\mathrm{TS_{RDS}}$ happens by addition of the proton (from hydrogen bromide) to one of the vinylic carbons, leaving the other vinylic carbon with a partial positive charge and the bromine atom with a partial negative charge. An instantaneous contact ion pair is formed subsequently, but the instantaneous carbocation is readily attacked by the other bromide ion to form the cyclohexyl bromide, also in accordance with experimental results (where the existence of carbocation was not observed).

Figure 17.10(C) shows the mechanism of the *syn* addition of hydrogen bromide to cyclohexene, which partly resembles that from *anti* addition, except for the absence of an expectant bromide ion to attack the other face of the instantaneous carbocation.

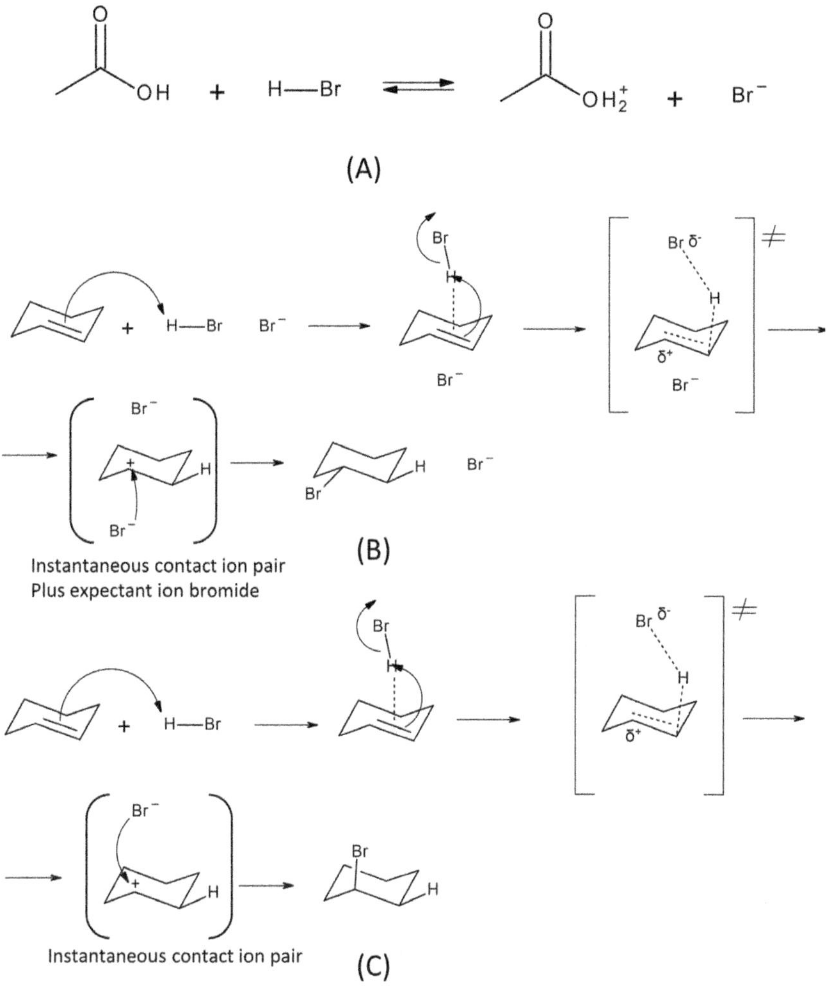

(A)

Instantaneous contact ion pair
Plus expectant ion bromide

(B)

Instantaneous contact ion pair

(C)

Figure 17.10 (A) Equilibrium reaction between bromide hydrogen and acetic acid; (B) *anti* mechanism; and (C) *syn* mechanism of hydrogen bromide addition to cyclohexene in polar solvent.

The *anti* and *syn* addition of HBr to cyclohexene were calculated at ωB97XD/6-31G++(d,p) and ωB97XD/6-311G++(d,p) levels of theory and the results are qualitatively the same (Firme 2019).

Figures 17.11(A) and (B) show the critical points of these reactions. Both π complexes are quite similar except for the absence of bromide ion in *syn* addition. Nonetheless, the TS$_{RDS}$ of *anti* addition is characteristic of an early TS (whose TS geometry is quite close to those from its reactants), and TS$_{RDS}$ of *syn* addition is characteristic of late TS whose geometry is closer to the subsequent intermediate or product. In fact, the energy barrier for the *anti* addition is very low (2.89 kcal mol^{-1}), which is another feature of an early TS, while the energy barrier for the *syn* addition is much higher (32.31 kcal mol^{-1}), which is associated with the late TS. In fact, there is a stronger intermolecular interaction between bromide ion (from opposite face) and positively charged carbon atom in the TS of the *anti* addition. On the other hand, in the TS of *syn* addition there is no interaction between bromine atom and positively charged atom (Firme 2019). The PES of both reactions are depicted in Fig. 17.12. Once again, there is no indication of carbocation formation in these reactions. From the TS$_{RDS}$ the subsequent instantaneous contact ion pair readily forms the alkyl bromide product without any previous intermediate or energy barrier.

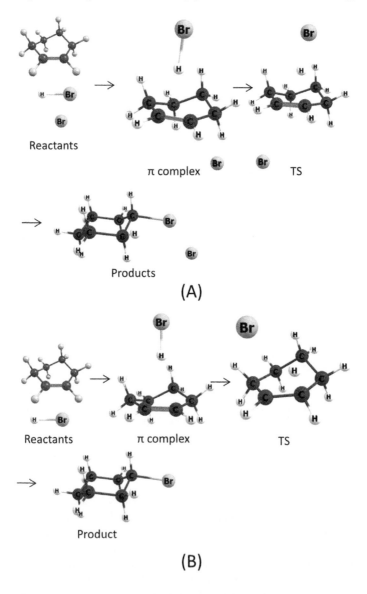

Figure 17.11 Calculated critical points of the PES from hydrogen bromide (A) *anti* addition and (B) *syn* addition to cyclohexene from ωB97XD/6-31G++(*d,p*) level of theory.

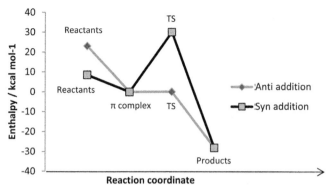

Figure 17.12   PES of the *anti* and *syn* addition of hydrogen bromide to cyclohexene.

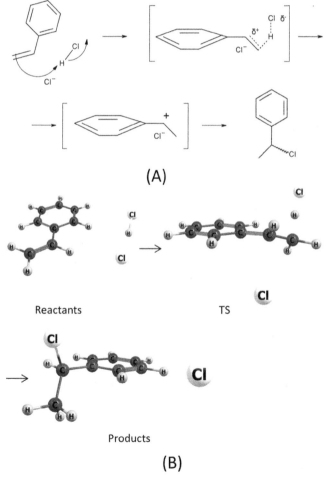

Figure 17.13   (A) Proposed mechanism of the hydrochlorination of styrene under fourth order kinetics; (B) critical points of PES of hydrochlorination of styrene under fourth order kinetics from ωB97XD/6-31G++(d,p) level of theory.

Fahey and McPherson alleged a possible existence of the carbocation intermediate for the hydrochlorination of styrene in polar solvent. However, they hadn't isolated and characterized this intermediate (Fahey and McPherson 1969). Our DFT calculations have shown that there is neither π-complex in this reaction nor the carbocation intermediate since there is only one transition state under the fourth order kinetics, as it is shown in Fig. 17.13. Like all other previous reactions of hydrobromination/hydrochlorination of alkenes, an instantaneous contact ion pair (which is not a critical point in the PES) becomes the product automatically (without second barrier and intermediacy). The energy barrier is 25.6 kcal mol^{-1}.

REARRANGEMENT IN ADDITION OF HX TO ALKENES: POLAR SOLVENT

For some alkenes it is possible to pass through rearrangement towards hydrogen halide addition. For example, the addition of HCl to 3,3-dimethylbut-1-ene using acetic acid as a solvent gives three products: 2-chloro-2,3-dimethylbutane (from rearrangement), 2-chloro-3,3-dimethylbutane (*anti* addition of HCl), and a third product from the *anti* addition of acetic acid (Fahey and McPherson 1969).

We have exchanged chlorine into bromine in our calculations, but the mechanism does not tend to change. The normal addition of HBr to form 2-bromo-3,3-dimethylbutane pass through π complex and a barrier of 7.15 kcal mol^{-1} by third order kinetics (first order in alkene, first order in hydrogen halide, and first order in halide ion) and 27.98 kcal mol^{-1} by a second order kinetics (first order in alkene and first order in hydrogen halide) from ωB97XD/6-311G++(d,p) level of theory, since both mechanisms are feasible in polar, protic media. From our IRC calculations, no carbocation exists from the corresponding transition state. The proposed mechanism for normal addition is shown in the upper part of Fig. 17.14(A).

We observed from IRC calculations that the rearrangement reaction (methyl shift) is only possible from 2-bromo-3,3-dimethylbutane as starting material (Firme 2019). Then, the product of normal addition of HX to 3,3-dimethylbut-1-ene needs to be formed so that a methyl shift from C3 towards C2 might occur subsequently. We found the transition state for this rearrangement (see Fig. 17.14(B)) and enthalpy barrier is 53.02 kcal mol^{-1} to form 2-bromo-2,3-dimethylbutane. The corresponding mechanism for this rearrangement is shown in the lower part of the Fig. 17.14(A).

ACID-CATALYZED HYDRATION OF ALKENES

Without acid catalysis, water alone cannot react with alkene. In this case both water and alkene in condensed phase form two heterogeneous mixtures. When an oxyacid (nitric acid or sulfuric acid) is added (usually at concentrations nearly 1M or higher), the hydration of alkene yields alcohol as a product. Nonetheless, be aware that the acid catalyst cannot be a hydrogen halide since it is a competing reagent for alkene's reaction.

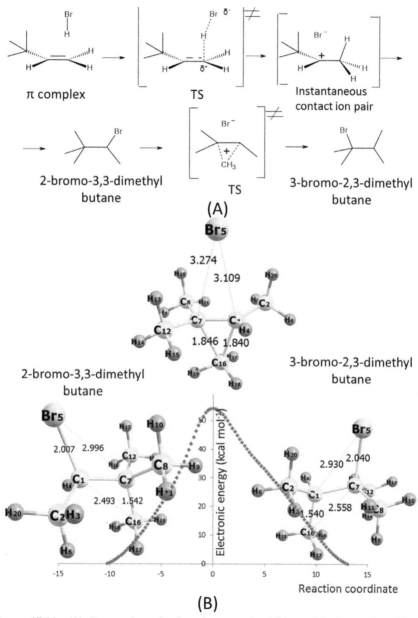

Figure 17.14 (A) Proposed mechanism for normal addition of hydrogen bromide to 3,3-dimethylbut-1-ene, followed by rearrangement from first product towards second product. (B) Calculated transition state from rearrangement reaction and corresponding IRC plot from ωB97XD/6-311G++(d,p) level of theory.

Relatively asymmetric alkenes follow Markovnikov regioselectivity. However, no carbocation was confirmed as an intermediate in the hydration of alkenes although some experimental results indicated a positively charged character in the reaction (Oyama and Tidwell 1976), which might be associated with the transition state, not to the carbocation. Chiang and Kresge alleged the existence of carbocation or a bridged hydro-carbocation in acid catalyzed hydration of 1-methyl-cis- and trans-cyclooctene (Chiang et al. 1984). However, their kinetic study lacks further investigation to prove the existence of the carbocation or bridged hydro-carbocation because: (i) it is an indirect observation; (ii) no consideration of a possible tunneling effect for proton transfer; (iii) no consideration of the π complex. (iv) no consideration of two reaction paths for this reaction. Then, it lacks a direct observation of the carbocation in acid-catalyzed hydration of alkenes.

Our quantum chemistry calculations indicate the non-existence of carbocation intermediate (see ahead). Theoretical results indicate that the transition state of the rate determining step is a late transition state with a developed partial atomic charge in one of the vinylic carbon, which could explain the above-mentioned experimental results about the positive character in TS.

Experimental evidences indicate that acid-catalyzed hydration of alkenes has a second order kinetics (first order in alkene and first order in the acid) and that the protonation of alkene is the rate determining step (Lucas and Liu 1934). These evidences are in agreement with quantum chemistry results (see ahead). Experimental and theoretical results also agree that acid-catalyzed hydration of alkenes passes through a π complex before the transition state of the rate determining step (see ahead).

Experimental results also indicate that this acid-catalysis is a general acid catalysis, which means that both hydronium ion and oxyacid concentrations influence the rate of the reaction (Kresge et al. 1971). In another words, there are two reaction paths for the hydration of alkenes depending on the protonating agent: one with hydronium ion prototating the alkene and the other with oxyacid protonating the alkene.

Let HA be the oxyacid catalyst, then the first step of the reaction has two distinguished paths. Here, we are regarding the protonation of water as an aside step.

$$HA + H_2O \rightleftarrows H_3O^+ + A^-$$
$$H_2C=CH_2 + H_3O^+ \rightleftarrows [H_3C-C^+H_2 \cdots H_2O]$$
$$H_2C=CH_2 + HA \rightleftarrows [H_3C-C^+H_2 \cdots A^-]$$

The rate law for acid-catalyzed hydration of alkene is:

$$r = k[HA][alkene]$$
$$[HA] = \frac{[H_3O^+][A^-]}{K_a}$$

Further experimental evidence indicates that the rate of hydration of alkene increases with the electron-donating ability of the groups attached to the vinylic carbon atoms. This is also in agreement with theoretical results.

The mechanism of the alkene's hydration is given in Fig. 17.15, where nitric acid is used as catalyst. There are two possible reaction paths. In the first path water is protonated by nitric acid, yielding hydronium ion which forms a π complex with alkene subsequently. It is succeeded by the transition state of the rate determining step, TS$_{RDS}$, giving protonated alcohol without precedent intermediate (such as carbocation and water). At last, protonated alcohol is deprotonated by nitrate ion, forming the corresponding alcohol. In the second path, the protonating agent is the nitric acid passing through similar critical points than those using hydronium ion until the TS$_{RDS}$. In the instantaneous intermediate, water preferably attacks the carbocation since it is more nucleophilic than nitric acid. Subsequent critical points are similar to those from the first path.

It is important to add that Chiang and Kresge alleged the existence of the tertiary carbocation as an intermediate for this reaction based on the comparison with the kinetic study of interconversion of 1-methylcyclooctyl system (Chiang et al. 1984). This kinetic study itself needs further investigation to prove the existence of carbocation/bridged-carbocation for this cyclic alkene. In fact they themselves added that "no direct information on the lifetimes of simple aliphatic carbocations in aqueous solution is available, but indirect, order-of-magnitude estimates can be made" (Chiang and Kresge 1985).

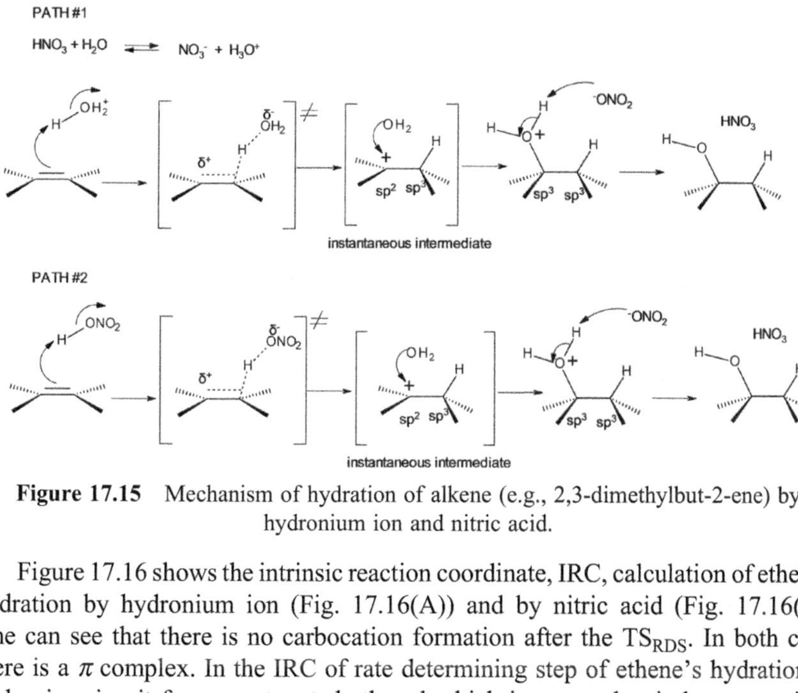

Figure 17.15 Mechanism of hydration of alkene (e.g., 2,3-dimethylbut-2-ene) by hydronium ion and nitric acid.

Figure 17.16 shows the intrinsic reaction coordinate, IRC, calculation of ethene's hydration by hydronium ion (Fig. 17.16(A)) and by nitric acid (Fig. 17.16(B)). One can see that there is no carbocation formation after the TS$_{RDS}$. In both cases there is a π complex. In the IRC of rate determining step of ethene's hydration by hydronium ion it forms protonated ethanol which in a next barrierless step yields ethanol. In the IRC of ethene's hydration by nitric acid it forms one product which is not the ethanol protonated since water is absent. However, in an actual reaction water is present in the media and since it is more nucleophilic than nitrate ion, it will preferably be added to ethyl moiety forming protonated ethanol. We can see that hydration by hydronium ion has 15 kcal mol^{-1} lower barrier than that from hydration by nitric acid. Then, hydration by hydronium ion is more reactive than by nitric acid and probably other oxyanions as well.

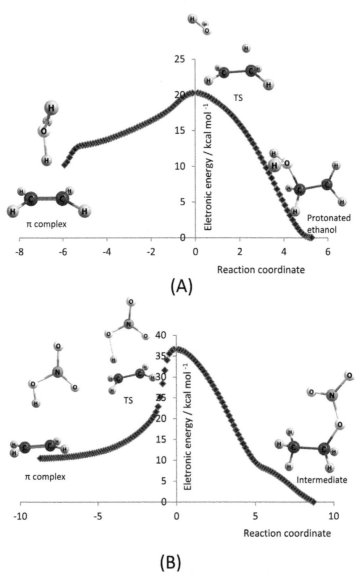

Figure 17.16 Intrinsic reaction coordinate, IRC, of rate determining step of acid-catalyzed hydration of ethene by (A) hydronium ion from MP2/6-311G++(*d,p*) and by (B) nitric acic from ωB97XD/6-311G++(*d,p*) level of theory along with the corresponding structures of π complex, transition state, and intermediate.

We have also performed the transition state calculation of rate determining step of alkene's hydration by hydronium ion and by nitric acid in propene, but-1-ene, *trans*-butene, and 2-methyl-but-2-ene. We have used both MP2/6-311G++(*d,p*) and ωB97XD/6-311G++(*d,p*) level of theories. Although we have obtained possible candidate for transition state in all cases (with just one imaginary frequency and its vibration which could be associated with the reaction coordinate – see chapter eight), only two transition states were confirmed by corresponding IRC calculation of the rate determining step. It is important to emphasize that IRC calculation is used to confirm or deny the possible transition state. These two confirmed transition states (besides those with ethene) are depicted in Fig. 17.17, showing their corresponding IRCs of the rate determining step. They are the IRC of hydration of propene by

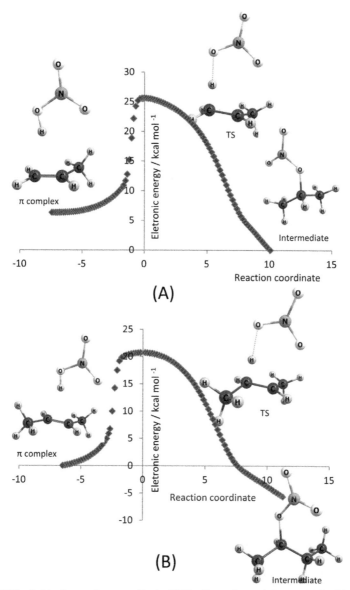

Figure 17.17 Intrinsic reaction coordinate, IRC, of rate determining step of acid-catalyzed hydration by nitric acid of (A) propene and (B) *trans*-butene from ωB97XD/6-311G++(d,p) level of theory along with the corresponding structures of π complex, transition state, and intermediate.

nitric acid (Fig. 17.17(A)) and the IRC of hydration of *trans*-butene by nitric acid (Fig. 17.17(B)). An important observation is the reactivity trend of alkenes. *As the alkene is more substituted it becomes more reactive, that is, the reactivity trend for alkene's hydration is: trans-butene (two substituents at vinylic carbons) > propene (one substituent at vinylic carbon) > ethene (zero substituent at vinylic carbon).*

From both Figs. 17.16 and 17.17, one can see that the vinylic carbon forming a new bond with proton from HX has a pseudo-pyramidal geometry as a characteristic of $sp^{2.5}$ hybridization at the TS_{RDS}. *Then, since TS_{RDS} of hydration of alkenes is a late transition state holding a partial positive charge in one carbon (former vinylic carbon), the stabilization features of the carbocations are similar to the corresponding TS_{RDS}.* One important stabilization factor for carbocation (and its precedent late TS) as stated before is: the more substituted the carbocation (and/or its precedent late transition state) the more stable it is (they are). This accounts for the reactivity trend shown in the last paragraph.

POLAR ADDITION OF HALOGEN TO ALKENES: THERMOCHEMISTRY

Polar addition of halogens to alkenes is an exothermic reaction for most of the halogen molecules. From ωB97XD/6-311G++(d,p) level of theory, fluorination, chlorination, and bromination of *trans*-butene have the following values: −130.65, −51.71, and −30.50 kcal mol^{-1} at gas phase. The experimental enthalpy change for iodine addition to propene is 11.51 and 14.70 kcal mol^{-1} at lowest and highest reaction temperatures, respectively, i.e., at 365.5 and 436.6 K (Benson and Amano 1962). Even through the endergonicity and endothermocity of iodination of alkenes, no oxidizing agent was used for the iodination of ethene, propene, and cyclopropane (Benson and Amano 1962). *Then, as the halogen atomic volume increases, the exothermicity from the corresponding addition reaction in alkenes decreases.* Due to the high exothermicity of fluorination in alkenes, the kinetics of this reaction is very difficult to be evaluated other than at very low temperatures (Merritt 1966). Due to fluorine's small atomic volume, fluoration undergoes a two-step reaction with β-fluorocarbocation as an intermediate or, at least, instantaneous intermediate instead of a fluoronium ion (see ahead), or a concerted mechanism in a *syn* addition (Carroll 2010).

POLAR ADDITION OF BROMINE TO ALKENES

Addition of bromine, Br_2, to alkene passes through a cationic three membered ring called bromonium ion in many reactions. The existence of this intermediate was confirmed by X-ray diffractometry from bromine reacting with adamantylideneadamantane (Slebocka-Tilk et al. 1985). Other experimental evidence (Rolston and Yates 1969) is the addition of bromine to *cis*-but-2-ene (yielding a racemic product) and to *trans*-but-2-ene (yielding a meso product), as depicted in Fig. 17.18.

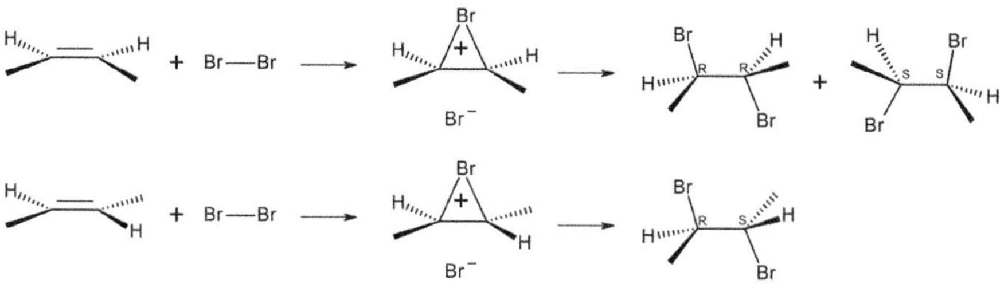

Figure 17.18   Electrophilic addition of bromine to *cis*-but-2-ene and *trans*-but-2-ene.

Experimental evidences also indicate the formation of bromine-alkene π complex (Bellucci et al. 1985a) prior to the TS_{RDS}. Our calculations (see ahead) also indicate the formation of this π complex.

The general rate law for bromination of alkenes is shown below:

$$r = (k_2[Br_2] + k_3[Br_2]^2 + k_{Br_3^-}[Br_3^-])[alkene]$$

The third term in the parentheses is only important when bromide salt is added in the reaction media (Bellucci et al. 1985b). Otherwise only second order kinetics (k_2) and third order kinetics (k_3) are important. *For apolar solvents, it has been shown that second order (k_2) and third order (k_3) mechanisms might occur simultaneously in bromination of alkene* (Fukuzumi and Kochi 1982, Schmidi and Toyonaga, 1984). In the bromination of cyclohexene in 1,2-dichloroethane, the reaction was exclusively a third order kinetics, second order in bromine, and first order in cyclohexene (Bellucci et al. 1985a).

$$r = (k_2[Br_2] + k_3[Br_2]^2)[alkene]$$

or

$$r = (k_2[Br_2])[alkene]$$

As Carroll noted: "Even though there is experimental evidence for the existence of bromonium ions, it need not be the case that all bromine addition reactions involve identical intermediates" (Carroll 2010). See ahead for more discussion about bromonium ion. The existence of open β-bromocarbocation for bromination of ethene has already been proved theoretically (Hamilton and Schaefer 1990). Our calculations from ωB97XD/6-311G++(*d,p*) level of theory also indicate that propene does not form bromine ion as well (see ahead). The inexistence of bromonium ion was also proved in bromination of styrene derivatives, where the intermediate is a derivative of benzylic carbocation (Yates et al. 1973).

The π complex of bromine in alkenes might involve one (Fig. 17.19(A)) or two (Fig. 17.19(B)) bromine molecules interacting with the alkene. The second bromine molecule interacts with the other bromine molecule as well as two hydrogen atoms from *trans*-butene and the bromine atom closer to the π bond interacts with it as depicted in the molecular graph of this complex (Fig. 17.19(C)).

From ωB97XD/6-311G++(*d,p*) level of theory, bromonium ion is not formed when bromine molecule reacts with propene under third order kinetics (second order in bromine), neither when tribromide ion (Br_3^-) is placed above (Fig. 17.20(A)) nor

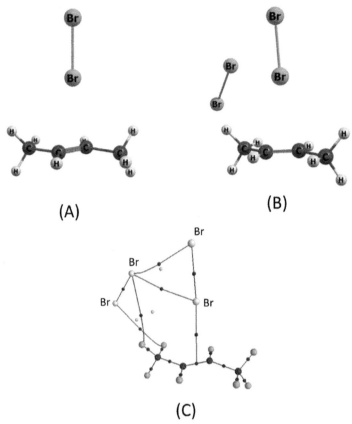

Figure 17.19 Optimized geometries of bromine and *trans*-butene π complex with (A) one and (B) two bromine molecules from ωB97XD/6-311G++(*d,p*) level of theory; (C) QTAIM molecular graph of two bromine and *trans*-butene π complex.

below (Fig. 17.20(B)) the bromine atom forming a three-membered ring as initial geometry prior to the optimization procedure. When these initial structures are optimized they form π complex or 1-bromo-2-tribromide propyl anion, respectively. Likewise, under second order kinetics (first order in bromine molecule), when the bromide ion is placed above the initial, random bromonium ion it forms a π complex and when the bromide ion is placed below the initial, random bromonium ion it forms the final *anti*-dibromide product (α,β-dibromide product). The unique situation where brominium ion is formed with propene is when it is isolated from its counterion (Fig. 17.20(C)).

On the other hand, *trans*-butene and *cis*-butene do form bromonium ion in third order kinetics at ωB97XD/6-311G++(*d,p*) level of theory when tribromide ion is placed above the bromonium ion (Fig. 17.21(B)). When the tribromide ion is placed below the initial, random bromonium ion it forms the final *anti*-product (α,β-dibromide product) after optimization without any barrier (Fig. 17.21(A)). A similar outcome occurs with bromide ion in second order kinetics. When the bromide

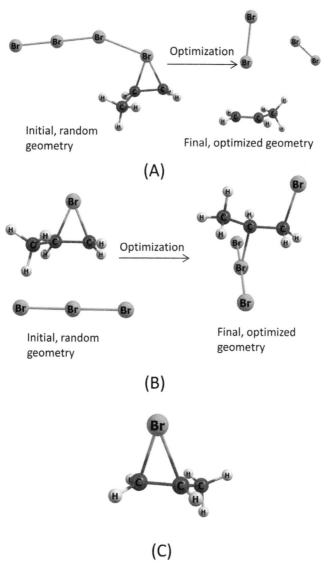

Figure 17.20 Initial, random geometry of tribromide ion placed (A) above or (B) below the bromonium ion (from propene adduct) along with their corresponding final, optimized geometries; (C) optimized geometry of bromonium ion from propene isolated from its counterion from ωB97XD/6-311G++(d,p) level of theory.

ion is placed above the initial, random bromonium ion it forms a π complex after optimization (Fig. 17.21(C)). Like the case of propene, but-2-ene forms bromonium ion when isolated from its counterion (Fig. 17.21(D)). Then, *depending on the position of the tribromide ion from third order kinetics, the final anti α,β-dibromide product can be formed without any further barrier or the bromonium ion is formed prior to the anti addition of the second bromine atom without any further barrier.*

In both cases of third order kinetics (plus the case of second order kinetics) of bromine addition to alkene, *anti* addition always happens. Probably, a dynamical process can give the barrier for the formation of bromonium ion, which is the TS_{RDS} of this reaction. Conversely, the TS_{RDS} relative to the formation of chloronium ion was obtained at ωB97XD/6-311G++(d,p) level of theory (see ahead).

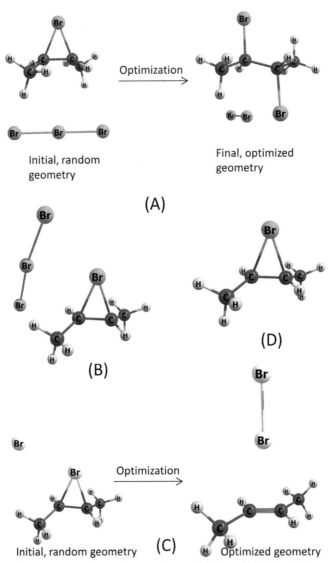

Figure 17.21 (A) Initial, random geometry of tribromide ion placed below the bromonium ion (from *trans*-butene adduct) along with their corresponding final, optimized geometries (minimum optimization); optimized geometries of (B) tribromide ion and bromonium ion from bromine addition to *trans*-butene; (C) initial, random geometry of bromide ion placed above the bromonium ion along with its final, optimized geometries; (D) optimized geometry of bromonium ion isolated from its counterion from ωB97XD/6-311G++(d,p) level of theory.

From second order kinetics, we observe that mono-, di-, tri-, and tetramethyl substituted alkenes undergo bromonium ion as intermediate. But-1-ene, 2-methylpropene, 2-methylbut-1-ene and 2,3-dimethylbut-1-ene form bromonium ion at ωB97XD/6-311G++(d,p) level of theory in Figs. 17.22 (A), (B), (C) and (D), respectively. In a relatively asymmetric alkene, the more substituted vinylic carbon has the longest C-Br bond length (i.e., the weakest C-Br bond) and it also has the highest positive charge density. Then, the most asymmetric alkene (2:0 substituted vinylic carbons) forms the bromonim ion closest to the geometry of a β-bromocarbocation.

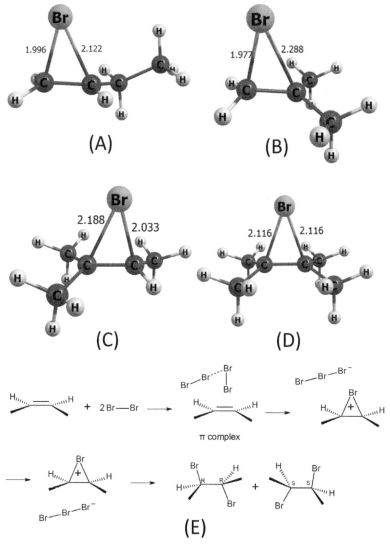

Figure 17.22 Optimized geometries of bromonium ion from (A) But-1-ene, (B) 2-methylpropene, (C) 2-methylbut-1-ene, (D) 2,3-dimethylbut-1-ene, and (E) bromination mechanism.

From the theoretical and experimental evidences, the mechanism of bromination from third order kinetics is given in Fig. 17.22(E). Firstly, the π complex is formed with two bromine molecules followed by the formation of bromonium ion and tribromide ion above where the barrier might be obtained dynamically. Afterwards, the tribromide ion migrates to be underneath the bromonium ion and a barrierless passage to the final products occurs.

The transition state of the first step of bromination to alkenes (propene and *trans*-butene) was not obtained in both third order and second order kinetics from ωB97XD/6-311G++(d,p) level of theory. *As mentioned above, polar addition of bromine to alkenes has just one barrier, since the second step is barrierless* (Fig. 17.21(A)).

POLAR ADDITION OF CHLORINE TO ALKENES

Polar addition of chlorine to alkenes has some similarities and also some differences with respect to the bromination of alkenes.

Like bromine addition to most alkenes, chloronium ion is an important intermediate in many reactions, and it was already experimentally characterized by NMR in the chlorination of pinacolone (Olah and Bollinger 1967). Similarly to bromination, chlorination of *trans*-butene gives anti α,β-dichloride meso product (Fig. 17.23(A)). From ωB97XD/6-311G++(d,p) level of theory, there is a chloronium ion from chlorine addition to *trans*-butene (Fig. 17.23(B)) under second order kinetics (first order in chlorine).

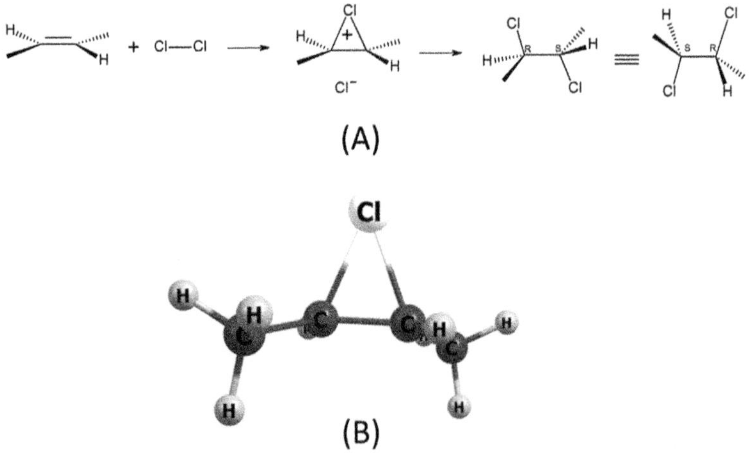

(A)

(B)

Figure 17.23 (A) Chlorination of *trans*-butene and (B) optimized geometry of corresponding chloronium ion from ωB97XD/6-311G++(d,p) level of theory.

On the other hand, unlike bromination of alkenes, chlorination of alkenes undergoes parallel substitution reaction (i.e., a reaction that interchanges the position of double bond and interchanges hydrogen atom by chlorine atom, simultaneously). This substitution reaction tends to increase for more alkyl-substituted alkenes (see ahead). Both reactions (addition and substitution) may undergo only chloronium ion as an intermediate (in an exclusive reaction path) or chloronium ion and β-chlorocarbocation as intermediates in two competing reaction paths as indicated by Poutsma (Poutsma 1965), although no experimental proof for β-chlorocarbocation as an intermediate was given in his work. Likwise, in Fahey's work, there is no experimental evidence for the existence of β-chlorocarbocation as an intermediate (Fahey 1965). Indeed, no theoretical evidence for the existence of β-chlorocarbocation as an intermediate was found in non-aromatic substituted alkenes. Then, the reaction between chlorine and 2,3-dimethylbut-2-ene might undergo two competing reaction paths, (1) and (2), yielding either addition product or substitution product (Fig. 17.24(A)).

On the other hand, experimental evidences indicate that β-chlorocarbocation as an intermediate might exist in the chlorination of *cis*-stilbene, forming 9:1 meso:racemic products (Buckles and Knaack 1960) and chlorination of acena-phthylene to give only the meso *cis*-dichloride product (Cristol et al. 1956), as depicted in Fig. 17.24(B). In these cases, the β-chlorocarbocation as an intermediate is probably stabilized by the phenyl ring(s). However, theoretical calculation using ωB97XD/6-31G++(2d,p) level of theory gave no β-chlorocarbocation and chloride anion as an intermediate. Probably, a different mechanism gives the *cis*-product passing by non-synchronous concerted *syn* addition (similar to that from hydrohalogenation of alkenes) rather than chloronium ion. No further investigation will be done here, so the hypothesis of β-chlorocarbocation as an intermediate will be considered. Then, *both chloronium ion and β-chlorocarbocation (with phenyl rings vicinal to the positively charged carbon) are possible intermediates in the chlorination of alkenes and two distinguished reactions might occur (addition and substitution).*

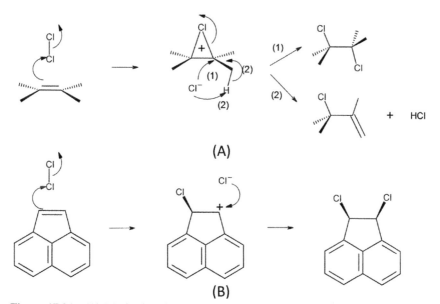

Figure 17.24 (A) Mechanism for addition and substitution reactions between chlorine and 2,3-dimethylbut-2-ene. (B) Mechanism of chlorination of acenaphthylene.

Chlorination of alkenes also has a π complex prior to the TS$_{RDS}$. The β-chlorocarbocation was not observed in chlorination of both propene and 2,3-dimethylbu-2-ene (see ahead) from ωB97XD/6-311G++(d,p) level of theory. Figures 17.25(A) and (B) show that β-chlorocarbocation and trichloride anion before optimization give an addition product. In Fig. 17.25(C), the trichloride anion is placed below the chloronium ion before the optimization and the final, optimized geometry is the 1,2-dichloride product. When the trichloride anion is placed above the chloronium ion it gives the optimized chloronium ion and trichloride ion after the optimization procedure (Fig. 17.25(D)).

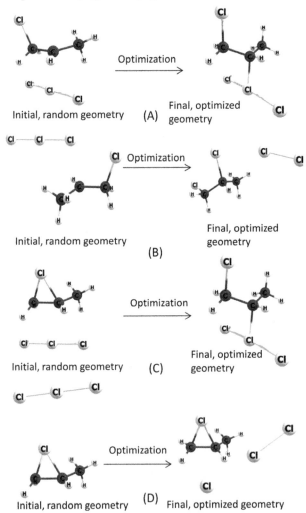

Figure 17.25 Initial geometries of trichloride ion placed (A) below the "possible" β-chlorocarbocation and (B) above the "possible" β-chlorocarbocation; (C) below the "possible" chloronium ion; (D) above the "possible" chloronium ion from chlorination of propene; under third order kinetics along with their corresponding final, optimized geometries using ωB97XD/6-311G++(d,p) level of theory.

As for the chlorination of 2,3-dimethylbu-2-ene under third order kinetics, when trichloride ion is placed below the initial, random geometry of its corresponding β-chlorocarbocation (which would be a tertiary carbocation), after optimization procedure, it yields a corresponding β-chloroalkene (Fig. 17.26(A)). This is the case of substitution reaction which was mentioned before (Fig. 17.24). When the trichloride ion is placed above the β-chlorocarbocation of the 2,3-dimethylbu-2-ene, after the optimization procedure, it gives the corresponding chloronium ion and trichloride ion as final, optimized geometry (Fig. 17.26(B)). Then, like the bromination of alkenes, *the second step of chlorination of alkenes is barrierless, giving the α,β-dichloride product. This product is preceded by the formation of chloronium ion as an intermediate or by the TS$_{RDS}$ followed by the instantaneous β-chlorocarbocation.* As a by-product, β-chloroalkene can be formed from substitution reaction (see Figs. 17.24 and 17.26(A)).

An important observation is that although there are two distinguished reaction paths being substitution and addition reactions of chlorination of alkenes, *there is only one intermediate (as a minimum of PES) in one reaction path (i.e., the chloronium ion) while in the other reaction path the β-chlorocarbocation is not a minimum in the PES but only an instantaneous entity.*

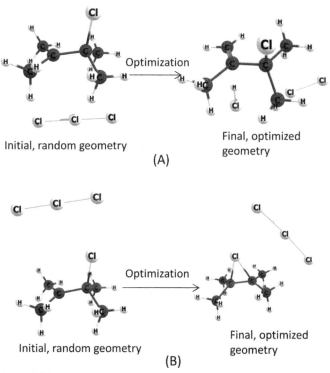

Figure 17.26 Initial geometries of trichloride ion placed (A) below the "possible" β-chlorocarbocation and (B) above the "possible" β-chlorocarbocation from chlorination of 2,3-dimethylbu-2-ene under third order kinetics along with their corresponding final, optimized geometries from ωB97XD/6-311G++(d,p) level of theory.

The transition state of chlorination of propane was obtained prior to the chloronium ion intermediate from third order kinetics (second order in chlorine) from ωB97XD/6-311G++(d,p) level of theory (Fig. 17.27(A)). Nonetheless, this transition state was not confirmed by further IRC calculation, although the displacement vectors are characteristic of the reaction coordinate indicated in Fig. 17.27(A). Then, one possible mechanism for the chlorination of alkenes (the addition reaction) passing through chloronium ion is given in Fig. 17.27(B).

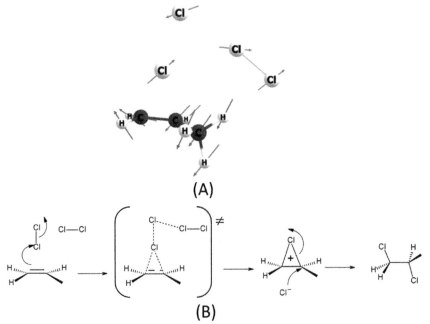

Figure 17.27 (A) Transition state of chlorination of propene under third order kinetics from ωB97XD/6-311G++(d,p) level of theory; (B) proposed mechanism for the chlorination of propene passing through chloronium ion.

HYDROBORATION

Hydroboration is a useful method to synthesize alcohol from relatively asymmetrical alkene using anti-Markovnikov regioselectivity. Whereas the acid-catalyzed addition of water to relatively asymmetrical alkene follows Markovnikov regioselectivity (where hydroxyl group is added to the most substituted vinylic carbon –the one which is more positively charged in the TS), in hydroboration the hydroxyl group is added preferably to the least substituted vinylic carbon (anti-Markovnikov regioselectivity). In most cases, anti-Markovnikov regioselectivity ranged from 80% to 99% for a large series of relatively asymmetrical alkenes (Brown and Zweifel 1960).

It is important to emphasize that no matter which reaction passes through Markovnikov or anti-Markovnikov regioselectivity, *in both cases (Markovnikov and anti-Markovnikov rule), the preferred, major product always comes from the smallest energy barrier.* In the case of Markvonikov regioselectivity, the electronic factor plays the major role, wherein the smallest barrier is ascribed to the most stabilized positively charged atom in the TS (the one attached to more alkyl groups). In the case of anti-Markovnikov rule for hydroboration, the steric factor plays the major role, where the addition of $-BH_2$ moiety from borane (BH_3) occurs at the least substituted vinylic carbon (which later will be exchanged into hydroxyl group).

The most efficient methods of hydroboration are: (a) mixture of sodium borohydride-aluminum chloride; (b) mixture of sodium borohydride-boron trifluoride; (c) diborane in ether, tetrahydrofuran, THF, or diglyme [bis(2-methoxyethyl) ether] as solvents (Brown and Rao 1959). Brown and Rao observed that "the ether functions to dissociate the diborane to some extent, producing in the reaction mixture a moderate concentration of a more active intermediate, the borane etherates". They observed that THF has several advantages as solvent (readily separated from products and high solubility of diborane in THF). *Later, it was proved that hydroboration is a second order kinetics (first order in alkene and first order in borane). This fact excluded diborane as a reactant* (Pasto et al. 1972), although borane (BH_3) is more stable in its dimer form: the diborane (B_2H_6). *They also found that borane-THF complex reacts with the alkene in syn addition instead of borane alone.*

$$r = k_2[alkene][borane]$$

In the reaction of borane and alkene, each B-H bond in borane is replaced by a B-R bond (where R represents alkyl group) by forming one equivalent of BR_3 (Fig. 17.28(A)). Without prior isolation of trialkylborane, BR_3, it can react with hydrogen peroxide in aqueous basic medium to produce the corresponding alcohol, that is, three equivalents of alcohol are formed.

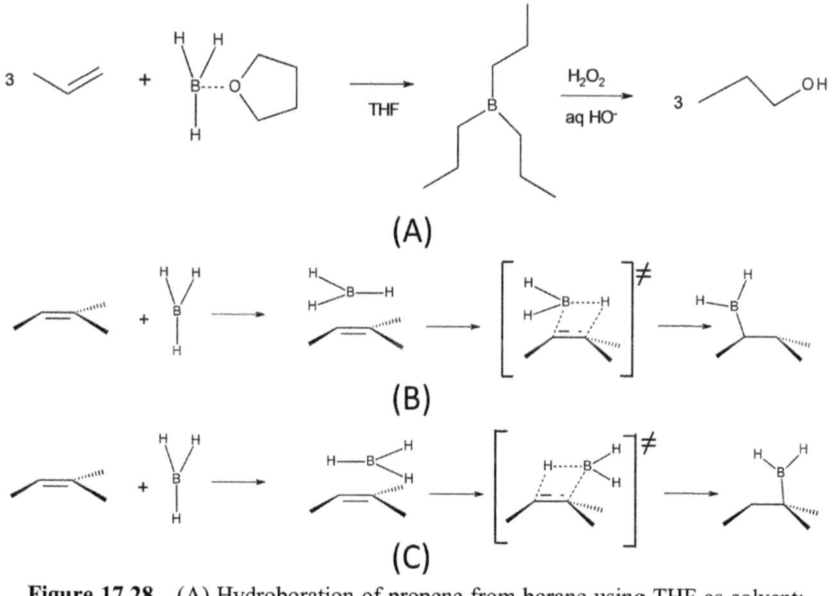

(A)

(B)

(C)

Figure 17.28 (A) Hydroboration of propene from borane using THF as solvent; (B) and (C) hydroboration of 2-methylbut-2-ene using apolar solvent.

Indeed, by using ωB97XD/6-31G++(d,p) level of theory, the optimization of mixture THF-BH$_3$-propene directly gave the *syn* adduct and THF (Fig. 17.29(A)). The addition of borane in propene also occurred in a barrierless manner after optimization of minimum in PES (Fig. 17.29(B)). Then, as already found by experimental evidences, the THF (and also acyclic ethers) catalyzes the borane addition in alkene. One possibility is that propene and borane reaction in THF does not follow transition state theory and the PES is determinied dynamically (Oyola and Singleton 2009).

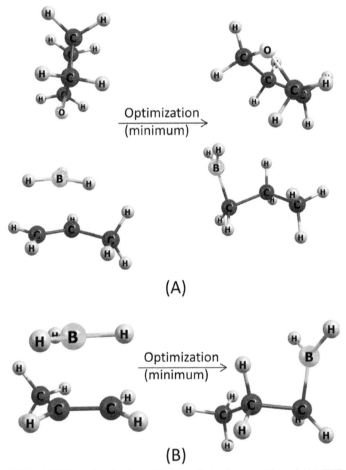

Figure 17.29   Initial, randomized, and final optimized geometries of (A) THF-BH$_3$-2-methylbut-2-ene and (B) BH$_3$-propene from ωB97XD/6-31G++(d,p) level of theory.

Likewise, the optimization of the mixture THF-BH$_3$-2-methylbut-2-ene also directly gave the *syn* adduct and THF (Fig. 17.30(A)). On the other hand, the optimization of borane and 2-methylbut-2-ene gave a complex prior to the TS according to the scheme in Fig. 17.28(B and C) and the optimized geometry shown in Fig. 17.30(B).

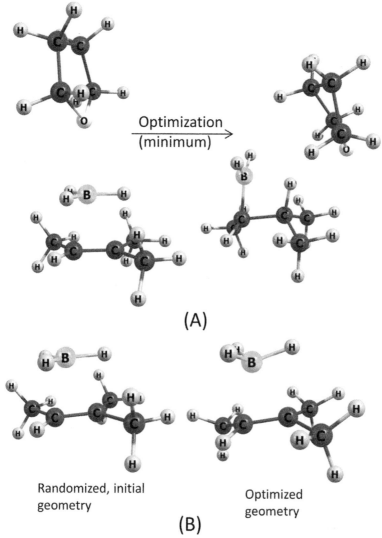

Figure 17.30 Initial, randomized, and final optimized geometries of (A) THF-BH$_3$-2-methylbut-2-ene and (B) BH$_3$-2-methylbut-2-ene from ωB97XD/6-31G++(d,p) level of theory.

After the optimization procedure from optimized BH$_3$-2-methylbut-2-ene complex to obtain the maximum of PES (i.e., the corresponding TS), it gave a transition structure for both additions, as indicated in the scheme in Figs. 17.28(B and C) and depicted in Fig. 17.31. The enthalpy difference between both transition states (anti-Markovnikov addition, at the left, and Markovnikov addition, at the right) is 2.44 kcal mol^{-1} which corresponds, according to Boltzmann distribution, to 62:1 anti-Markovnikov:Markovnikov ratio at 298 K, that is, 1.6% is Markovnikov product and 98.4% is anti-Markovnikov product.

$$\frac{N_i}{N_j} = \exp\left[-\left(E_i - E_j\right)\Big/RT\right] \therefore R = 8.314\,\mathrm{JK}^{-1}\mathrm{mol}^{-1} \therefore T = 298\,\mathrm{K}$$

$$\Delta H = 2.44\,\mathrm{kcal}\cdot\mathrm{mol}^{-1} = 10.20\,\mathrm{kJ}\cdot\mathrm{mol}^{-1}$$

$$\frac{N_i}{N_j} = \exp\left[10200/2477\right] = e^{4.12} = 61.72 \approx 62$$

This theoretical result is strikingly close to the experimental result for 2-methylbut-2-ene (Brown and Zweifel 1960). *This agreement between theoretical and experimental results indicates that the reaction between 2-methylbut-2-ene and borane in THF does follow transition state theory.* Moreover, our theoretical results do not take into account the direct influence of the solvent (Fig. 17.31). Then, probably, more experimental studies about the influence of solvent on the kinetics of hydroboration are needed.

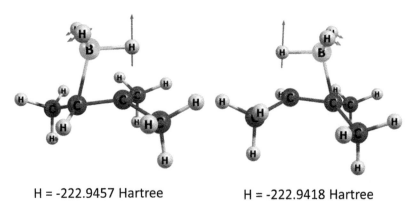

H = -222.9457 Hartree   H = -222.9418 Hartree

ΔH = 2.44 kcal mol^{-1}

Figure 17.31   Optimized structures of TS involving BH$_3$ and 2-methylbut-2-ene from ωB97XD/6-31G++(d,p) level of theory.

After the formation of the trialkylborane, its oxidation occurs with hydrogen peroxide in aqueous base. According to the transition states found with ωB97XD/6-31G++(d,p) level of theory, the proposed mechanism is somewhat different from that in Carrol's book (Carrol 2010). Firstly, hydroxide removes one proton from hydrogen peroxide forming hydroperoxide anion which is subsequently added to the trialkylborane. In Fig. 17.32, the optimized structure of tripropylborane and its tetrahedral anionic intermediate is depicted. We have performed TS calculation of this intermediate and its corresponding IRC to confirm the TS and to obtain the structures of initial and final points. The final point of the IRC is another tetrahedral anionic intermediate shown at the bottom right of Fig. 17.32. The TS structure is depicted in the bottom left of Fig. 17.32 along with the vectors of its motion from its imaginary frequency, where one C-B and O-O bonds are being broken, while C-O and B-O bonds are being formed.

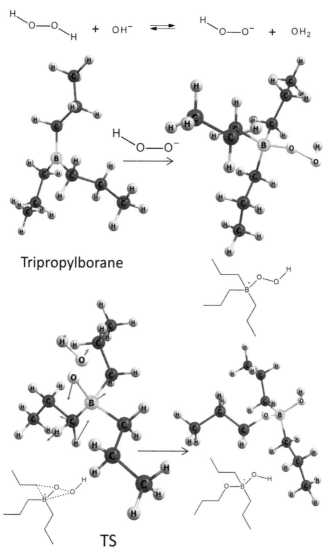

Figure 17.32 Optimized structures of tripropylborane, its anionic intermediate after addition of hydroperoxide anion, its subsequent transition state, and its posterior intermediate from ωB97XD/6-31G++(d,p) level of theory.

Afterwards, one water molecule approaches the previously formed anionic intermediate (at the bottom right of Fig. 17.32) and a hydrogen bond is formed between water and the oxygen atom bonded to the propyl group as depicted at the top of Fig. 17.33. Another transition state is found where one hydrogen atom from water is transferred to the oxygen atom bonded to the propyl group and simultaneous O-B bond cleavage occurs in TS (at the bottom of Fig. 17.33) to form propan-1-ol, hydroxide, and hydroxydipropylborane. This reaction repeats twice to form three equivalents of propan-1-ol and trihydroxyborane (boric acid).

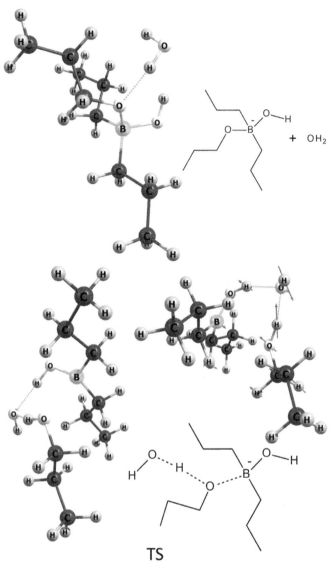

Figure 17.33 Optimized structure of hydroxydipropylpropanoxyborane anion and its subsequent TS from ωB97XD/6-31G++(d,p) level of theory.

RADICAL ADDITION OF HYDROGEN HALIDE OR HALOGEN MOLECULE

Radical addition of hydrogen halide to alkene may occur by the presence of solvent impurities, air, peroxides, and some other reactants. It follows anti-Markovnikov regioselectivity for relatively asymmetrical alkenes. For example, addition of hydrogen bromide to propene gives 2-bromopropane if the reaction is done in an inert condition free of impurities. On the other hand, in the presence of radicals, the main product is 1-bromopropane – an anti-Markovnikov product (Kharasch et al. 1933). Kharasch and coworkers named this the "peroxide effect" and similar results of anti-Markvonikov regioselectivity were found for pent-1-ene (Kharasch et al. 1934).

The peroxide effect is associated with the formation of oxide radical. One of the sources of oxide radical is benzoyl peroxide, which forms benzoyloxyl radical under heating at 60°C or UV incidence (Fig. 17.34(A)), which in turn abstracts the hydrogen atom from hydrogen bromide forming bromine radical whose transition state is shown in Fig. 17.34(B). This is the initiation step of this radical reaction. The bromine radical is added to the alkene. When the alkene is relatively asymmetrical it is added preferably so as to form the most stable radical. For example, the bromine

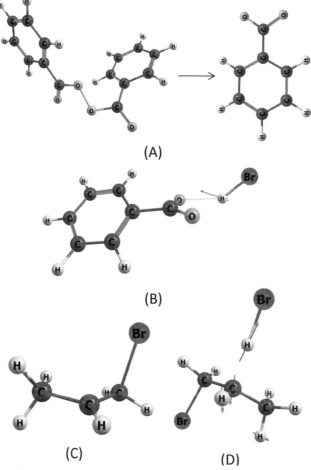

Figure 17.34 (A) Optimized geometries of benzoyl peroxide and benzoyloxyl radical; (B) transition state of hydrogen abstraction of hydrogen bromide from benzoyloxyl radical; (C) tentative TS search of bromine radical addition in C1 of propene; (D) TS of hydrogen abstraction from hydrogen bromide from β-bromo-1-propyl radical from ωB97XD/6-31G++(d,p) level of theory.

radical addition to propene occurs preferably in C1 so that β-bromo-2-propyl radical is formed (a secondary radical which is more stable than β-bromo-1-propyl radical). This is an anti-Markovnikov regioselectivity. However, no transition state for this reaction involving propene with bromine radical nor chlorine radical was found for B3LYP, MP2, and ωB97XD methods (Fig. 17.34(C)). In the next step, the β-bromo-1-propyl radical abstracts the hydrogen from a second hydrogen bromide molecule to yield the alkyl halide anti-Markovnikov product and another bromine radical to return to propagation step of the radical reaction. The corresponding optimized transition state is given in Fig. 17.34(D). The termination step might occur when two radicals react with each other (see Exercise 2).

Halogen molecule under mild heating or UV incidence may also form halogen radicals, one of which is added to alkene in the same manner described in last paragraph to form β-halogen alkyl radical. Afterwards, the second halogen radical is added to the β-halogen alkyl radical (termination step) or the β-halogen alkyl radical abstracts one halogen atom from a second halogen molecule to form vicinal alkyl dihalide product and another halogen radical (propagation step) (see exercise 3).

ADDITION OF HALOGEN AND WATER

The aqueous solution of halogen (bromine water, chlorine water) forms hypobromous acid or hypochlorous acid which is an oxidizer along with hydrogen bromide or hydrogen chloride, respectively, in equilibrium with water and halogen (Fig. 17.35). In the past, hypochlorous or hypobromous acid was thought to be the reactant (Bartlett and Rosenwald 1934). Nonetheless, chlorine/bromine and water are the reactants in the addition to alkene instead. The product is chlorohydrin or bromohydrin, respectively. The halogen is the first to react, forming bromonium ion or chloronium ion, which is subsequently attacked by the water molecule forming a protonated halohydrin (a β-halo-alcohol). The water is added preferably at the most substituted vinylic carbon of the asymmetrical alkene because of the asymmetrical chloronium ion with partial positive charge at the most substituted former vinylic

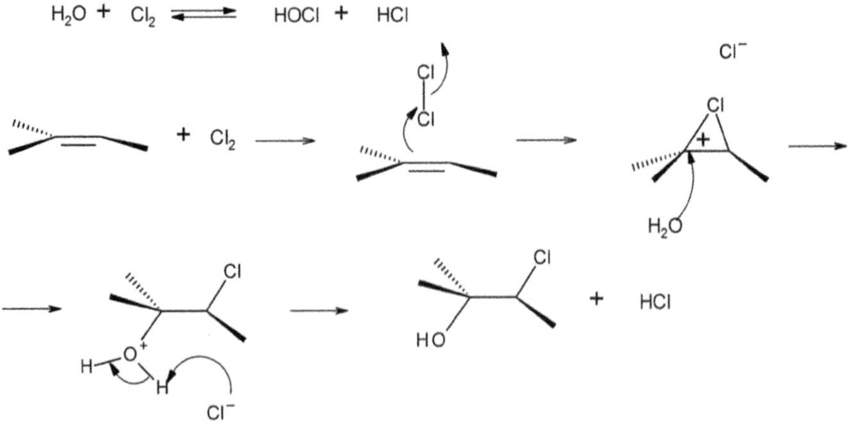

Figure 17.35 Mechanism of chlorine and water addition to 2-methylbut-2-ene.

carbon (Mare and Salama 1956). After deprotonation, the halohydrin major product comes from anti-addition, which is also a consequence of the formation of the halonium ion (Bartlett 1935).

EPOXIDATION

Epoxidation is the reaction of alkene with peroxy acid (peracid), RCO_3H, to form substituted epoxide (epoxy, oxirane, ethylene oxide are other acceptable names). The reaction has a concerted mechanism, where the peracid oxygen (from the hydroxyl group) is directed to the C=C double bond. Hereafter, the alkene's π electron pair migrates to this oxygen inducing (O=C-)O-O bond cleaveage, which in order to form another C=O double bond move the actual C=O π electron pair to abstract the proton from the former hydroxyl group. The σ electron pair of this former hydroxyl group moves to the second vinyl carbon to form the epoxide and carboxylic acid (Fig. 17.36(A)). The transition state of this reaction is shown in Fig. 17.36(B), which proves the concerted mechanism for this reaction and the spiro geometry of the TS instead of the planar TS.

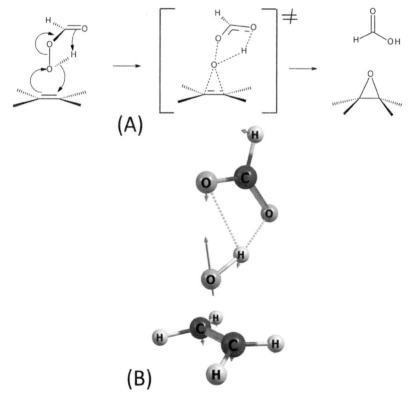

Figure 17.36　(A) Mechanism for the epoxidation of 2,3-dimethylbut-2-ene from peroxyformic acid; (B) transition state for the epoxidation of ethene by peroxyformic acid from ωB97XD/6-311G++(d,p) level of theory.

DIELS-ALDER REACTION (INTRODUCTION)

Pericyclic reactions can be divided into four classes, *cycloadditions, electrocyclic reactions, σ-tropic rearrangements,* and the less common *group transfer reactions.* All these categories share a common feature, that is, a concerted mechanism involving a cyclic transition state with a concerted movement of electrons. Concerted periclyclic cycloaddition involves the reaction of two unsaturated molecules, resulting in the formation of a cyclic product with two new σ-bonds. Examples of cycloaddition reactions include the 1,3-dipolar cycloaddition (also referred to as Huisgen cycloaddition), nitrone-olefin cycloaddition[5], and cycloadditions such as the Diels-Alder reaction. The number of π-electrons involved in the components characterizes the cycloaddition reactions, and for the three above-mentioned cases this would be [2+3], [3+2], and [4+2] respectively.

Otto Diels and Kurt Alder described an important reaction that involves the formation of C-C bond in a six membered ring with high stereochemical control, named Diels-Alder reaction. Since 1928, the Diels-Alder reaction has been used in the synthesis of new organic compounds, providing a reasonable way for forming a 6-membered system.

The simplest Diels-Alder reaction involves the reagents butadiene and ethane (Fig. 17.37(A)), but 200°C under high pressure is needed to give a low yield, though (Joshel and Butz 1941). Butadiene reacts more easily with acrolein (Fig. 17.37(B)). A methyl substituent on the diene, on C1 or C2, the reaction rate increases, e.g., *trans*-piperylene reacting with acrolein[16, 17] at 130°C (Fig. 17.37(C)). These last two reactions achieved a yield close to 80%.

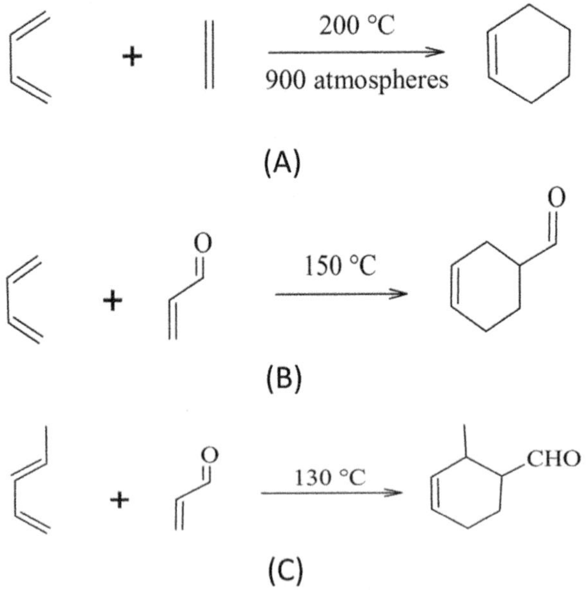

(A)

(B)

(C)

Figure 17.37   Examples of Diels-Alder reaction between (A) ethene and butadiene, (B) acrolein and butadiene, (C) acrolein and *trans*-piperylene.

In Diels-Alder reaction, butadiene and derivatives are the diene while ethene and derivatives are the dienophile. In these reactions, ethene and its derivatives react as electrophile by means of their LUMO (see chapter two), using Frontier Molecular Orbital theory (FMO). The electron withdrawing group (EWG) decreases their energy and increases their reactivities. On the other hand, butadiene and derivatives react as nucleophiles by means of their HOMO (see chapter two). The electron donating group (EDG) attached to carbon 1 or carbon 2 increases the reactivity of the diene by increasing the energy of its HOMO, according to FMO.

Figure 17.38 shows the optimized geometry of ethene and butadiene along with their LUMO and HOMO orbitals, respectively. One can see that the lobes of the same sign (same color) of HOMO and LUMO overlap in order to form two new σ bonds passing through the transition state, which shows the ethene

HOMO$_{\text{diene}}$

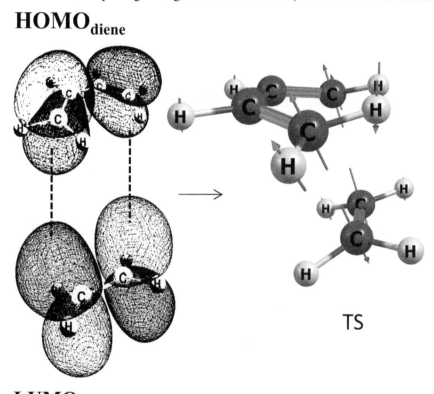

LUMO$_{\text{dienophile}}$

Figure 17.38 Optimized geometries of butadiene (along with its HOMO orbital), ethene (along with its LUMO orbital), and their corresponding transition state from B3LYP/6-311G++(d,p) level of theory.

approaching below the butadiene for the effective overlap of their orbitals. In fact, nitroethene reacting with butadiene has a lower enthalpy barrier ($15.7\,\text{kcal mol}^{-1}$) than that from ethene and butadiene ($23.03\,\text{kcal mol}^{-1}$), and nitroethene and 1-aminobutadiene also have a lower enthalpy barrier ($17.9\,\text{kcal mol}^{-1}$), where nitro and amino groups are EWG and EDG, respectively. Nonetheless, the reaction between aminoethene and 1-nitrobutadiene (with substituents that deactivate the reaction leading to the high increase of the barrier with respect to the reference reaction) only has a slightly increase in enthalpy barrier by 0.5 kcal mol^{-1} (23.53 kcal mol^{-1}) and the reaction between ethene and 1-aminobutadiene has a higher enthalpy barrier (25.04 kcal mol^{-1}) with respect to reference reaction. These barriers were obtained from B3LYP/6-311G++(d,p) level of theory. Then, FMO is not applicable to all cases of Diels-Alder reaction.

Regarding the thermodynamics of Diels-Alder reaction, its exothermic nature is the result of converting two weaker π-bonds into two stronger σ-bonds. Because it is an addition reaction it has negative entropy change. However, the Diels-Alder reactions are exergonic due to the high negative enthalpy change.

POLYMERIZATION (INTRODUCTION)

Alkenes might also undergo several types of addition polymerization reactions, i.e., radical polymerization, cationic polymerization, anionic polymerization, and catalyzed polymerization. Several olefin polymers are industrially manufactured: polyethylene, polypropylene, polystyrene, etc. In this book, we will briefly focus on the catalyzed polymerization of olefins.

The group 4 (zirconium, titanium, and hafnium) bent metallocene is used in conjunction with alkyalumoxanes (MAO or BAO) to produce a highly active catalytic system for olefin polymerization (Firme et al. 2005). Metallocenes contain transition metal and two cyclopentadienyl anions (or substituted cyclopentadienyl anions) as π-bonded ligands. The main representative is the bis(cyclopentadienyl) iron, Cp_2Fe (Fig. 17.38(A)), where its QTAIM analysis indicates five bond paths from each carbon atom of each cyclopentadienyl ligand towards iron metal (Firme et al. 2010). However, only group 4 bent metallocenes can be successfully used for olefin polymerization, for example dimethyl-bis(cyclopentadienyl)titanium, $Cp_2Ti(CH_3)_2$, and dimethyl-bis(indenyl)titanium, $Ind_2Ti(CH_3)_2$, depicted in Fig. 17.39(B) and Fig. 17.39(C), respectively. The QTAIM analysis indicates that they have less Ti-Cp/Ind bond paths. Indeed, they are less stable (Firme et al. 2010). Barron group was the first to synthesize the alkyalumoxanes, $(RAlO)_n$ from the hydrolysis of trialkylaluminium, and also showed they are also cocatalysts for olefin polymerization (Harlan et al. 1995). Figure 17.29(D) shows the optimized structure of methylalumoxane (BAO) which was characterized by X-ray crystallography.

The reaction between the group 4 bent-metallocene (catalyst) and MAO/BAO (co-catalyst) molecules imparts the true polymerization active species: alumoxane anion and metallocene cation (Sishta et al. 1992). Figure 17.40(A) shows the mechanism for the reaction between $Cp_2Ti(CH_3)_2$ and MAO. After forming the metallocene (zirconocene, titanocene, or hafnocene) cation, it reacts with alkene (e.g., propene) to pass to a four-membered transition state, giving another cationic intermediate which reacts with another propene molecule, and so on until the end of the reaction where polypropylene is produced (Fig. 17.40(B)).

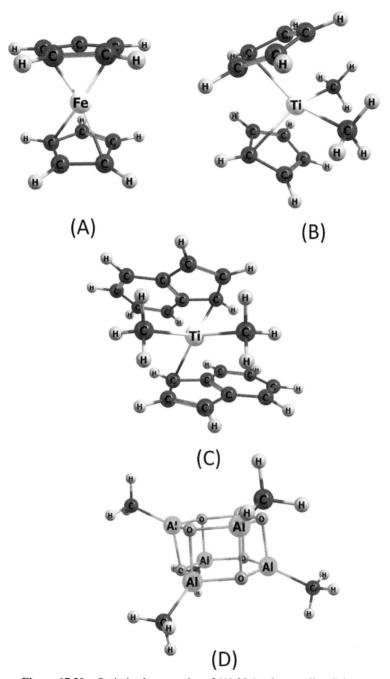

Figure 17.39   Optimized geometries of (A) bis(cyclopentadienyl) iron,
(B) dimethyl-bis(cyclopentadienyl)titanium, (C) dimethyl-bis(indenyl)titanium,
(D) methylalumoxane from B3LYP/6-311G++(*d,p*) level of theory.

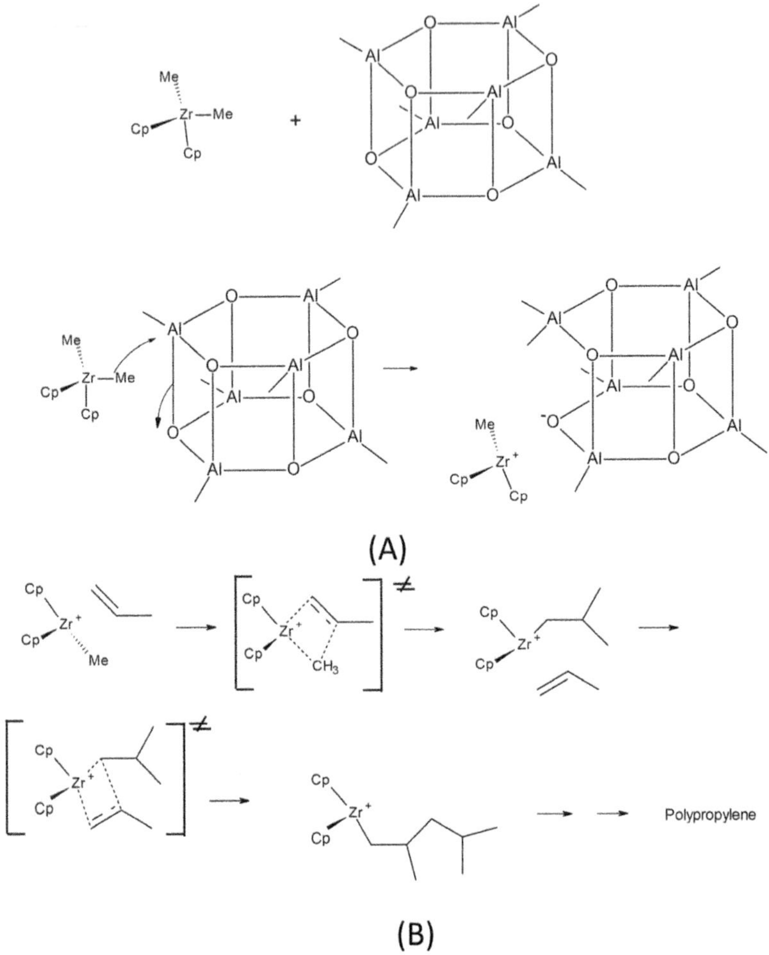

(A)

(B)

Figure 17.40 Mechanism of the reaction between (A) $Cp_2Ti(CH_3)_2$ and MAO, and (B) cationic $Cp_2Ti(CH_3)^+$ and propene.

The π-bonded ligands in group 4 bent-metallocene catalysts strongly influence the stereo- and regioregularity of the produced polyolefins (Blom et al. 2001). Some σ-bonded ligands can also play a role in olefin polymerization. Voluminous σ-bonded alcoholato ligands influence the molecular weight of ethylene polymerization (Grafov et al. 2005). The mixture of group 4 bent-metallocene, MAO, and voluminous alcohols change the regioregularity of polypropylene as well (Firme et al. 2005).

REDUCTION OF ALKENES BY CATALYTIC HYDROGENATION

See the end of chapter ten in the section "Fundamentals of heterogeneous catalysis. Study case: hydrogenation of alkenes".

Examples

1. The mechanism of bromination of cyclopropane is shown in Fig. 17.41(A), where it is considered only second order kinetics. Bromine molecule approaches either upper face or lower face of cyclopropane forming two π complexes. Afterwards, the two bromonium ions are formed. The bromide anion attacks the opposite side where the bromine atom in the bromonium ion is located. Two products are formed. In order to assign their asymmetric carbons, the easiest way is to draw the products into lateral view. Afterwards, change the lateral view into upper view. The lateral view is just used to facilitate the correct drawing of the upper view. The assignment can be easily done in upper view which can also be checked out in any drawing software which gives the name (*R* or *S*) of the asymmetric carbon after drawing its corresponding molecule. One should observe that, as in the case of *trans*-dimethylcyclopropane, there is not center of symmetry or inversion center for the *trans*-disubstituted- cyclopropane (see chapter thirteen), *trans*-disubstituted-cyclohexane (see chapter fifteen), or other *trans*-disubstituted cycloalkane. *Then, two enantiomers are formed from any trans-disubstituted cycloalkane as long as there is no plane of symmetry* (Smith and March 2007).

2. The mechanism of bromination of 1-methylcycloprone where it is considered only second order kinetics is shown in Fig. 17.41(B).

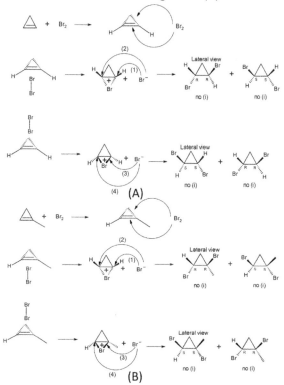

Figure 17.41 Mechanism of bromination of (A) cyclopropane and (B) 1-methylcyclopropane, considering only second order kinetics.

3. The mechanisms of the bromination of propene and *trans*-butene are depicted in Fig. 17.42, considering only second order kinetics. In the bromination of propene, one can see bromine approaching propene's double bond upwards and downwards, and two bromonium ions as intermediates are obtained. Henceforth, the attack of bromide ion (1-4) gives four products, where (1) equals (4) and (2) equals (3). Then, one bromonium ion is necessary to obtain the products.

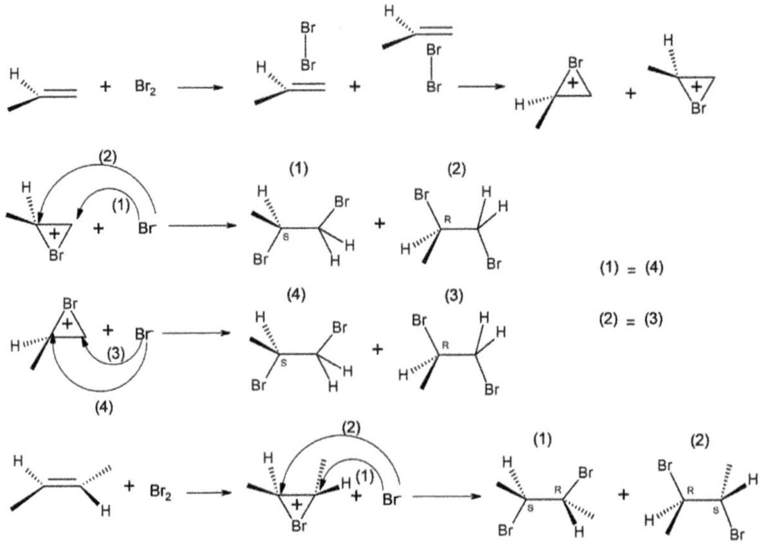

Figure 17.42 Mechanism of bromination of propene and *trans*-butene.

4. The mechanism of hydrochlorination of 1-methylcyclopropene by considering both *syn* and *anti* additions and only second order kinetics is shown in Fig. 17.43. The proton from hydrogen chloride can be added in carbon (C1) or carbon (C2) of 1-methylcylopropene. In C1 it can be added by path #1 and #2. Let's consider first the addition in carbon C1 from path #1. The instantaneous carbocation is formed and it can be attacked from both sides by the chloride anion: path #1 and path #2 give two diastereoismores as products. Afterwards, let us consider the addition of the proton in C1 through path #2. Again, the instantaneous carbocation is formed and the chloride ion can be added by paths #1 and #2 to give another pair of diastereoisomers. Finally, the proton is added in C2, whose corresponding carbocation is attacked by chloride anion by paths #1 and #2 to give the same product (1-chloro-1-methyl-cyclopropane).

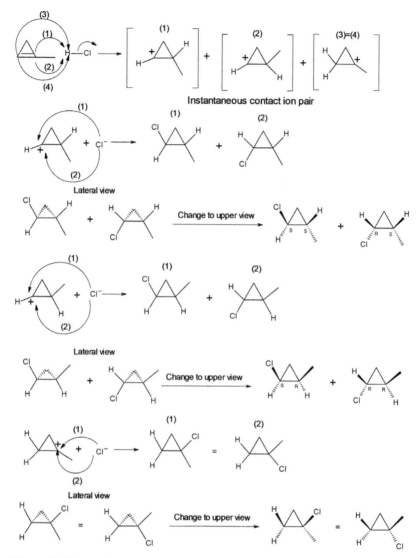

Figure 17.43 Mechanism of hydrochlorination of 1-methylcyclopropene, where *syn* and *anti* addition is considered in a second order kinetics.

5. The mechanism of hydrobromination of *trans*-2-methyl-hex-3-ene is shown in Fig. 17.44 by considering only second order kinetics and both *syn* and *anti* addition. The addition of bromide anion in C3 or C4 forms a pair of enantiomers each. Rearrangement from hydrogen shift might also occur to give a third product 2-bromo-2-methylhexane.

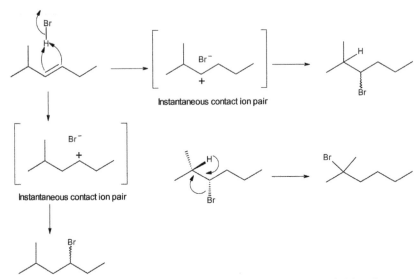

Figure 17.44 Mechanisms of hydrobromination of *trans*-2-methyl-hex-3-ene considering only second order kinetics and *syn/anti* additions.

EXERCISES

1. For the iodination of ethene at 432.2 K the equilibrium constant is 12.28 atm. By knowing its entropy change (31.5 eu), give the corresponding enthalpy change.
2. Give the mechanism of the radical addition reaction of hydrogen chloride in propene initiated by benzoyl peroxide, including initiation, propagation, and possible termination steps.
3. Give the mechanism of the radical addition reaction of bromine in propene initiated by light, including initiation, propagation, and termination steps.
4. Do the first two steps of polymerization of cyclopropene from zirconocene cation as a catalyst and give the Diels-Alder product of the reaction between cyclopentadiene and cyclopropene (Wiberg and Bartley 1960).
5. Give the mechanism of chlorination of cyclohexene. Remember chapter fifteen, where the *trans*-disubstituted cyclohexanes have two enantiomers.
6. Give the mechanisms of the hydrobromination of 2-methylbut-2-ene by considering only fourth order kinetics.
7. Give the mechanism of chlorination of 3-methylcyclopropene considering third order kinetics.
8. Give the mechanism for radicalar chlorination of 1-methylcyclopentene.
9. Give the mechanism for polar chlorination of 1-methylcyclopentene considering third order kinetics and chloronium ion.
10. Give the transition states of Example 5 from Fig. 17.44.
11. Give the products for the reactions in Fig. 17.45.

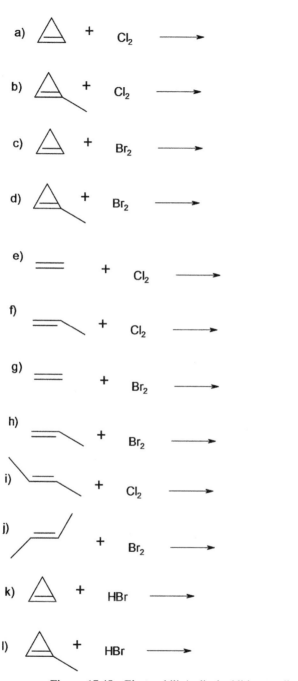

Figure 17.45 Electrophilic/radical addition to alkenes.

12. Give the products for the reactions in Fig. 17.46.

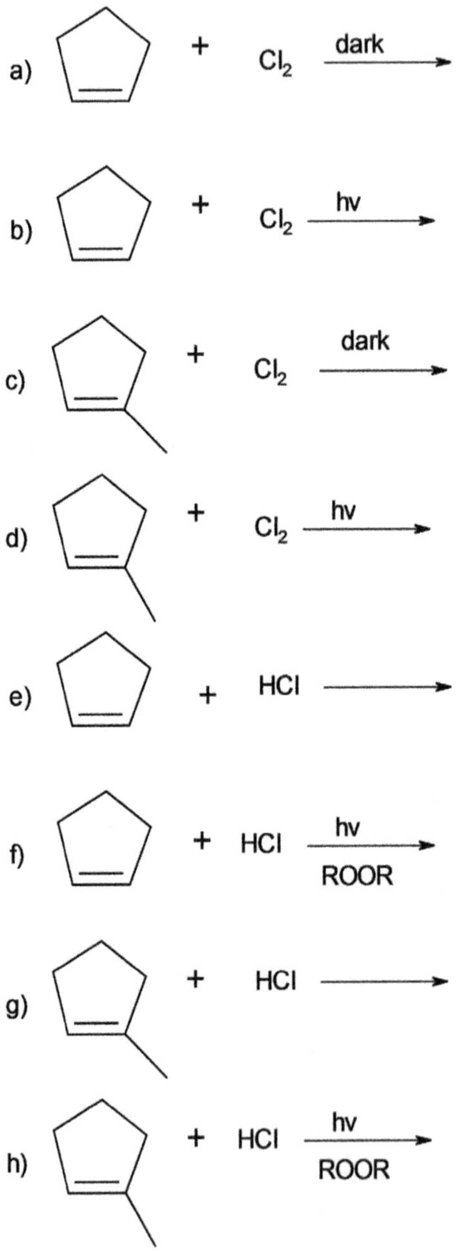

a) [cyclopentene] + Cl_2 →(dark)

b) [cyclopentene] + Cl_2 →(hv)

c) [methylcyclopentene] + Cl_2 →(dark)

d) [methylcyclopentene] + Cl_2 →(hv)

e) [cyclopentene] + HCl →

f) [cyclopentene] + HCl →(hv / ROOR)

g) [methylcyclopentene] + HCl →

h) [methylcyclopentene] + HCl →(hv / ROOR)

Figure 17.46   Electrophilic/radical addition to alkenes.

13. Give the mechanisms for addition of hydronium ion to methylpropene and borane/hydrogen peroxide in basic medium to methylpropene.

REFERENCES CITED

Bartlett, P.D. and Rosenwald, R.H. 1934. Cis- and trans-chlorohydrins of Δ^1-methyl-cyclohexene. J. Am. Chem. Soc. 56: 1990-1994.

Bartlett, P.D. 1935. Cis- and trans-chlorohydrins of cyclohexene. J. Am. Chem. Soc. 57: 224-227.

Bellucci, G., Bianchini, R. and Ambrosetti, R. 1985a. Direct evidence for bromine-olefin charge transfer complexes as essential intermediates of the fast ionic addition of bromine to cyclohexene. J. Am. Chem. Soc. 107: 2464-2471.

Bellucci, G., Bianchini, R., Ambrosetti, R. and Ingrosso, G. 1985b. Comparison of molecular bromine and tribromide ion as brominating reagents. 1. Kinetic evidence for different mechanisms of addition to cyclohexene. J. Org. Chem. 50: 3313-3318.

Benson, S.W. and Amano, A. 1962. Thermodynamics of iodine addition to ethylene, propylene and cyclopropane. J. Chem. Phys. 36: 3464-3471.

Blom, R. Follestad, A., Rytter, E., Tilset, M. and Ystenes, M. (Eds.). 2001. Organometallic catalysts and olefin polymerization – catalysts for a new millennium. Springer-Verlag, Berlin.

Brown, H.C. and Rao, B.C.S. 1959. Hydroboration. II. A remarkably fast addition of diborane to olefins – scope and stoichiometry of the reaction. J Am. Chem. Soc. 81:6428-6434.

Brown, H.C. and Zweifel, G. 1960. Hydroboration. VII. Directive effects in the hydroboration of olefins. J. Am. Chem. Soc. 82: 4708-4712.

Buckles, R.E. and Knaack, D.F. 1960. Tetrabutylammonium iodotetrachloride as a chlorinating agent. J. Org. Chem. 25:20-24.

Carroll, F.A. 2010. Perspectives on Structure and Mechanism in Organic Chemistry. John Wiley & Sons Inc. Hoboken.

Chiang, Y., Chwang, W.K., Kresge, A.J., Powell, M.F. and Szilagyi, S. 1984. Acid-catalyzed olefin-alcohol interconversion in the 1-methylcyclooctyl system. Strain-relief acceleration of the hydration of 1-methyl-trans-cyclooctene. J. Org. Chem. 49: 5218-5224.

Chiang, Y. and Kresge, A.J. 1985. Mechanism of hydration of simple olefins in aqueous solution. *Cis-* and *trans*-cyclooctene. J. Am. Chem. Soc. 107: 6363-6367.

Cristol, S.J., Stermitz, F.R. and Ramey, P.S. 1956. Mechanism of elimination reactions. XVII. The *cis-* and *trans*-1,2-dichloroacenaphthenes; *trans*-1,2-dibromoacenaphthene. J. Am. Chem. Soc. 78: 4939-4941.

Fahey, R.C. and Schubert, C. 1965. Polar additions to olefins. I. The chlorination of 2-butene and 1-phenylpropene. J. Am. Chem. Soc. 87, 5172-5179.

Fahey, R.C. and McPherson, C.A. 1969. The mechanism of the hydrochlorination of t-butylethylene and styrene in acetic acid. J. Am. Chem. Soc. 91:3865-3869.

Fahey, R.C. and McPherson, C.A. 1971. Kinetics and stereochemistry of the hydrochlorination of 1,2-dimethylcyclohexene. J. Am. Chem. Soc. 93: 2445-2453.

Fahey, R.C., Monahan, M.W. and McPherson, C.A. 1970. Hydrochlorination of cyclohexene in acetic acid. Kinetic and product studies. J. Am. Chem. Soc. 92: 2810-2815.

Fahey, R.C. and Monahan, M.W. 1970. Stereochemistry of the hydrochlorination of cyclohexene-1,3,3,-d_3 in acetic acid. Evidence for termolecular *anti* addition of acids to olefins. J. Am. Chem. Soc. 92: 2816-2820.

Firme, C.L., Grafov, A.V. and Dias, M.L. 2005. Ethylene and propylene polymerization with bis(indenyl)zirconium/MAO catalytic systems modified by sterically demanding alcohols. J. Polymer Sci. A 43: 4248-4259.

Firme, C.L., Pontes, D.L. and Antunes, O.A.C. 2010. Topological study of bis(cyclopenta-dienyl)titanium and bent titanocenes. Chem. Phys. Lett. 499: 193-198.

Firme, C.L. 2019. Revisiting the mechanism of polar hydrochlorination of alkenes. J. Mol. Model. 25: 128.

Fogler, H.S. 2011. Essentials of Chemical Reaction Engineering. Prentice Hall, Pearson Education, Inc. Westford.

Fukuzumi, S. and Kochi, J.K. 1982. Transition-state barrier for electrophilic reactions. Solvation of charge-transfer ion pairs as the unifying factor in alkene addition and aromatic substitution with bromine. J. Am. Chem. Soc. 104: 7599-7609.

Harlan, J.C., Bott, S.G. and Barron, A.R. 1995. Three-coordinate aluminum is not a prerequisite for catalytic activity in the zirconocene-alumoxane polymerization of ethylene. J. Am. Chem. Soc. 117: 6465-6474.

Haugh, M.J. and Dalton, D.R. 1975. The gas phase addition of hydrogen chloride to propylene. J. Am. Chem. Soc. 97: 5674-5678.

Hamilton, T.P. and Schaefer III, H.F. 1990. Structure and energetics of C2H4Br+: ethylenebromonium ion vs. Bromoethyl cations. J. Am. Chem. Soc. 112: 8260-8265.

Grafov, A.V., Firme, C.L., Grafova, I.A., Benetollo, F., Dias, M.L., Abadie, M.J.M. 2005. Olefin polymerization with hafnocenes: a bridged alicyclic alcohol as a ligand and as the hafnocene modifier. Polymer 46: 9626-9631.

Kharasch, M.S., McNab, M.C. and Mayo, F.R. 1933. The peroxide effect in the addition of reagents to unsaturated compounds. III. The addition of hydrogen bromide to propylene. J. Am. Chem. Soc. 55: 2531-2533.

Kharasch, M.S., Hinckley, J.A. and Gladstone, M.M. 1934. The peroxide effect in the addition of reagents to unsaturated compounds. J. Am. Chem. Soc. 56: 1642-1644.

Kresge, A.J., Chiang, Y., Fitzgerald, P.H., McDonald, R.S. and Schmid, G.H. 1971. General acid catalysis in the hydration of simple olefins. Mechanism of olefin hydration. J. Am. Chem. Soc. 93: 4907-4908.

Joshel, L.M. and Butz, L.W. 1941. The synthesis of condensed ring compounds. VII. The successful use of ethylene in the Diels—Alder reaction 1. J. Am. Chem. Soc. 63:3350–3351.

Lucas, H.J. and Liu, Y.-P. 1934. The hydration of unsaturated compounds.III. The hydration rate of trimethylethylene in aqueous solutions of acids. J. Am. Chem. Soc. 56: 2138-2140.

Mare, P.B.D.de la and Salama, A. 1956. The kinetics and mechanisms of addition to olefinic substances. Part III. The carbonium ionic intermediate involved in addition of hypochlorous acid to isobutene. J. Chem. Soc. 3337-3346.

Mayo, F.R. and Katz J.J. 1947. The addition of hydrogen chloride to isobutylene. J. Am. Chem. Soc. 69: 1339-1348.

Mayo, F.R. and Savoy, M.G. 1947. The addition of hydrogen bromide to propylene. J. Am. Chem. Soc. 69: 1348-1351.

Merritt, R.F. 1966. Direct fluorination of 1,1-diphenylethylene. J. Org. Chem. 31: 3871-3873.

O'Connor, S.F, Baldinger, L.H., Vogt, R.R. and Hennion, G.F. 1939 Solvent Effects in Addition Reactions. I. Addition of Hydrogen Bromide and Chloride to Cyclohexene and 3-Hexene, J. Am. Chem. Soc. 61: 1454–1456.

Olah, G.A. and Bollinger, J.M. 1967. Stable carbonium ions. XLVII. Halonium ion formation via neighboring halogen participation. Tetramethylethylene halonium ions. J. Am. Chem. Soc. 89: 4744-4752.

Oyama, K. and Tidwell, T.T. 1976. Cyclopropyl substituent effects on acid-catalyzed hydration of alkenes. Correlation by σ^+ parameters. J. Am. Chem. Soc. 98: 947-951.

Oyola, Y. and Singleton, D.A. 2009. Dynamics and the failure of transition state theory in alkene hydroboration. J. Am. Chem. Soc. 131: 3130-3131.

Pasto, D.J., Lepeska, B. and Cheng, T.-C. 1972. Measurement of the kinetics and activation parameters for hydroboration of tetramethylethylene and measurement of isotope effects in the hydroboration of alkenes. J. Am. Chem. Soc. 94: 6083-6090.

Poutsma, M.L. 1965. Chlorination studies of unsaturated materials in nonpolar media. IV. The ionic pathway for alkylated ethylenes. Products and relative reactivities. J. Am. Chem. Soc. 87: 4285-4293.

Rolston, J.H. and Yates, K.J. 1969. Polar additions to styrene and 2-butene systems. I. Distribution and stereochemistry of bromination products in acetic acid. J. Am. Chem. Soc. 91: 1469-1476.

Sergeev, G.B., Stepanov, N.F., Leensov, I.A., Smirnov, V.V., Pupyshev, V.I., Tyurina, L.A., Mashyanov, M.N. 1982. Molecular mechanism of hydrogen bromide addition to olefins. Tetrahedron 38: 2585–2589.

Schmidi, G.H. and Toyonaga, B. 1984. Nonlinear least-square method of separating the second- and third-order rate constants for the ionic bromination of alkenes in CCl_4 at 25°C. J. Org. Chem. 49: 761-763.

Sishta, C., Hathorn, R.M. and Marks, T.J. 1992. Group 4 metallocene-alumoxane olefin polymerization catalysts. CPMAS-NMR spectroscopic observation of "cation-like" zirconocene alkyis. J. Am. Chem. Soc. 114: 1112-1114.

Slebocka-Tilk, H., Ball, R.G. and Brown, R.S. 1985.The question of reversible formation of bromonium ions during the course of electrophilic bromination of olefins. 2. The crystal and molecular structure of the bromonium ion of adamantylideneadamantane. J. Am. Chem. Soc. 107: 4504-4508.

Smith M.B. and March, J. 2007. March's Advanced Organic Chemistry. Reactions, Mechanisms and Structure. John Wiley and Sons Inc. Hoboken, New Jersey.

Wiberg, K.B. and Bartley, W.J. 1960. Cyclopropene. V. Some reactions of cyclopropane. J. Am. Chem. Soc. 82:6375-6380.

Yates, K., McDonald, R.S. and Shapiro, S.A. 1973. Kinetics and mechanisms of electrophilic addition. I. A comparison of second- and third-order brominations. J. Org. Chem. 38: 2460-2464.

Chapter Eighteen

Alkynes (properties and reaction)

INTRODUCTION

The alkynes, whose general formula is C_nH_{2n-2}, have two *sp*-hybridized carbon atoms relative to the CC triple bond, which is made up of two π bonds and one σ bond. The four atoms in the triple bond and bonded to the atom of the triple bond (two *sp* carbon atoms and each substituent bonded to them) are in a linear geometry. See the optimized geometries of pent-1-yne and oct-4-yne in Figs. 18.1(A) and (B), respectively.

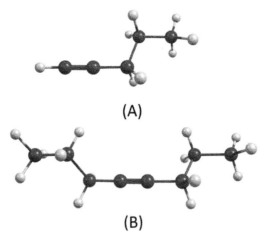

(A)

(B)

Figure 18.1 Optimized geometries of (A) pent-1-yne and (B) oct-1-yne from B3LYP/6-31G(*d,p*) level of theory.

Since the triple CC bond is stronger than double CC bond (in alkene), the former has smaller CC bond length than the latter. Figure 18.2(A) shows both singly-occupied GVB orbitals of CC σ bond and one singly-occupied GVB orbital of C-H bond in ethyne. One can see that the shape of both σ bonds is different. Figure 18.2(B) shows both GVB π bonds (two singly-occupied orbitals for each

π bond). Figure 18.2(C) shows all GVB singly-occupied GVB bonds of triple bond in ethyne, which roughly resembles an elliptical shape which is confirmed in ethyne's 3-D electron density (Fig.18.2(D)). Only single and triple bonds have elliptical shapes in their local electron density, which corresponds to zero value in ellipticity of QTAIM with respect to these bonds.

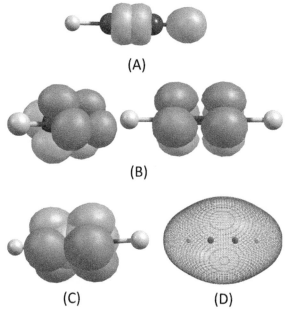

Figure 18.2   GVB singly-occupied orbital of (A) σ CC bond and one orbital of σ CH
bond, (B) both π CC bond, (C) all CC triple bond in ethyne from
6-31G++(d,p) basis set, (D) 3-D electron density of ethyne.

The 1-alkyne is referred to as the terminal alkyne, while the triple bond in any other position is referred to as internal alkynes. Similar to alkenes, the numbering scheme in alkynes gives the smallest number to the first carbon of the triple bond.

POLAR ADDITION OF HALOGEN

The electrophilic additions also occur in alkynes with some differences with respect to those in alkenes.

Contrary to the halogenation of alkenes, the polar addition of halogens does not lead to the halonium ion as an intermediate, as far as is determined by the level of theory used here.

In the bromination of hex-3-yne, only *anti* product is formed (Carroll 2010), as shown in Fig. 18.3(A). However, this is not because of the formation of bromonium ion as an intermediate – a critical point at the PES. Probably, the bromonium ion is formed as an instantaneous species, which is readily attacked by its counterion, as indicated in Fig. 18.3(B).

The bromonium ion after addition of bromine in hex-3-yne only exists isolated from its counterion (Fig. 18.3(B)). When the counterion (bromide anion) is present and close to the positive charge of the instantaneous bromonium ion, corresponding to the real situation, it leads to the bromination of hex-3-yne. This can be shown after the optimization of the bromonium ion (obtained from previous optimization disregarding the counterion) along with its counterion (Fig. 18.3(C)).

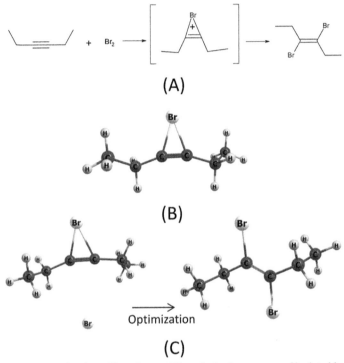

Figure 18.3 (A) Bromination of hex-3-yne; (B) optimized geometry of isolated bromonium ion from addition of bromine in hex-3-yne; (C) initial geometry of bromonium ion and its counterion and its final geometry after optimization procedure from ωB97XD/6-31G++(2d,p) level of theory.

Except for the products of acetic acid addition, the two products of the bromination of phenylprop-1-yne in acetic acid medium are derived from *anti* and *syn* addition forming dibromide alkenes (Pincock and Yates 1968), as depicted in Fig. 18.4(A). They are *trans*- and *cis*-1,2-dibromophenylprop-1-ene. The formation of both *anti* and *syn* products excludes the possibility of bromonium ion as an intermediate (where only *anti* addition could occur). Similar results (formation of both *anti* and *syn* products) were found in another work (Bianchini et al. 1999).

In an apolar medium (where bromide ion is not present as initial reactant as in polar, protic medium), we show the optimized geometries of the π complex of bromination of phenylprop-1-yne, along with the corresponding transition state and *anti* product (Fig. 18.4(B)). In Fig. 18.4(C), one can see that the enthalpy barrier is 11.3 kcal mol^{-1} and that the reaction is very exothermic (-41.4 kcal mol^{-1}).

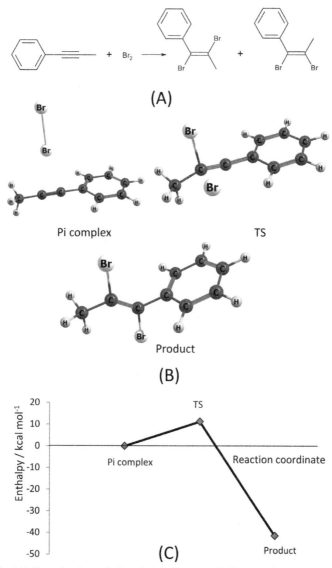

Figure 18.4 (A) Bromination of phenylprop-1-yne excluding products from acetic acid addition; (B) optimized structures of π complex, transition state and product from bromination of phenylprop-1-yne in apolar medium; (C) potential energy (in terms of enthalpy) of the corresponding reaction calculated from ωB97XD/6-31G++(2d,p) level of theory.

POLAR ADDITION OF HYDROGEN HALIDES

Alkynes also react with hydrogen halides from both polar and radical mechanism. Likewise in alkenes, radical mechanism can occur when impurities are present. Then, in order to assure only polar mechanism, solvent and reactant must be purified

(Fahey and Lee 1968). The reaction hydrochlorination of phenylprop-1-yne passes through the π complex and the transition state where chloride ion is at the same side with respect to the added hydrogen atom in the alkyne to form the *syn* product (Fig. 18.5(A)). The syn product is the main product in this reaction (81%) which is in accordance with our theoretical calculations (Fahey et al. 1974). The enthalpy barrier for hydrochlorination is 40 kcal mol^{-1} and the reaction is exothermic by 20.58 kcal mol^{-1} (Fig. 18.5(B)).

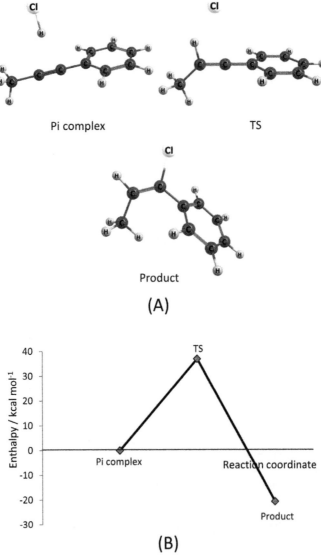

Figure 18.5 (A) Optimized structures of π complex, transition state, and product from bromination of phenylprop-1-yne in apolar medium; (B) potential energy (in terms of enthalpy) of the corresponding reaction calculated from ωB97XD/6-31G++(2d,p) level of theory.

ADDITION OF WATER

The terminal and internal alkynes undergo acid catalyzed hydration to yield enol as an intermediate, which is in tautomeric equilibrium with its corresponding aldehyde or ketone, respectively.

Figure 18.6(A) shows the IRC plot, the optimized TS along with structures of initial and final points of the IRC calculation from the reaction between hydronium ion and ethyne. One of the acids which reacts with the alkyne is the hydronium ion.

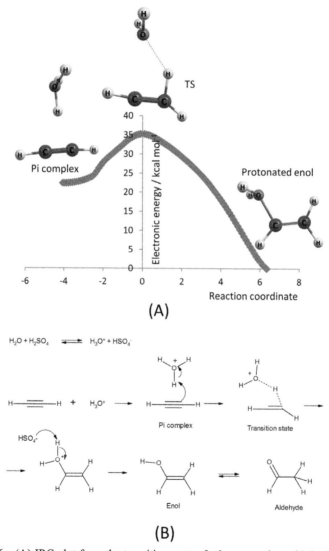

(A)

(B)

Figure 18.6 (A) IRC plot from the transition state of ethyne reacting with hydronium ion from ωB97XD/6-31G++(d,p) level of theory; (B) proposed mechanism according to the IRC calculations.

It forms a π complex with the alkyne. In the transition state, the water molecule stands inwards to the alkyne's face and not outwards, where there is an interaction between the positive charge of the acetylenic carbon in the TS and the water. Next, it forms the corresponding protonated enol. This is the rate determining step. Once again, no carbocation is formed as an intermediate. Afterwards, the conjugate base of the mineral acid (used as a catalyst) removes one of the hydrogen (as proton) bonded to oxygen atom which yields the enol, and the enol is in equilibrium with the corresponding aldehyde (Fig. 18.6(B)). Usually, this equilibrium tends to form a greater amount of aldehyde/ketone than enol.

REFERENCES CITED

Bianchini, R., Chiappe, C., Moro, G.L., Lenoir, D. 1999. Spectroscopic and theoretical investigation of electrophilic bromination reactions of alkynes: the first evidence for π complexes as reaction intermediates. Chem. Eur. J. 5: 1570-1580.

Carroll, F.A. 2010. Perspectives on Structure and Mechanism in Organic Chemistry. John Wiley & Sons Inc., Hoboken.

Fahey, R.C. and Lee, D.-J. 1968. Polar additions to olefins and acetylenes. V. Bimolecular and termolecular mechanisms in the hydrochlorination of acetylenes. J. Am. Chem. Soc. 90: 2124-2131.

Fahey, R.C., Payne, M.T.and Lee, D.-J. 1974. Reaction of acetylines with hydrogen chloride in acetic acid. Effect of structure upon AdR2 and Ad3 reaction rates. J. Org. Chem. 39: 1124-1130.

Pincock, J.A. and Yates, K. 1968. On the mechanism of bromination of acetylenes. J. Am. Chem. Soc. 90: 5643-5644.

Chapter Nineteen

Aromaticity and Aromatic Compounds

BRIEF HISTORY OF BENZENE

Benzene is the iconic example of an aromatic compound being the most aromatic of the benzenoid series (see ahead). The name "benzene" has its origin from incense commerce in Indonesian islands to Euroupe, where it was named "gum benzoin" from which it was isolated to an odorless liquid called "benzin" by German chemist Eilhard Mitscherlich (Mitscherlich 1834). In fact, the gum benzoin, a balsamic resin, had benzoic acid as one of its components. In 1825, Michael Faraday was the first to isolate and identify the benzene, originally named bicarburet of hydrogen from compressed illuminating gas (Faraday 1825). In 1849, large scale production of benzene had begun from coal tar - a by-product from coke and coal gas (Mansfield 1849).

In 1858, Archibald Couper developed the theory of chemical structure (Couper 1858), and simultaneously Kekulé did similar work (Kekulé 1858). Kekulé spread his ideas in his books as well, which gave him more popularity than Couper. In 1861, Joseph Loschmidt was the first to draw a cyclic structure for benzene (Rocke 2014). Another notorious contribution to the structural chemistry came from Loschmidt. He made the first representation of: (1) allyl and vinyl moieties; (2) cyclopropane; (3) double and triple bonds; (4) the first line-notation formula; and (5) benzene as well. These findings were published in 1861 in a 47-page booklet which was the first book containing graphical molecules, and it was republished in 1913 (Loschmidt 1861, Wiswesser 1989).

By the mid 19th century, chemical composition of benzene was known, in which the low content of hydrogen per carbon was a challenge for chemists at that time. Probably, Kekulé hadn't read Loschmidt's booklet, because the latter was nearly anonymous to scientific society, although there is a brief mention in Kekulé's paper of Loschmidt. In 1865, Kekulé presented in this work the hexagonal structure of benzene based on the properties of symmetry and his vision of a snake biting its own tail (Kekulé 1865). Later, other propositions to the chemical structure of benzene were done (Kolb 1979), most of which were based on the Kekulé hexagonal structure (Fig. 19.1(A)). However, the benzene structure which is still most used nowadays (although not completely correct– see ahead) is Kekulé's representation.

After Kekulé's proposal for benzene structure, it was found that 1,2-disubstituted and 1,3-disubstituted derivatives are the same. Then, in 1872, Kekulé proposed a rapid interconversion between two Kekulé's structures (Fig. 19.1(B)). Then, the most appropriate representation of benzene is shown in Fig. 19.1(B) at the right, where the circle represents the delocalization of the π electrons.

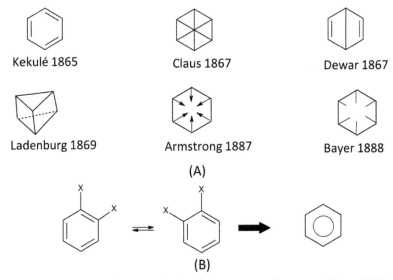

Figure 19.1 (A) Propositions to the chemical structure of benzene; (B) two Kekulé's resonance structures and benzene (resonance hybrid)

BENZENE'S STRUCTURE AND ELECTRONIC NATURE

Benzene is a planar and highly symmetrical molecule. All CC bonds are uniform (1.394 Å), all CH bonds have the same bond length (1.085 Å), and all CCC bond angles are equal (120°) (Fig. 19.2(A)). It has D_{6h} point group symmetry, which means it has one horizontal plane of symmetry, six vertical planes of symmetry, six C2 axes of symmetry, one C6 axis of symmetry, and one inversion center. Some of its symmetry elements are shown in Fig. 19.2(B). The CC bond in benzene is 0.044 Å longer than double CC bond in alkenes (1.35 Å in average) and 0.16 Å smaller than single CC bond in alkanes (1.54 Å in average). By using QTAIM DI (see chapter two) as indicative of bond order (Firme et al. 2009), the CC bond in benzene has a formal bond order 1.5, while single and double CC bonds have 1.0 and 2.0 formal bond orders, respectively.

According to Kekulé's representation of benzene, it has three pairs of π electrons. Then, there are three molecular orbitals to represent all π electrons, since each molecular orbital is doubly occupied. Four molecular orbitals of benzene are shown in Fig. 19.3. The molecular orbitals associated with the three pairs of π electrons are HOMO, HOMO-1, and HOMO-4. Surprisingly, the HOMO-2 which was supposed to represent π electrons instead of the HOMO-4 represents σ electrons. The HOMO and HOMO-1 has two orbital nodes, but HOMO encompasses all carbon atoms, while HOMO-1 encompasses only four carbon atoms. On the other hand, the HOMO-4 have only one orbital node. According to CCSD/6-31G+(d,p) level of theory, the energies, in atomic units, of HOMO, HOMO-1, HOMO-2, and HOMO-4 are: –0.33701, –0.33705, –0.49519, and –0.50267, respectively. Then, the first two HOMOS (HOMO and HOMO-1) are degenerate (same energy) and the third MO associated with π electrons (HOMO-4) have a smaller energy. *From MO theory, the molecular orbitals associated with the three pairs of π electrons are not degenerate and it seems to be an expected result due to the high symmetry of benzene.*

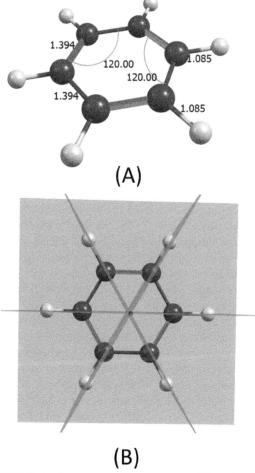

Figure 19.2   (A) Optimized geometry of benzene from G4 method along with some geometrical parameters; (B) some symmetry elements of benzene.

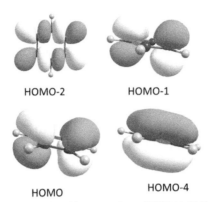

HOMO-2   HOMO-1

HOMO   HOMO-4

Figure 19.3   Molecular orbitals of benzene from CCSD/6-31G+(*d,p*) level of theory.

In chapter two, we have mentioned that spin-coupled valence bond, SCVB, is similar to generalized valence bond, GVB, when all resonance structures are included in the GVB calculation. A SCVB input does not need to specify the resonance structures because all spin-paired possibilities are already included. Then, SCVB is recommended to delocalized systems. As for GVB, it uses only one spin-pairing in its calculation and it is only recommended to localized systems. Nonetheless, when all possible resonance structures are included in the input of GVB calculation (by using flag $02VBSTR), the outcome is similar to that from SCVB for a delocalized system. See in Appendix 5 the inputs for GVB (for one resonance structure), GVB (for all five resonance structures), and SCVB calculations in VB2000 package of previously optimized benzene.

The only difference between the input of GVB for one resonance structure and GVB including all resonance structures of benzene is the flag $02VBSTR along with the number of resonance structures (5) and the spin-pairing scheme of the *p* electrons from carbon atoms 1 to 6. The first two lines (after the line containg number 5) are related to the spin-pairing scheme of Kekulé's resonance structures, and the last three lines are related to Dewar's resonance structures. For instance the first line after number five has 1 2 3 4 5 6 characters, which means spin-pairing between carbon atoms 1 and 2 (vicinal atoms), between 3 and 4 (vicinal atoms), and 5 and 6 (vicinal atoms as well). See in Fig. 19.4(A) a part of GVB input of benzene for all resonance structures. Figure 19.4(B) shows the five resonance structures of benzene according to the order of appearance in the input below.

It is important to remember (from chapter two) and emphasize that both GVB and SCVB yield singly occupied orbitals, while restricted MO theory yields doubly occupied orbitals, where restricted means the imposition of double occupation, which is the default calculation of singlet molecules in MO theory.

The singly occupied $C(2p_z)$ orbitals of GVB calculation using only one resonance structure (localized GVB) depicted in Fig. 19.4(C) are asymmetrically delocalized in order to overlap with other singly occupied $C(2p_z)$ GVB orbitals, for example between C1 and C2. As to the singly occupied $C(2p_z)$ orbitals from GVB calculation of all five resonance structures (delocalized GVB) and from SCVB, they are similar and they are symmetrically delocalized to both sides. *In fact, each* *p singly occupied C(2p$_z$) orbital from delocalized GVB and SCVB are localized in its corresponding carbon with a slight delocalization to its vicinal carbon atoms in benzene.* Since wave function from modern VB has permutation symmetry and it is univocal (unlike MO wave function), it presents a more reliable result (according to this author's opinion) of the electronic nature of a given molecule. As a consequence, one can state that there is neither π bond nor π electron in benzene, but six p electrons localized in each carbon atom with a slight delocalization towards its vicinal carbon atoms (Gerrat et al. 1987).

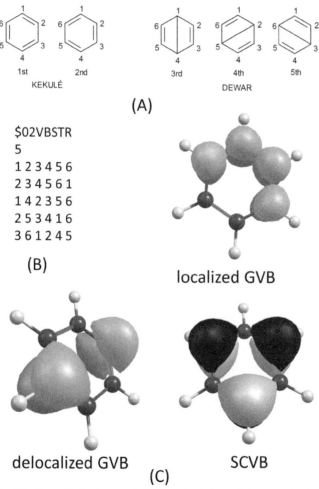

Figure 19.4 (A) Part of the input of GVB calculation of benzene including all five resonance structures; (B) Resonance structures of benzene; (C) Some singly occupied GVB C(2p$_z$) orbitals from one resonance structure (localized GVB) and all resonance structures (delocalized GVB) along with some singly occupied SCVB orbitals of benzene.

The delocalization index, DI, which is the amount of shared electron(s) between two atomic basins (nearly similar to atoms), is 1.0 in single CC bond of ethane. In benzene, the DI in each CC bond is 1.39. When subtracting 1.0 from 1.39, that is, by removing the amount of shared electrons of a single CC bond from the CC bond in benzene, there remains 0.39 shared electron between each CC bond in benzene. Since each p electron is a unit, when subtracting 0.39 from 1.0, there remains 0.61 p electron localized in each carbon atom of benzene. Then from QTAIM, there is 0.61 p electron localized in each carbon atom and 0.39 p shared electron in each CC bond. Although QTAIM uses charge density and VB uses wave function, the results from SCVB and QTAIM are in agreement for benzene (Firme et al. 2007a).

AROMATICITY AND RESONANCE ENERGY

The aromaticity is an extra stability which aromatic molecules have in comparison with reference non-aromatic molecules. Nonetheless, aromaticity is also related to electronic, geometrical, topological, and magnetic features, which will be discussed ahead.

As already stated, stability is a thermodynamic property which is evaluated in a specific reaction (e.g., hydrogenation, combustion, etc.). There are several ways to evaluate the aromaticity. One of them is the hydrogenation reaction:

$$\text{benzene} + H_2 \rightarrow \text{cyclohexa} - 1,3 - \text{diene} \therefore \Delta H = 10.1 \text{ kcal mol}^{-1}$$

Where the resonance energy is $10.1 \text{ kcal mol}^{-1}$.

All hydrogenation reactions of benzene (to cyclohexene or to cyclohexane), cyclohexadiene, and cyclohexene are exothermic (due to the π bond – σ bond change), except for the hydrogenation of benzene to cyclohexadiene. Then, benzene (plus one hydrogen molecule) is more stable than cyclohexadiene, although one π bond turns into a σ bond.

The adiabatic resonance energy, ARE, is the difference in energy between the real aromatic molecule (e.g., benzene) and the corresponding virtual most stable resonance structure (e.g., virtual cyclohexatriene). The ARE in terms of enthalpy from G4 method is 39.1 kcal mol^{-1} (see Fig.19.5). For comparison, the resonance energy of cylohexa-1,3-diene is only 1.7 kcal mol^{-1}. *Then, there is a huge increase in the stability of benzene due to its aromaticity, which does not exist in cyclohexa-1,3-diene.*

Resonance energy can also be obtained from isodesmic and homodesmotic reactions (Suresh and Koga 2002). These reactions are hypothetical and theoretical. In isodesmic reactions there is equivalence in the number of single, double, and triple bonds in reactants and products. One isodesmic reaction associated with the resonance energy of benzene is:

$$\text{benzene} + 6 \text{ methane} \rightarrow 3 \text{ ethene} + 3 \text{ ethane}$$

In a homodesmotic reaction there is an equal number of each type of C-C bond and equal number of each type of hybridized carbon atom in reactant and product. For example,

$$\text{benzene} + 3 \text{ ethene} \rightarrow 3 \text{ butadiene}$$

The resonance energy for this homodesmotic reaction is 23.9 kcal mol^{-1}.

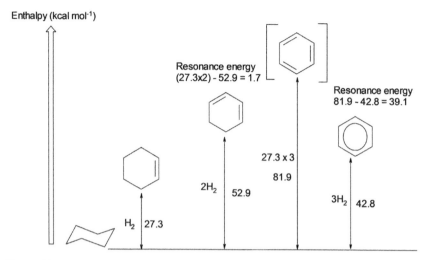

Figure 19.5 Hydrogenation enthalpy of cyclohexene, cyclohexa-1,3-diene, and benzene in kcal mol^{-1} from G4 method along with the corresponding resonance energies.

Aromaticity can also be estimated from the delocalization energy (DE), which is the energy difference between the resonance hybrid (Ψ) and its most stable resonance structure (φ). The vertical resonance energy (VRE) is the energy difference between resonance hybrid and its most stable resonance structure in the same geometry. The adiabatic resonance energy (ARE) is the corresponding energy difference from optimal hybrid and optimal resonance structure, which are different from each other (Jia et al. 2012).

$$DE = E(\varphi) - E(\Psi)$$

Although aromaticity is associated to a huge increase in the stability of the aromatic compound in comparison with a corresponding non-aromatic compound, the reason for this high stability is still in debate and perhaps will never reach a consensus. One reason for the high stability of benzene is associated with its five resonance structures (Fig. 19.4(A)). Nonetheless, other reasons for the high stability of benzene are attributed to a multi-center bonding (6c-6e) involving all p electrons of benzene (Nascimento 2008) or to the σ bonds of benzene (see ahead).

AROMATICITY CRITERIA AND HUCKEL'S RULE

There are four criteria associated with aromaticity: geometric, energetic, electronic, and magnetic. The energetic criterion was discussed in the previous section. The geometric criterion takes into account the regularity of CC bond length in benzene and it is the reference for an aromaticity index called HOMA (see ahead).

The electronic criterion of aromaticity arises from Huckel's work, known as Huckel method or Huckel Molecular Orbital, HMO. The HMO is based on the square Huckel matrix, M, whose matrix elements are 1 (if the atoms are the nearest neighbors) or 0 (otherwise). The number of lines/columns is the number of carbon atoms. One improvement of HMO is the inclusion of the alternation matrix, A, in the Huckel matrix. Then, for annulenes (conjugated monocyclic hydrocarbons) having $4n+2$ p/π electrons, the expectation value of A vanishes, i.e., there is no bond alternation. On the other hand, for annulene having $4n$ p/π electrons, the expectation value of A is non-zero, which means there is a bond alternation, where n is a non-negative integer number (Kutzelnigg 2006). These observations from HMO are known as Huckel's rule. There are some exceptions to Huckel's rule, which will be discussed ahead.

MAGNETIC AND ELECTRIC FIELDS AND MAGNETIC CRITERION OF AROMATICITY

When a magnetic field (B) is applied to the static charge (Q) over a given length (L) of a conducting wire which is in an oscilating movement in an orthogonal direction to the B (with velocity v), it generates a magnetic force (F) and an electromotive force (EMF) which induces the movement of the charge Q under EMF.

$$\left|\vec{F}\right| = Q \cdot \left|\vec{v}\right| \cdot \left|\vec{B}\right|$$

$$W = \left(Q \cdot \vec{v} \times \vec{B}\right) \cdot L$$

$$\text{EMF} = L \cdot \vec{v} \times \vec{B}$$

On the other hand, the electric current (I) in a conducting wire generates a circulating magnetic field around the wire with radius (r), whose modulus is given by:

$$\left|\vec{B}\right| = \frac{\mu I}{2\pi r}$$

Aromatic molecules (cyclic conjugated molecules with $4n+2$ π electrons) contain delocalized electrons moving in a closed circuit with a specific electric current, which generates an induced magnetic field and ring current (the circulating magnetic field around the CC bonds of the aromatic molecule). The induced magnetic field is in the same direction of the external field outside the aromatic ring. Then, for example, the hydrogens of benzene which are outside the ring experience a strong downfield (chemical) shift. Other types of magnetic criteria for aromaticity can be found elsewhere (Gershoni-Poranne and Stanger 2015).

ROLE OF SIGMA AND PI ELECTRONS IN AROMATICITY

Based on Nascimento's work (Nascimento 2008) on quantum-mechanical interference phenomenon as the main driving force for the stability of a bond, Cardozo and Nascimento used an energy partitioning scheme from GVB wave functions (dividing into covalent and non-covalent contributions) to show that the bond stabilization is due to the lowering of the kinetic energy term from the so-called interference phenomenon (Cardozo and Nascimento 2009). The non-covalent contributions came partly from the exchange energy from the antisymmetrization of electrons in different groups – σ and π groups. Afterwards, they selected five out of 30 normal modes of vibration of benzene to evaluate its VB wave function during these five vibrations from its optimized ground-state, its D_{6h} structure. The VB wave function of benzene was divided into three main groups: core electrons in HF method, σ electrons in GVB method, and π electrons in SCVB method, and its energy was partitioned according to Cardozo and Nascimento's partitioning scheme (Cardozo et al. 2014). They concluded that the resilience of benzene to preserve its D_{6h} structure (the *de facto* benzene's optimized structure) is due to the non-covalent contributions from the interaction between σ and π electrons and the nuclear potential energy. On the other hand, using stockholder partitioning scheme (S) from molecular orbital wave functions, Kovacevi et al. found that σ electrons stabilize benzene while π electrons destabilize it, which is known as π distorcivity (Kovacevic et al 2004).

$$E_{HF}^{\sigma}(S) = E(T)_{HF}^{\sigma} + V_{ne}^{\sigma} + V_{ee}^{\sigma\sigma} + \left(n_{\sigma}/N\right)V_{ee}^{\sigma\pi} + V_{nn}^{\sigma}(S)$$

$$E_{HF}^{\pi}(S) = E(T)_{HF}^{\pi} + V_{ne}^{\pi} + V_{ee}^{\pi\pi} + \left(n_{\sigma}/N\right)V_{ee}^{\sigma\pi} + V_{nn}^{\pi}(S)$$

Where $E(T)$ is the kinetic energy term, V is the potential energy, n stands for nucleus, e is the electron, and HF is the Hartree-Fock method where they found negligible influence of correlation energy term in the stability of benzene.

Then, the role of π electrons on the stability of benzene is unclear from the studies of energy partitioning shown above. However, as can be seen from both analyses, *the σ electrons play a stabilizing role in the aromaticity of benzene.*

AROMATICITY AND STABILITY FROM ELECTROSTATIC FORCE MODEL

From the electrostatic force model we have used throughout this book we can also show the stability of benzene in comparison with hexatriene. We will focus only on π electrons in this analysis. Hexatriene and benzene have only two resonance structures involving only π electrons. In hexatriene there is one neutral resonance structure and one zwitterionic resonance structure. The overall electrostatic force for these structures are:

$$F_{\pi(hexatriene)}^{neutral} = 36\left(K\frac{q_{e\pi}\left(Z_{eff(C)}q_e\right)}{r_{C1\pi}^2}\right)$$

$$F_{\pi(hexatriene)}^{zwitterion} = 16\left(K\frac{q_{e\pi}\left(Z_{eff(C)}q_e\right)}{r_{C1\pi}^2}\right)$$

The number 36 indicates that each of the six carbon atoms interacts with each of the six π electrons of neutral hexatriene resonance structure. The number 16 indicates that each of the four atoms with double bond interacts with each of the four π electrons of the zwitterion hexatriene resonance structure.

As for the two Kekulé structures of benzene, the total electrostatic force regarding only π electrons is:

$$F_{\pi\,(Kekulé\text{-}benzene)}^{overall} = 72\left(K\,\frac{q_{e\pi}\left(Z_{eff\,(C)}q_e\right)}{r_{C1\pi}^2}\right)$$

The number 72 means 2×36 for each Kekulé's resonance structure of benzene. Then, π bonding in benzene is stronger than that in hexatriene by sixteen times the electrostatic force between a carbon atom and a π electron. This model does not take into account the difference between the stability of both the resonance structures of hexatriene. If this is taken into account, then the difference between the electrostatic force of benzene and hexatriene is higher than sixteen times the electrostatic force between a carbon atom and a π electron. Moreover, this model does not take into account the three Dewar resonance structures of benzene for simplicity.

ANTI-AROMATICITY, ANTI-AROMATIC, AND NON-AROMATIC MOLECULES

Hereafter, it is important to differentiate three classes of molecules: aromatic, anti-aromatic, and non-aromatic molecules.

The aromatic molecules are cyclic, conjugated, flat (exceptions in cyclophanes– see ahead), have $(4n+2)$ π electrons (exceptions are shown ahead in ionic aromatic molecules), fully delocalized electronic circuit and are unsually stable.

The anti-aromatic molecules are cyclic, conjugated, flat, have $(4n)$ π electrons, and are extremely unstable.

The non-aromatic molecules are those which do not fulfill all the conditions of the aromatic or anti-aromatic molecules. They have saturated carbon (sp^3) carbon in the ring or their electronic circuit is not completely delocalized as evidenced by resonance structures

The anti-aromatic molecules are very difficult to synthesize and they are isolated under very restricted conditions.

While oxirane is very easy to obtain from the reaction of alkene with peroxy acid (peracid), the epoxidation of alkyne has never been successful. The corresponding oxirene (oxirane with a CC double bond) has never been obtained so far. Oxyrene is the simplest case of (virtual) anti-aromatic molecule with a lone pair of electrons from oxygen and two π electrons from double CC bond (Fig. 19.6(A). Its nitrogen analog, 1H-azirene, has never been obtained either (Fig. 19.6(B)).

The iconic example of anti-aromatic molecule is cyclobutadiene. Some derivatives of cyclobutene (with 4 π electrons) were synthesized, but they are unstable in contact with air. The cyclobutadiene dimerizes and it is converted to cyclooctatetratrane above 35 K. The cyclobutadiene has only been trapped in the interior of hemicarcerand macrocyclic molecule (Cram et al. 1991). The cyclobutadiene has a rectangular geometry instead of a square geometry whose CC bond lengths are 1.347 and 1.569 Å in an optimized geometry (a minimum at PES) from CCSD/6-31G+(d,p) level of theory (Fig. 19.6(B).

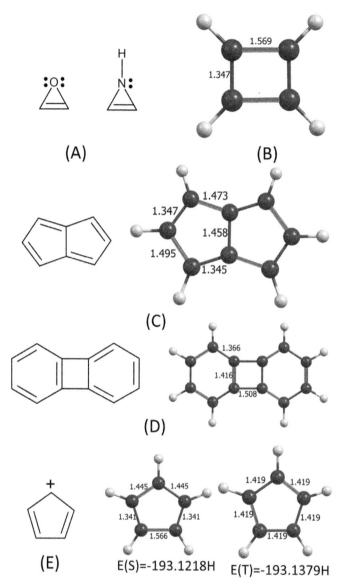

Figure 19.6 (A) Chemical structures of oxirene and 1H-azirene; (B) optimized geometry of cyclobutadiene from CCSD/6-31G+(*d,p*) level of theory; chemical structure, and optimized geometries of (C) pentalene; (D) biphenylene; and (E) singlet and triplet cation cyclopentadienyl obtained from ωB97XD/6-311G++(*d,p*).

The pentalene (a two fused cyclopentadiene rings) is an example of anti-aromatic molecule with 8 π electrons. Like butadiene, its CC bond lengths are not uniform with alternating double and single CC bonds (Fig. 19.6(C)). Pentalene was first synthesized from photocleavage of its corresponding dimer in argon matrices (Bally et al. 1997). It is stable only below minus a hundred degrees Celsius (−100°C).

The biphenylene is made up of two benzene rings attached by σ bonds from two vicinal carbons of each ring. The middle ring is an ($4n$) anti-aromatic ring with alternate (1.416 and 1.508 Å) CC bond lengths (see Fig. 19.6(D)). It was first synthesized in 1941 (Lothrop 1941). Biphenylene has drawn the scientists' attention because it has an ambiguous behavior as both an anti-aromatic and aromatic molecule (Dauben et al.1969).

The cyclopentadienyl cation also has $4n$ ($n=1$) π electrons. It has three distinguished CC bond lengths (1.445, 1.347, and 1.566 Å) and it is a planar structure (Fig. 19.6(E)). Then, it is an anti-aromatic molecule. However, its triplet is aromatic. All CC bond lengths in triplet cyclopentadienyl cation are uniform (1.419 Å), and it is 10.1 kcal mol^{-1} more stable than its singlet analog (see Fig. 19.6(E)).

The cyclooctatetrane has $4n$ ($n=2$) π electrons and a conjugated π system. However, it is a non-aromatic molecule because it has a non-planar structure. (Fig. 19.7(A)). Other examples of non-aromatic molecules are those with a non-conjugated π system (where there is a sp^3 carbon atom in the ring (Fig. 19.7(B)).

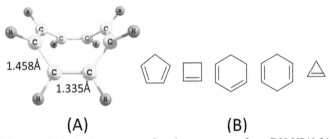

(A)   (B)

Figure 19.7 (A) Optimized geometry of cyclooctatetraene from B3LYP/6-31G++(d,p) level of theory and; (B) other examples of non-aromatic molecules.

AROMATICITY INDICES

As we have already stated: "An aromaticity index represents an attempt to quantify aromaticity based on a specific criterion (energetic, geometric, magnetic, reactivity, or electronic). However, all the aromaticity indices are somewhat arbitrary, which leads to a lack of agreement towards the best aromaticity index. As a consequence, it is advisable to use two or more aromaticity indices for the evaluation of a set of aromatic molecules" (Araújo et al. 2015).

Based on the geometric criterion, Krygowski and Kruszewski created HOMA (Harmonic Oscillator Model of Aromaticity), which became a well-known geometric aromaticity index that quantifies the aromaticity in comparison with benzene as a reference molecule (Kruszewski and Krygowski 1972).

$$HOMA = 1 - \frac{\alpha}{n} \sum_i \left(R_{opt} - R_i \right)^2$$

Where n is the number of chemical bonds; α is an empirical constant ($\alpha=257.7$); R_i is the individual bond length; and R_{opt} is 1.388 Å (CC bond length in benzene). One limitation of HOMA is that it needs a reference molecule.

Several electronic criteria were developed for aromaticity quantification based on quantum theory of atoms in molecules, QTAIM. The most famous are PDI, FLU, and D3BIA(D2BIA). All of them have the delocalization index, DI (the amount of shared electrons between each atomic pair derived from the Fermi hole density) in their formulas.

Based on the assumption that the delocalization indices between *para*-related carbon atoms are higher than those between *meta*-related carbon atoms in benzene, Poater et al. created the *para*-DI, PDI aromaticity index, which is an arithmetic average of all *para*-DI in six membered-rings (Poater et al. 2003). One limitation of PDI is that it can only be applied to six-membered rings.

From the same research group, the aromatic fluctuation index, FLU, was developed (Matito et al. 2005). Like HOMA, FLU also depends on an aromatic molecule as reference.

$$FLU = \frac{1}{n} \sum_{A-B}^{ring} \left[\left(\frac{\sum_{A \neq B} \delta(B, A)}{\sum_{B \neq A} \delta(A, B)} \right)^{\alpha} \left(\frac{\delta(A, B) - \delta_{ref}(A, B)}{\delta_{ref}(A, B)} \right) \right]^2$$

$$\alpha = \begin{array}{l} 1 \Leftrightarrow \sum_{A \neq B} \delta(B, A) > \sum_{B \neq A} \delta(A, B) \\ -1 \Leftrightarrow \sum_{A \neq B} \delta(B, A) \leq \sum_{B \neq A} \delta(A, B) \end{array}$$

Where the summation involves all atomic pairs in the aromatic site and $\delta(A,B)$ is the DI value between atoms A and B. This aromaticity index is also dependent on the reference molecule.

This author has developed the D3BIA – an aromaticity index based on three electronic factors: the electronic density within the aromatic ring; the uniformity of electronic delocalization in the aromatic ring; and the degree of degeneracy of the atoms in the aromatic ring (Firme et al. 2007a).

$$D3BIA = \left[100 - \left(\frac{100\sigma}{\overline{DI}} \right) \right] \cdot \left[(1 + \overline{\lambda_2}) \cdot \rho_{RCP} \right] \cdot \delta \therefore \delta = \frac{n}{N}$$

Where $\overline{DI}$ is the average of all DIs from the atoms of the aromatic ring and σ is the mean deviation of all DIs from the atoms in the aromatic ring; $\overline{\lambda_2}$ is the average of all second eigenvalues from the Hessian matrix of the charge density, λ_2, of all bond critical points in the aromatic ring, whose eigenvector, $\overline{u_2}$, points from BCP belonging to the aromatic ring towards the center of the ring; n is the number of atoms in the aromatic circuit that follows the rules for maximum degree of degeneracy shown below and N is the total number of atoms in the aromatic circuit.

The first term in the D3BIA equation is associated with delocalization index uniformity (DIU), the second term is associated with the ring density formula (RDF), and δ is related to degeneracy of the atoms in the aromatic site (degree of degeneracy). The DIU factor was empirically obtained from the analysis of several aromatic molecules, including benzenoid, hetero, and ionic aromatic systems, with varying ring sizes.

The D3BIA can also be used for the analysis of homoaromatic molecules (see ahead) as well. In that case, RDF formula is changed into:

$$RDF = \rho_{RCP/CCP}$$

Where ρ_{RCP} is the charge density in the ring critical point of cyclic homoaromatic molecule and ρ_{CCP} is the charge density in the cage critical point of caged homoaromatic molecule.

From QTAIM, it is possible to obtain the atomic energy according to the virial theorem (see chapters two and three). As a consequence, it is feasible to derive the degree of degeneracy for all atoms, δ, for an aromatic, homoaromatic, or any other aromatic system. The reason of δ in D3BIA formula is to evaluate the influence of heteroatom(s) in the aromatic site. For instance, Si_6H_6, N_6, and benzene have the same degree of degeneracy ($\delta = 1$), while the values of degree of degeneracy in pyridine, pyrazine, 1,3,5-triazine, 1,2,4-triazine (less symmetric) are 0.83, 0.67, 0.50, and 0.33, respectively.

The rules for maximum degree of degeneracy for aromatic systems are shown below. Although these rules are arbitrary, they were empirically obtained from a long range of aromatic (neutral and ionic aromatic molecules, heteroaromatic molecules, aromatic molecules with varying ring size) and non-aromatic systems so that heteroaromatic systems have $\delta < 1$.

I)   If all atoms of the ring are degenerate and if all of these atomic pairs have the same delocalization indices, the molecule has the maximum degree of degeneracy ($\delta = 1$).

II) If all atoms of the aromatic ring are near-degenerate (within the range of 0.3 au.) and if they have near delocalization indices (where $\Delta DI \leq 0.50e$), the system also has a maximum degree of degeneracy ($\delta = 1$).

In the case of homoaromatic systems, the degree of degeneracy was empirically obtained so that the influence of heteroatoms (and/or atoms of same chemical element but with very different atomic energies) in the aromatic circuit could be taken into account for evaluating the homoaromaticity. The general formula for degree of degeneracy is also the same for aromatic molecules where n is the number of atoms in the homoaromatic circuit whose $\Delta E(\Omega) \leq 0.009$ au. and N is the total number of participating atoms in the multicenter bond of the homoaromatic circuit. The minimum value for n is 1, where there is no atomic pair whose $\Delta E(\Omega) \leq 0.009$ au.

The D2BIA was an improvement of D3BIA where the δ (degree of degeneracy) was removed. This term has some arbitrary rules that are not so straightforward for new users of D3BIA (Firme and Araújo 2017). Then D2BIA is:

$$D2BIA = \left[100 - \left(\frac{100\sigma}{\overline{DI}} \right) \right] \cdot \left[\left(1 + \overline{\lambda_2} \right) \cdot \rho_{RCP} \right]$$

The aromaticity index most used from magnetic criterion is the nuclear independent chemical shift (Schleyer et al 1996), NICS, which a theoretical NMR calculation obtains from the absolute chemical shifts of all atoms of the molecule and all ghost atoms which are artificially placed in the center of the ring in the plane of the ring (NICS(0)) and/or one Angstrom above the center of the ring (NICS(1)).

The NICS can also be successfully used for caged aromatic or homoaromatic molecules in the center of the cage. The ghost atom is a probe that senses the absolute chemical shift at any region where it is placed after the NMR calculation.

ACENES

Acenes are fused benzenoid rings in one direction (linear acenes) or in more than one direction (non-linear acenes). It is an important class of organic semiconducting materials. The pentacene (with five fused benzenoid rings) has potential applications in the electronic industry.

As the size of the acene increases (i.e., the number of fused benzenoid rings increases), the acene becomes less aromatic. In Table 19.1, there is a collection of experimental resonance energies (Baird 1971) which increases from benzene to tetracene (four fused benzenoid rings). The increase in experimental resonance energy means the decrease of the aromaticity of the corresponding acene so that heptacene (seven fused benzenoid rings) has only been obtained in a matrix condition until 2017, where it was isolated for the first time from thermal cleavage of diheptacene (Einholz et al. 2017).

Table 19.1   Experimental resonance energy, in kcal mol^{-1}, of benzene and some acenes.

Molecule	Number of benzenoid rings	Experimental RE (kcal mol^{-1})[a]
Benzene	1	+21
Naphthalene	2	+33
Anthracene	3 (linear)	+43
Phenanthrene	3 (non-linear)	+49
Tetracene	4 (linear)	+59

(a) From Baird 1971.

From the resonance model, one can understand the reason why naphthalene is less aromatic than benzene. The naphthalene and benzene have two Kekulé resonance structures. However, in the former only one benzenoid ring in both resonance structures is an aromatic ring, that is, it has six p electrons in the ring (Fig. 19.8(A)).

As for anthracene, it has four resonance structures (Fig. 19.8(B)) where, for simplicity, it is assumed they have the same value of resonance coefficients ($c_i = 0.25$). Only in two resonance structures, each benzenoid ring of anthracene is aromatic. However, in one of these two resonance structures, each benzenoid ring has a shared aromatic ring (i.e., two p electrons are shared between two vicinal benzenoid rings). As we will show ahead, two shared benzenoid aromatic rings are less stable than two isolated benzenoid aromatic rings. On the other hand, in each resonance structure of naphthalene, it has one isolated benzenoid ring. Then, in two resonance structures (out of four) of anthracene (let's say φ_1 and φ_2) there are less than two aromatic rings for each benzenoid ring, while in naphthalene one (out of two) resonance structure has one isolated aromatic ring. This means that in the hybrid of naphthalene, each benzenoid ring is more aromatic than in the hybrid of anthracene.

$$\Psi_{naphthalene} = 0.5\varphi_1 + 0.5\varphi_2$$
$$\Psi_{anthracene} = 0.25\varphi_1 + 0.25\varphi_2 + 0.25\varphi_3 + 0.25\varphi_4$$

As for phenanthrene, there are five resonance structures where the internal benzenoid ring is less aromatic than the external benzenoid rings (Fig. 19.8(C)). Again, let us assume that all resonance structures have the same resonance coefficient ($c_i=0.20$). The external benzenoid rings are aromatic in four out of five resonance structures, while the internal benzenoid ring is aromatic in only three out of five resonance structures.

$$\Psi_{phenanthrene} = 0.20\varphi_1 + 0.20\varphi_2 + 0.20\varphi_3 + 0.20\varphi_4 + 0.20\varphi_5$$

Figure 19.8 also shows the structure of tetracene (Fig. 19.8(D)) and two of its isomers, pyrene and chrysene (Figs. 19.8(E and F)).

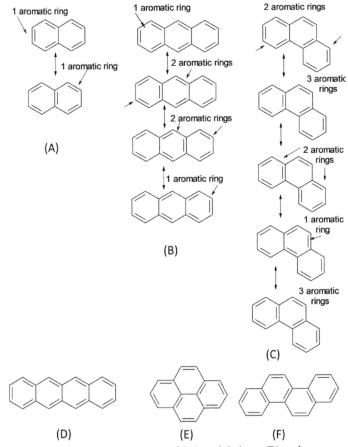

Figure 19.8 Resonance structures of (A) naphthalene, (B) anthracene, and (C) phenanthrene; chemical structures of (D) tetracene, (E) pyrene, (F) chrysene.

The values of D3BIA/D2BIA (in the case of acenes where $\delta = 1$, D3BIA = D2BIA) of some acenes, naphthalene, phenanthrene, anthracene, pyrene, tetracene, pentacene, are depicted in Figs. 19.9(A), (B), (C), (D), (E) and (F), respectively. along with their CC bond lengths, in Angstroms. We can see that all acenes have non-uniform CC bond lengths which are indicative of their smaller stability (or even instability for higher acenes) in comparison with benzene. The D2BIA/D3BIA in external rings decreases linearly as the size of the linear acenes increases (Araújo et al. 2015). The internal rings of linear acenes are more aromatic than the external rings of the corresponding acene. On the other hand, the interal rings of non-linear acenes are less aromatic than external rings. Let us compare the D3BIA values of anthracene and phenanthrene and then we can see that it is in accordance with the resonance model we have used for anthracene and phenanthrene (Fig. 19.8).

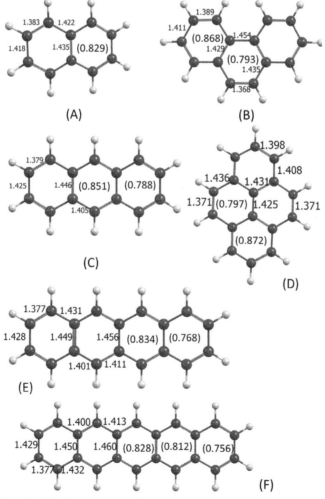

Figure 19.9 Optimized geometries of some acenes along with their CC bond lengths, in Angstroms, and D3BIA/D2BIA values (Araújo et al. 2015).

Although the internal rings of linear acenes are more aromatic than corresponding external rings (as indicated by all aromaticity indices), the Diels-Alder cycloaddition reaction of ethene and a linear acene (e.g., anthracene) tends to occur in the internal ring than in external ring. The cycloaddition reaction in the internal ring leads to two isolated aromatic rings, while the cycloaddition in the external ring leads to two shared aromatic rings (Fig. 19.10(A)). The Diels-Alder adduct from addition

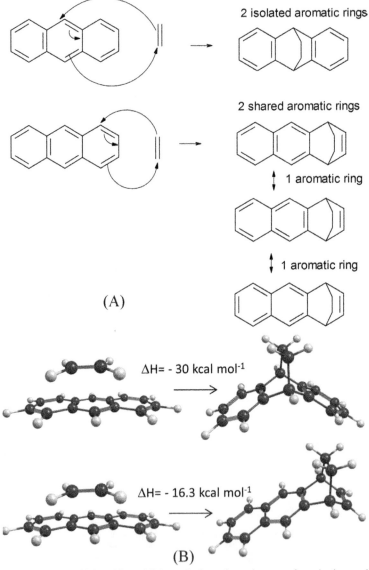

Figure 19.10 (A) Diels-Alder addition of ethene in anthracene from its internal and external ring; (B) optimized geometries and enthalpy changes, in kcal mol^{-1}, from these corresponding reactions using ωB97XD/6-31G+(d,p) level of theory.

in the external ring of anthracene has three resonance structures, where each benzenoid ring is aromatic in only two out of the three resonance structures. On the other hand, the Diels-Alder adduct from addition in internal ring always has two aromatic rings in any Kekulé resonance structure. Then, from the resonance model, *the adduct of Diels-Alder addition in the external ring of a linear acene is less stable than the corresponding adduct whose addition was in the internal ring*. This statement is confirmed by thermodynamic values in Fig. 19.10(B) using ωB97XD/6-31G+(d,p) level of theory. The Diels-Alder cycloaddition reaction of ethene in anthracene is more exothermic (by 14 kcal mol^{-1}) when it occurs in the internal ring than in the external ring. Moreover, *one can assume that two isolated aromatic rings are more stable than two shared aromatic rings.*

HETEROARENES

The IUPAC's definition of heteroarenes is: "heterocyclic compounds formally derived from arenes by replacement of one or more methine (-C=) and/or vinylene (-CH=CH-) groups by trivalent or divalent heteroatoms, respectively, in such a way as to maintain the continuous π-electron system characteristic of aromatic systems (...) corresponding to the Huckel rule".

Most heteroarenes containing a single ring are six- or five-membered rings. *Any six-membered heteroarene is less aromatic than benzene. As the number of heteroatoms increase in the six-membered ring, the hereroarene becomes less aromatic.*

From the definition of vertical resonance energy, VRE – the energy difference between resonance hybrid and its most stable resonance structure in the same geometry – we have used the localized GVB energy and the corresponding SCVB with the same optimized geometry of each aromatic molecule in order to obtain their VREs.

$$VRE = E(GVB) - E(SCVB)$$

Table 19.2 shows the valence bond VREs of benzene and some six-membered heteroarenes with increasing number of nitrogen atoms in the ring. As this number increases, the aromaticity, according to the valence bond VRE, decreases.

Table 19.2 GVB and SCVB energies of previously optimized heteroarenes, in Hartree, along with the corresponding VRE, in kcal mol^{-1}. using 6-31G++(d,p) basis set.

Molecule (number of heteroatoms)	GVB energy (Hartree)	SCVB energy (Hartree)	VRE (kcal mol^{-1})
Benzene (0)	−230.7712	−230.7854	8.9
Pyridine (1)	−246.7650	−246.7795	9.1
Pyrazine (2)	−262.7554	−262.7707	9.6
1,3,5-Triazine (3)	−278.7516	−278.7776	16.3

Figure 19.11 shows the optimized geometry of several heteroarenes (Firme and Araújo 2017). Their names and corresponding aromaticity measured from D2BIA, FLU, and NICS are given in Table 19.3. It is important to emphasize that the higher

the D2BIA (and D3BIA) value, the higher the aromaticity. On the other hand, the smaller the FLU value, the higher the aromaticity. As for NICS, the more negative the NICS value, the higher the aromaticity.

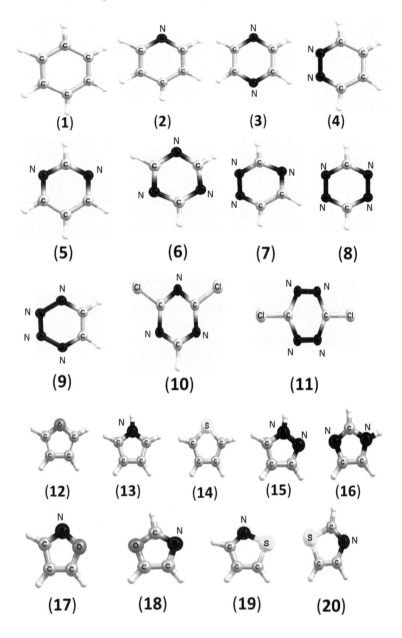

Figure 19.11 Optimized geometry of benzene and several heteroarenes (Firme and Araújo 2017) (see the Acknowledgment section).

Table 19.3 D2BIA, FLU, and NICS, in ppm, values of heteroarenes given in Fig. 19.11.

Molecule	D2BIA	FLU	NICS[a]
Benzene (1)	1.0187	0.0000	−8.05
Pyridine (2)	0.9972	0.0022	−6.83
Pyrazine (3)	0.9946	0.0050	−5.33
Pyridazine (4)	0.8461	0.0038	−5.33
Pyrimidine (5)	0.9525	0.0070	−5.52
1,3,5-triazine (6)	0.8956	0.0133	−4.05
1,2,4-triazine (7)	0.7904	0.0422	−3.75
1,2,4,5-tetrazine (8)	0.6038	0.1116	−1.72
1,2,3,4-tetrazine (9)	0.5373	0.1116	−2.62
Dichloro-1,3,5-triazine (10)	0.9219	0.0166	−4.88
Dichloro-1,2,4,5-tetrazine (11)	0.6325	0.0156	−1.91
Furan (12)	2.2035	0.1504	−11.85
Pyrrole (13)	2.3413	0.0489	−13.59
Thiophene (14)	2.1110	0.0947	−12.94
Pirazole (15)	2.1866	0.0337	−13.62
Imidazole (16)	2.3066	0.0589	−13.09
Isoxazole (17)	1.9398	0.1477	−12.29
Oxozole (18)	2.1572	0.1670	−11.44
Izothiazole (19)	2.2787	0.0756	−13.32
Thiazole (20)	2.1060	0.1069	−12.91

(a) In ppm.

From all aromaticity indices in Table 19.3, regarding the six membered rings, as the number of hetereoatoms increases, the aromaticity decreases. In addition, for the heteroarenes with the same number of heteroatoms (nitrogen), the less symmetric the less aromatic. For example, 1,3,5-triazine (more symmetric) is more aromatic than 1,2,4-triazine. As for the five-membered heteroarenes analysis from D2BIA, the aromaticity decreases as the number of heteroatoms increases, when considering the same type(s) of heteroatoms. For example, pyrrole is more aromatic than pirazole and imidazole. Furan and pyrrole are more aromatic than isoxazole and oxozole.

IONIC AROMATIC MOLECULES AND EXCEPTIONS TO HUCKEL'S RULE

The ionic aromatic molecules are cyclic, conjugated carbocation, or carbanion molecules having usual aromatic features: extra stability and uniformity of CC bond lengths in a planar, cyclic geometry similar to benzene. Most of the ionic aromatic molecules follow Huckel's rule, but there are a couple of exceptions. Like benzene, they can be represented by resonance structures. For example, cyclopentadienyl anion has five resonance structures where the negative charge is located in every carbon atom in each resonance structure (Fig. 19.12). Figure 19.12 also depicts the corresponding resonance hybrid.

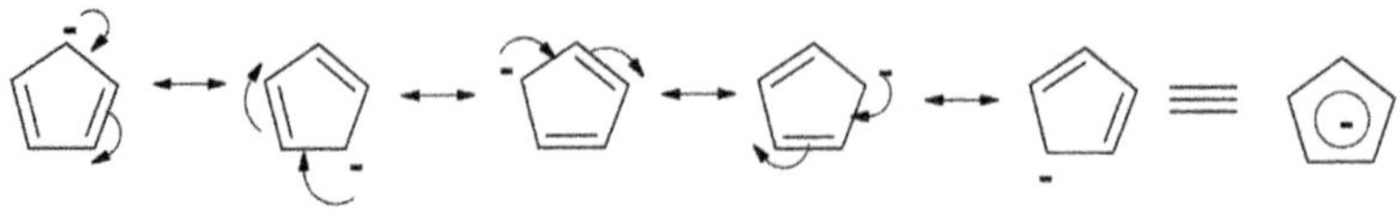

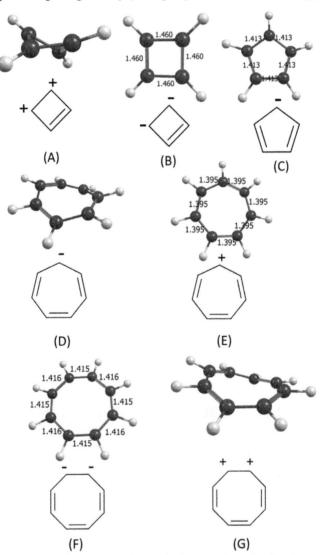

Figure 19.12   Resonance structures and resonance hybrid of cyclopentadienyl anion.

It is important to emphasize that a carbanion has a carbon with pseudo-tetrahedral geometry (with nearly sp^3 hybridization) and there are two electrons in the p orbital corresponding to the negative charge. On the other hand, the carbocation has a carbon with planar trigonal geometry (with sp^2 hybridization) and an empty p orbital.

Figure 19.13   Optimized structures of some ionic aromatic molecules along with their CC bond lengths, in Angstroms, from ωB97XD/6-31G+(d,p) level of theory.

Figure 19.13 shows some ionic aromatic molecules. All of them are planar geometries with uniform CC bond length and follow Huckel rule. They are: cyclobutenyl dianion ($4n+2=6$ and CC bond length 1.460 Å), cyclopentadienyl anion ($4n+2=6$ and CC bond length 1.413 Å), cycloheptatrienyl cation or tropylium cation ($4n+2=6$ and CC bond length 1.395 Å), and cyclooctatrienyl dianion ($4n+2=10$ and CC bond length 1.415(6) Å).

There are some exceptions to Huckel's rule. The cycloheptatrienyl anion has $4n=8$ π electrons and it is not planar, which means it is not an anti-aromatic ion, but rather a non-aromatic ion. The cyclobutenyl dication and cyclooctatrienyl dication have $4n+2=2$ and $4n+2=6$ π electrons, respectively, but they are not planar and they are not aromatic. However, their dianion parents are planar and aromatic. As we stated "its 6π electron resonance (from cyclobutenyl dianion) compensates its carbon-to-carbon repulsive interaction in which each carbon atom has a charge of –0.5 au Its parent dication (cyclobutenyl dication) probably has a puckered structure because the 2π-electron resonance does not compensate its carbon-to-carbon repulsive interaction (with 0.5 au positively charged carbons)" (Firme et al. 2007a). This reasoning is similarly applied to cyclooctatrienyl dication and dianion.

Nonetheless, some substituents lead the cyclobutenyl dication to its planarity. Two or more phenyl rings as substituents in cyclobutenyl are able to delocalize the positive charges in the cyclobutenyl moiety, and the corresponding derivatives become planar and aromatic. Curiously, some electron withdrawing groups are also able to yield planar cyclobutenyl dications because they reduce the π distorcivity of the ring (Firme et al. 2007b).

ANNULENES

Annulenes are completely conjugated monocyclic hydrocarbons which can be aromatic, anti-aromatic, or non-aromatic. They have C_nH_n general formula (for n as an even number). Cyclobutadiene, benzene, and cyclooctatetraene are the simplest cases of annulenes.

Although [10]annulene (Fig. 19.14(A)) and [14]annulene (Fig. 19.14(B)) have 10 and 14π-electrons theoretically, and follow Huckel's rule, they are not aromatic since they are not planar. The planar [10]annulene has three imaginary frequencies. As the interatomic distance between internal hydrogen atoms increase, the number of imaginary frequencies decrease (Fig. 19.14(C)). The most stable [10]annulene structure has no imaginary frequency and the interatomic hydrogen-hydrogen distance is 2.135 Å. However, it is a non-planar geometry and it is non-aromatic annulene. Vogel et al. solved the problem of the puckered geometry and the high ring strain in [10]annulene by replacing internal hydrogen atoms into a methylene bridge, which yielded the stable 1,6-methano[10]annulene (Hill et al. 1988). Although it distorts from the planarity (Fig. 19.14(C), its ^{1}H NMR data revealed its aromatic character. This is an important example of quasi-planar aromatic molecule. See the case of cyclophanes ahead.

The [14]annulene, $4n+2=14$ ($n=3$), would be an aromatic molecule, but due to its high ring strain has no planar structure (Fig. 19.14(D)) and it is quite unstable (Sondheimer and Gaoni 1960). Then, [14]annulene is a non-aromatic molecule.

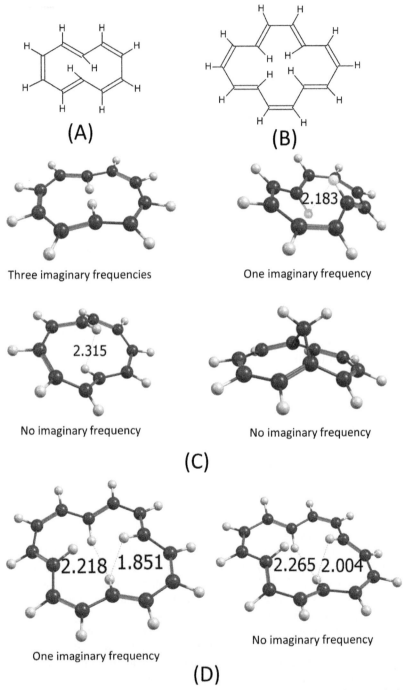

(A)

(B)

Three imaginary frequencies

One imaginary frequency

2.183

2.315

No imaginary frequency

No imaginary frequency

(C)

2.218 1.851

One imaginary frequency

2.265 2.004

No imaginary frequency

(D)

Figure 19.14   Chemical structure of (A) [10]annulene and; (B) [14]annulene; optimized geometries of (C) [10]annulene and 1,6-methano[10]annulene; and (D) [14]annulene from B3LYP/6-31G(*d,p*) level of theory.

CYCLOPHANES

Cyclophanes have aromatic rings bonded to aliphatic chains in a closed circuit involving non-adjacent bonds (*meta* or *para*) of the benzenoid rings. Most cyclophanes have distorted phenyl (or polycyclic aromatic) rings. We showed that benzenoid geometric parameters are responsible for 70% of the (in)stability of cyclophanes while 20% percent comes from one aliphatic parameter, global aromaticity and the intermolecular interactions (Firme and Araújo 2018).

In the nomenclature of cyclophanes there are numbers in squared brackets and they mean the number of methylene carbons bonded to each side of the phenyl ring. For example, *anti*-[2.2]metacyclophane (Fig. 19.15(A)), [2.2]paracyclophane (Fig. 19.15(B)), *syn*-[2.2]metacyclophane (Fig. 19.15(C)), [2.2]metaparacyclophane Fig. 19.15(D)), the multi-layered paracyclophane (Fig. 19.15(E)) and metacyclophane (Fig. 19.15(F)), [2]naphathaleno[2]paracyclophane (Fig. 19.15(G)) and pyrenophane (Fig. 19.15(H)).

There are two geometric factors that influence the aromaticity of benzenoid rings in cyclophanes: the CC bond length root mean square deviation (CC bond RMSD) and the deviation from planarity (DFP). The reference for RMSD is benzene's CC bond length and the DFP formula is an average of three dihedral angles (DA) involving carbon atoms in the ring (Firme and Araújo 2018).

$$DFP = \frac{\left|DA_{1'2'4'5'}\right| + \left|DA_{2'3'5'6'}\right| + \left|DA_{3'4'6'1'}\right|}{3}$$

The DFP is given in degrees where $1', 2',...$ are the carbon atoms in the benzenoid ring. The smaller the DFP, the closer is the bezenoid ring from planarity. Likewise, the smaller the CC bond RMSD, the closer is the benzenoid ring to benzene in terms of CC bond length. The DFP ranges from 0 to $16.2°$ in the benzenoid rings of the cyclophanes, and the CC bond RMSD ranges from 0.5 to 105×10^{-6}. The D2BIA values of the benzenoid rings ranged from 1.1077 to 0.9316 (Firme and Araújo 2018). *Then, aromaticity may exist even in quasi planar benzenoid or higher annulene rings.*

In general, benzenoid rings in cyclophanes with smaller DFP and CC bond RMSD have higher D2BIA (aromaticity), while benzenoid rings with higher DFP and CC bond RMSD have lower D2BIA. Nonetheless, a product of DFP and CC bond RMSD is necessary for a better comparison with D2BIA of the corresponding benzenoid ring (Firme and Araújo 2018).

Some cyclophanes have intramolecular interactions (π-stacking between parallel-displaced phenyl rings) which give small contributions to the stability of cyclophanes. These π-stacking are characterized in QTAIM by C-C intramolecular interactions from the molecular graph (Figs. 19.15(I and J)). Cyclophanes with parallel (not displaced), sandwich phenyl rings have repulsion between these rings which can be observed from NCI calculation (Firme and Araújo 2018).

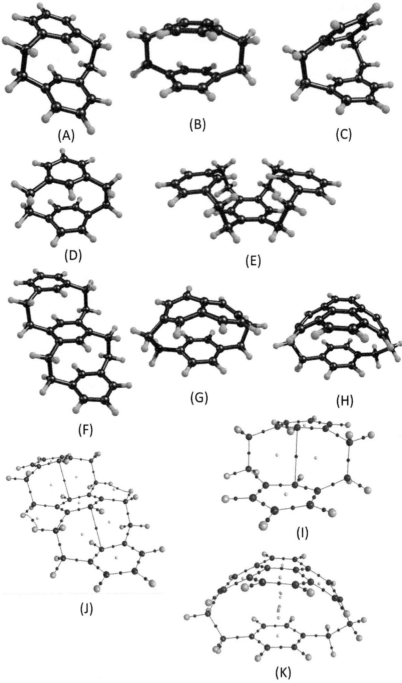

Figure 19.15 Optimized geometries of cyclophanes (A) to (H) from from
ωB97XD/6-311G++(d,p) level of theory. Molecular graph of (I) cyclophane (F),
(J) cyclophane (A), and (K) cyclophane (H).

SIGMA AROMATICITY

Tetrahedrane (Fig. 19.16(A)) is one of the most strained organic molecule, with a highly symmetrical structure, unusual bonding, and it is thermodynamically unstable. Conversely, tetrahedrane has σ-aromaticity because it has a large negative NICS value. Some substituents lead to a minimum at PES with the tetrahedrane structure (Figs. 19.16(B)-(D)), while other substituents lead to the cage opening (Fig. 19.16(E)).

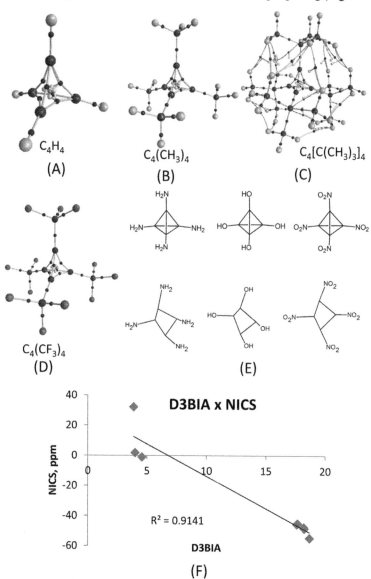

Figure 19.16 Optimized geometries of tetrahedrane and derivatives (A) to (D) and (E) open structures from initial tetrahedrane cage geometries; (F) plot of D3BIA versus NICS of corresponding tetrahedrane derivatives.

In our previous work we have stated that: "according to NICS and D3BIA and their relations with the corresponding formation enthalpy, the aromaticity is an important factor for the stability of the tetrahedrane cage. The σ-EDGs increase both aromaticity and stability of the tetrahedrane cage, whereas the EWGs decrease both aromaticity making and cage stability, except for tetrakis (*tert*-butyl)tetrahedrane, whose stability is due to the intramolecular hydrogen-hydrogen bonds" (Monteiro et al. 2014). In Fig. 19.16(F) one can see a linear relation between NICS and D3BIA. Then, tetrahedrane and derivatives represent a special case of aromaticity where there are only σ bonds.

HOMOAROMATICITY AND HOMOAROMATIC MOLECULES

The nonclassical ions (carbonium ions) have multi-center bonding (three-center two-electron, 3c-2e, or four-center two-electron, 4c-2e, bonding) from σ or π delocalized electrons towards the positively charged atom(s) so that this charge is delocalized over the multicenter bonding (see chapter twelve).

Homoconjugation occurs when a conjugation between adjacent centers is interrupted by insertion of a saturated unit, such as methylene carbon. Then, homoaromaticity is the energy lowering of the cyclic delocalization of homoconjugative interaction(s) of $(4n+2)$ electrons resulting in aromatic properties (chemical and magnetic aromatic properties).

Most of the carbonium ions are homoaromatic. Homoaromaticity is the energy lowering from cyclic delocalization of homoconjugative interaction(s) of $(4n+2)$ electrons resulting in aromatic properties. Winstein introduced the term homoaromatic to describe systems that have aromaticity in spite of one or more saturated linkages (sp^3 hybridized atoms in the ring) interrupting the formal cyclic conjugation (Winstein 1959). Most of the homoaromatic molecules are ions, but neutral homoaromatic molecules also exist.

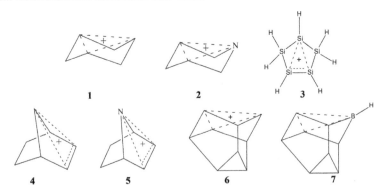

Figure 19.17   Chemical structures of 3-bicyclo[3.1.0]hexyl cation, its 3-N analog, silicon analog of cyclopentenyl cation, 7-norbornenyl cation, its 7-N analog, 9-pentacyclo[4.3.0.0.0]nonyl cation, and its 9-BH neutral analog.

A derivative of homocyclopropenium cation was the first system proposed to be homoaromatic. The term "homo-conjugation" is refered to a sp^3 center interrupting the conjugation. Then when there are two or three sp^3 centers interrupting the conjugation, it is termed "bishomo" and "trishomo", respectively. Then, 3-bicyclo[3.1.0]hexyl cation (1) is a tris-homocyclopropenyl cation, and 7-norbornenyl cation (4) is a bishomocyclopropenyl cation. In (2) and (5) there appear derivatives of (1) and (4), respectively where –CH- group is replaced by –N- atom (Fig. 19.17). Figure 19.17 also shows the silicon analog of cyclopentenyl cation (3) and 9-pentacyclo[4.3.0.0.0]nonyl cation (6) and its derivative changing –CH- group into –BH- (7).

In Fig. 19.18, the optimized geometries of the homoaromatic molecules (1) to (7) are depicted, along with their DI values from the multicenter bonding. Nearly half an

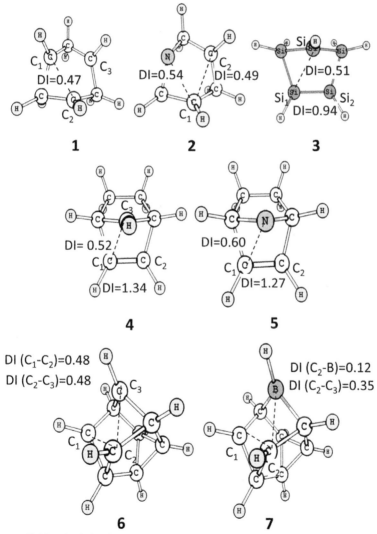

Figure 19.18   Optimized geometries along with selected DI values of (1) to (7) from B3LYP/6-311G++(d,p) level of theory.

electron charge density is shared between atoms of the multicenter bonding which are not directly bonded by a single or double bond. This is a significant amount of shared electron for through-space interaction. Nitrogen atom replacing –CH– group gives the same homoaromatic character in (2) and (5). Moreover, the 9-BH neutral analog of (6), that is (7), proves that homoaromatic molecules can also be neutral.

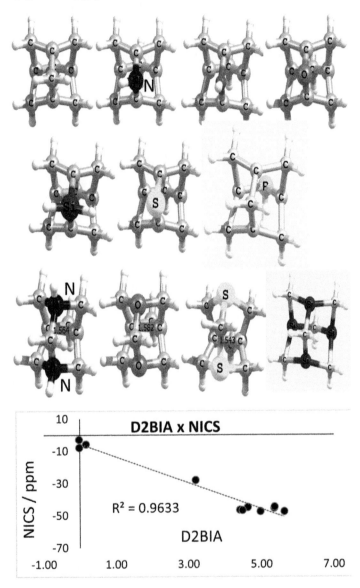

Figure 19.19 Optimized geometry of 1,3-dehydro-5,7-adamantanediyl dication and its derivatives changing one or more carbon atoms into nitrogen, boron, oxygen, silicon, sulfur, and phosphorus atoms along with the D2BIA versus NICS plot from these molecules. See the Acknowledgment section.

Homoaromatic molecules also exist through a tri-dimensional homoconjugation instead of the bi-dimensional homoconjugation shown before. The iconic example is the 1,3-dehydro-5,7-adamantanediyl dication, which has a $4c$-$2e$ bonding system. Figure 19.19 shows its optimized geometry and its analogs changing one or more carbon atoms in the multicenter bonding by nitrogen, boron, oxygen, silicon, sulfur, and phosphorus. Figure 19.19 also shows the excellent correlation between D2BIA and NICS from these homoaromatic molecules. There is a 99.99% correlation between D2BIA and D3BIA for this series. The 1,3-diazo, 1,3-dioxo, and 1,3-ditio-substituted analogs of the 1,3-dehydro-5,7-adamantanediyl dication are not homoaromatic as indicated by their low or negative values of D2BIA (−0.0130, 0.1728, and −0.0160, respectively) and small negative values of NICS (−7.88, −5.79 and −2.88 ppm, respectively), while the homoaromatic molecules have D2BIA ranging from 3.1935 to 5.6252 and NICS ranging from −28.1 to −47.56ppm (Firme and Araújo 2017).

EXERCISES

1. Give the seven resonance structures of cycloheptatrienyl cation where the positive charge is located in every carbon atom in each resonance structure. Afterwards, give its resonance hybrid.
2. Indicate the aromatic, non-aromatic, and anti-aromatic molecules from Fig. 19.20 by using Huckel rule and resonance structure (to prove the closed electronic delocalization or not).

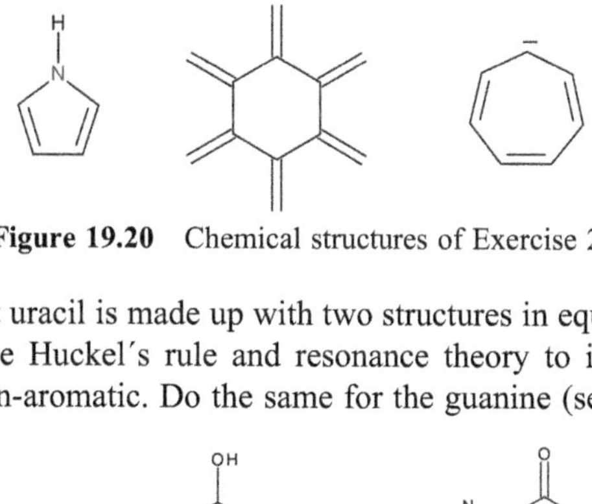

Figure 19.20 Chemical structures of Exercise 2.

3. By knowing that uracil is made up with two structures in equilibrium (lactam and lactim), use Huckel's rule and resonance theory to indicate which is aromatic and non-aromatic. Do the same for the guanine (see Fig.19.21).

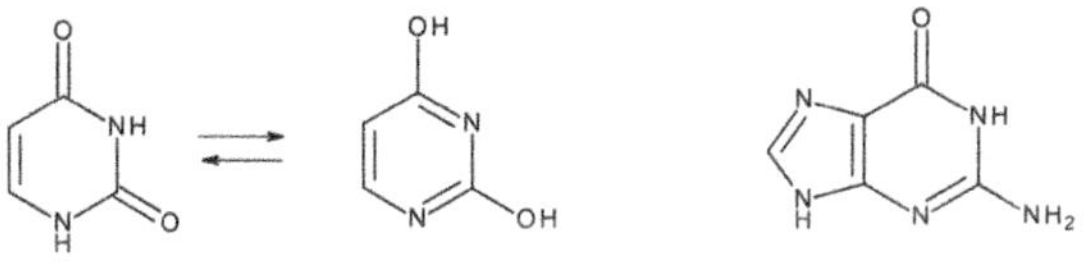

Uracil (lactam)   Uracil (lactim)   Guanine

Figure 19.21 Chemical structures of Exercise 3.

Acknowledgment

Figures 19.11 and 19.19 were published in our own article, whose copyrights belong to Springer-Verlag, and we appreciate the publisher's permission to print these figures in this book.

REFERENCES CITED

Araújo, D.M., da Costa, T.F. and Firme, C.L. 2015. Validation of the recently developed aromaticity index D3BIA for benzenoid systems. Case study:acenes. J Mol. Model. 21: 248.

Baird, N.C. 1971. Dewar resonance energy. J. Chem. Educ. 48: 509-513.

Bally, T., Chai, S., Neuenschwander, M. and Zhu, Z. 1997. Pentalene: formation, electronic, and vibrational structure. J. Am. Chem. Soc. 119: 1869-1875.

Cardozo, T.M. and Nascimento, M.A.C. 2009. Energy partitioning for generalized product functions: the interference contribution to energy of generalized valence bond ans spin coupled wave functions. J. Chem. Phys. 130: 104102-1.

Cardozo, T.M., Fantuzzi, F. and Nascimento, M.A.C. 2014. The non-covalent nature of the molecular structure of the benzene molecule. Phys. Chem. Chem. Phys. 16: 11024-11030.

Couper, A.S. 1858. XII. On a new chemical theory. The London, Edinburgh, and Dublin Phil. Mag. J. Sci. 16:104-116.

Cram, D.J., Tanner, M.E. and Thomas, R. 1991. The taming of cyclobutadiene. Angew. Chem. Int. Ed. 30: 1024-1027.

Dauben, H.J., Wilson, J.D. and Laity, J.L. 1969. Diamagnetic susceptibility exaltation in hydrocarbons. J. Am. Chem. Soc. 91: 1991-1998.

Einholz, R., Fang, T., Berger, R., Grüninger, P., Früh, A., Chassé, T., Fink, R.F. and Bettinger, H.F. 2017. Heptacene: characterization in solution, in the solid state, and in films. J. Am. Chem. Soc. 139: 4435-4442.

Faraday, M. 1825. XX. On new compounds of carbon and hydrogen, and on certain other products obtained during the decomposition of oil by heat. Phil. Trans. R. Soc. Lond. 115: 440-466.

Firme, C.L, Antunes, O.A.C. and Esteves, P.M. 2007a. Density, degeneracy, delocalization-based index of aromaticity (D3BIA). J. Braz. Chem. Soc. 18: 1397-1404.

Firme, C.L, Antunes, O.A.C. and Esteves, P.M. 2007b. Electronic nature of planar cyclobutenyl dication derivatives. J. Phys. Chem. A 111: 11904-11907.

Firme, C.L., Antunes, O.A.C. and Esteves, P.M. 2009. Relation between bond order and delocalization index of QTAIM. Chem. Phys. Lett. 468: 129-133.

Firme, C.L. and Araújo, D.M. 2017. D2BIA - flexible, not (explicitly) arbitrary and reference/ structurally invariant - a very effective and improved version of D3BIA aromaticity index. J. Mol. Model. 23: 253.

Firme, C.L. and Araújo, D.M. 2018. Revisiting electronic nature and geometric parameters of cyclophanes and their relation with stability – DFT, QTAIM and NCI study. Comp. Theor. Chem. 1135: 18-27.

Gerratt, J., Raimondi, M. and Cooper, D.L. 1987. The electronic structure of the benzene molecule. Nature 329: 492-493.

Gershoni-Poranne, R. and Stanger, A. 2015. Magnetic criteria of aromaticity. Chem. Soc. Rev. 44: 6597-6615.

Hill, R.K., Giberson, C.B. and Silverton, J.V. 1988. Forfeiture of the aromaticity of a bridged [10]annulene by benzannelation. J. Am. Chem. Soc. 110: 497-500.

Jia, J.-F., Wu, H.-S. and Mo, Y. 2012. The generalized block-localized wave function method: a case study of the conformational preference and C-O rotational barrier of formic acid. J. Chem. Phys. 136: 144315.

Kekulé, A. 1858. Ueber die constitution und die metamorphosen der chemischen verbindungen und über die chemische natur des kohlenstoffs. Ann. Chem. Pharm. 106: 129-159.

Kekulé, A. 1865. Sur la constitution des substances aromatiques. Bull. Soc. Chimique de Paris 3: 98-110.

Kolb, D. 1979. The aromatic ring. J. Chem. Educ. 56: 334-337.

Kovacevic, B., Baric, D, Maksic, Z.B. and Muller, T. 2004. The origin of aromaticity: important role of the sigma framework in benzene. Chem. Phys. Chem. 5: 1352-1364.

Kruszewski, J. and Krygowski, T.M. 1972. Definition of aromaticity basing on harmonic oscillator model. Tetrahedron Lett. 13: 3839-3842.

Kutzelnigg, W. 2006. What I like about Huckel theory. J. Comp. Chem. 28: 25-34.

Lothrop, W.C. 1941. Biphenylene. J. Am. Chem. Soc. 63:1187-1191.

Loschmidt, J. 1861. Chemische studien, A. Constitutions-formalen der organischen chemie in geographischer darstellung, B. Das Mariotte´sche gesetz. Vienna.

Mansfield, C.B. 1849. Untersuchung des steinkohlentheers. Annalen der Chemie Chemie und Pharmacie. 69: 162-180.

Matito, E., Duran, M. and Solà, M. 2005. The aromatic fluctuation index (FLU): a new aromaticity index based on electron delocalization. J. Chem. Phys. 122: 014109.

Mitscherlich, E. 1834. Uber das benzol und die säuren der oel- und talgarten. Annalen der Pharmacie 9: 39-48.

Monteiro, N.K.V.; Oliveira, J.F. and Firme, C.L. 2014. Stability and electronic structures of substituted tetrahedranes, silicon and germanium parents – a DFT, ADMP, QTAIM and GVB study. New J. Chem. 38: 5892-5904.

Nascimento, M.A.C. 2008. The nature of the chemical bond. J. Braz. Chem. Soc. 19: 245-256.

Poater, J., Fradera, X., Duran, M. and Solà, M. 2003. The delocalization index as an electronic aromaticity criterion: application to a series of planar polycyclic aromatic hydrocarbons. Chem. Eur. J. 9: 400-406.

Rocke, A.J. 2014. It began with a daydream: The 150th anniversary of the Kekulé benzene structure. Angew. Chem. Int. Ed. 53: 2-7.

Schleyer, P.v.R., Maerker, C., Dransfeld, A., Jiao, H., Hommes, N.J.R.v.E. 1996. Nucleus-independent chemical shift: a simple and efficient aromaticity probe. J. Am. Chem. Soc. 118: 6317-6318.

Sondheimer, F. and Gaoni, Y. 1960. Unsaturated macrocyclic compounds. XV. Cyclotetra-decaheptaene. J. Am. Chem. Soc. 82: 5765-5766.

Suresh, C.H. and Koga, N. 2002 Accurate calculation of aromaticity of benzene and antiaromaticity of ciclobutadiene: new homodesmotic reactions. J. Org. Chem. 67: 1965-1968.

Winstein, S. 1959. Homo-aromatic structures. J. Am. Chem. Soc. 81: 6524-6525.

Wiswesser, W.J. 1989. Johann Josef Loschmidt (1821-1895): a forgotten genius. Aldrichimica Acta 22: 17-19.

Chapter Twenty

Substituent Groups and Electrophilic Aromatic Substitution

INTRODUCTION

Benzene, as the most important representative of aromatic compounds, normally does not undergo addition reaction unless under very drastic conditions. While cyclohexene and 1,3-cyclohexadiene undergo bromination (normal addition reaction in darkness and impurity-free solvent), bromine cannot react with benzene even under light exposure by addition reaction (Fig. 20.1(A)). However, in a catalyzed synthesis, benzene can react with bromine in a substitution reaction rather than an addition reaction. Then, aromaticity in benzene changes its reaction behavior drastically in comparison with other unsaturated molecules.

Reactants such as bromine, chlorine, nitric acid, sulfur trioxide, and alkyl/acyl halides, using catalysists (except for sulfur trioxide) can react as electrophiles with benzene (besides other benzenoid and five-membered heteroaromatic systems). In this case, benzene (and other benzenoid/five-membered heteroaromatic compounds) reacts as a nucleophile. As mentioned above, in this reaction, a substitution occurs instead of an addition reaction, i.e., the hydrogen atom from benzene is exchanged by the electrophile, E. In another words, the net result of this reaction is that the E replaces H in benzenoid ring without eliminating its aromaticity. This reaction is known as electrophilic aromatic substitution, EAS.

Sometimes, the mechanism for electrophilic aromatic substitution is known as Wheland mechanism because it passes through the Wheland intermediate or σ-complex (Fig. 20.1(B)). In 1942, G.W. Wheland was the first to propose theoretically the existence of the arenium ion (the σ-complex) as an intermediate in the EAS (Wheland 1942).

In this mechanism, benzene and an electrophile interact with each other as a π complex, and henceforth two p electrons (see the next paragraph) from benzene attack the electrophile (here represented as $E\text{-}Y$), which passes through the activated complex of the rate determining step (Fig. 20.1(C)). After that, it goes to the intermediate so-called σ complex, followed by the barrierless removal of the hydrogen from benzene to yield the substituted aromatic compound.

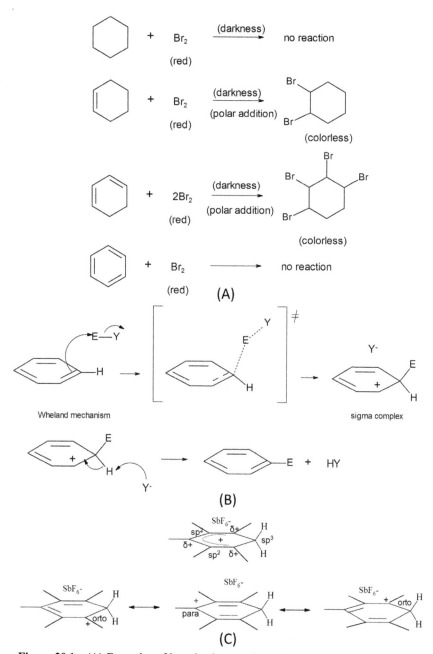

Figure 20.1   (A) Examples of bromination reaction; (B) Wheland mechanism for electrophilic aromatic substitution; (C) resonance hybrid and resonance structures of an isolated arenium ion (Olah 1965).

Let us recall that benzene does not have a typical π bond, but p electrons localized in each carbon atom with slight distortion to vicinal carbon atoms according to SCVB result (see chapter nineteen). However, for simplicity, in EAS mechanism, we will usually represent benzene (and other benzenoid systems) as one of its Kekulé structure where one of its "π bond" attacks the electrophile. Although the representation of benzene is simplified to one Kekulé structure, we will always refer to p electrons instead of π electrons of benzene when mentioning EAS mechanism.

The arenium ion is the structure of the σ complex in EAS mechanism. In the arenium ion, one carbon has sp^3 hybridization (the one attacked by the electrophile). The remaining 4 p electrons (or 2 π bonds) in the arenium ion are delocalized among five sp^2 carbon atoms by means of three resonance structures (Fig. 20.1(C)). The formal positive charge is delocalized in orto and para positions with respect to the sp^3 carbon atoms. The arenium ion was already isolated as a stable species (Olah 1965) which reinforces the existence of this intermediate in EAS mechanism. Nonetheless, ahead we will see that some EAS reactions do not undergo Wheland intermediate.

AROMATIC NITRATION

The aromatic nitration is the EAS using nitronium ion as an electrophile. This reaction is done under reflux at 55°C. The sources of nitronium ion can be nitronium salts, NO_2^+ ClO_4^-/BF_4^-/etc. (Guk et al. 1983), or nitric acid catalyzed by sulfuric acid. The sulfuric acid protonates the nitric acid, yielding nitronium ion (NO_2^+) and water. The sulfuric acid is regenerated at the end of the reaction (Fig. 20.2(A)).

The formal positive charge of nitronium ion (which has a linear geometry) is located in the sp-hybridized nitrogen atom. In the π complex, the nitronion ion is positioned vertically in the middle of the benzene ring. One oxygen atom is interacting with the six carbon atoms through dipole-quadrupole interaction represented by six bond paths, according to QTAIM, in the π complex (Fig. 20.2(B)).

According to B3LYP/6-31G++(d,p) level of theory, aromatic nitration has enthalpy change –27.74 kcal mol^{-1} and entropy change –0.004 kcal mol^{-1} K^{-1}. Then, nearly no entropy change occurs and the exothermic reaction is because the C-N bond is stronger than the N-O bond.

Two mechanisms are proposed for the aromatic nitration in gas phase: the polar mechanism (Fig. 20.2(C)) and the SET mechanism (Fig. 20.2(D)). The former was proposed by Hughes et al. (Hughes et al. 1950) and the latter was proposed by Weiss (Weiss 1946). The polar mechanism was theoretically supported by Parker et al. (Parker et al. 2013) and the SET mechanism was experimentally supported by Kochi and collaborators (Kochi 1992) using time-resolved spectroscopy. In both polar and SET mechanisms, the π complex is formed. In the polar mechanism, an energy barrier (and an activated complex) exists between π complex and σ complex where two electrons are transferred from benzene to form C-N bond (polar mechanism meaning electron-pair movement). In the SET mechanism, there is a charge transfer between the π complex and intimate pair (a sort of pre-σ complex) before forming the σ complex.

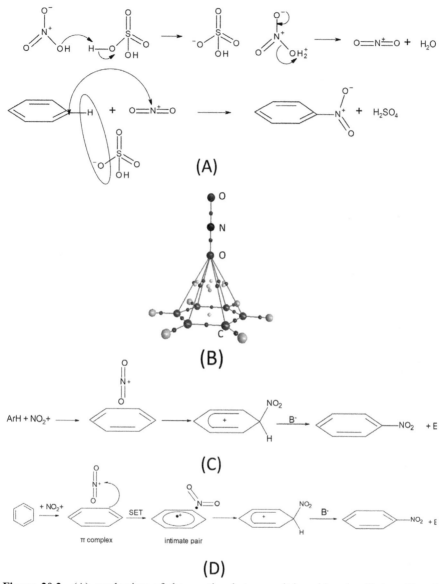

Figure 20.2 (A) mechanism of the reaction between nitric acid and sulfuric acid and overall scheme of the aromatic nitration; (B) molecular graph of the nitronium ion-benzene π complex; (C) general aromatic substition mechanism of aromatic nitration; (D) SET mechanism for aromatic nitration.

Besides the experimental evidences for the SET mechanism in the aromatic nitration, it was also theoretically supported by Gwaltney et al. using MP2/6-31G** and CCSD(T)/6-31G** levels of theory (Gwaltney et al. 2003), and by Chen and Mo using the BLW method and ab initio VB method, where strictly electron-localized diabatic states for π complex and intimate pair were obtained (Chen and

Mo 2013). One important conclusion of Chen and Mo's theoretical work where they compared the reactivity of nitronium ion and nitrosonium ion (NO^+) towards benzene: "(...) though NO_2^+ has a seemingly lower binding energy than NO^+ to benzene, it has a much higher charge transfer stabilization energy then higher chemical reactivity than NO^+; the so-called π complex $[C_6H_6NO_2]^+$ is more like a pre-σ complex, and its change to the σ complex is thus nearly barrierless" (Chen and Mo 2013). This pre-σ complex is the intimate pair depicted in Fig. 20.2(D). Another important consequence is that aromatic nitration in gas phase takes place without any energy barrier.

Then, the intimate pair is an important critical point in the potential energy surface (PES) in order to prove the SET mechanism. One can see that the only difference in Fig. 20.2(C) and Fig. 20.2(D), where transition states are omitted, is the intimate pair in Fig. 20.2(D). In our own calculations using PBE1PBE/6-311G++(d,p) level of theory, we have found the intimate pair. See the geometric differences (interatomic distances) comparing π complex intimate pair and σ complex in Fig. 20.3(A). When using a functional with corrections in dispersion interaction (e.g., ωB97XD), the intimate pair could not be found (Fig. 20.3(B)). When using MP2/6-311G++(d,p), the same method used by Parker et al., we found the intimate pair (Fig. 20.3(C)). In order to be completely sure about the SET in gas phase, we used the CCSD(T) method, which is one of the most accurate post-Hartree-Fock methods (See acknowlegment section). When using CCSD(T)/aug-cc-vdz level of theory, we found the intimate pair as critical point at the PES using benzene, nitrobenzene, and toluene (Fig. 20.3(D)).

Table 20.1 depicts the QTAIM atomic charge of each atom in nitronium ion (or its moiety) along the reaction coordinate (passing through π complex, intimate pair, and σ complex). Then, by substracting a unit positive charge (from the isolated nitronium ion) from the overall charge in the complexed nitronium ion ($\Sigma q(NO_2^+)$), one finds the charge transfer towards the nitronium ion along the reaction coordinate, $CT(NO_2^+)$. One can see that in the intimate pair, there is 80% of an electron transfer towards the nitronium ion.

Table 20.1 Atomic charge (q) of nitrogen and oxygen atoms of nitronium ion along with its overall charge [$\Sigma q(NO_2^+)$] and the amount of charge transferred to it [$CT(NO_2^+)$], in au., in all three intermediates of benzene.

Intermediate	QTAIM atomic charge (q) (au.)				$CT(NO_2^+)$ (au)
	$q(N)$	$q(O1)$	$q(O2)$	$\Sigma q(NO_2^+)$	
π complex	+0.921	−0.089	−0.112	+0.720	0.280
intimate pair	+0.749	−0.285	−0.265	+0.199	0.801
σ complex	+0.530	−0.368	−0.368	−0.206	1.206

Then, it is nearly certain that the SET mechanism occurs for aromatic nitration in gas phase. Nonetheless, no theoretical work has been done so far in condensed phase using static quantum chemistry and metadynamics. Whatever is the mechanism for aromatic nitration in condensed phase, it has to take into account the kinetics for aromatic nitration either using the complete expression:

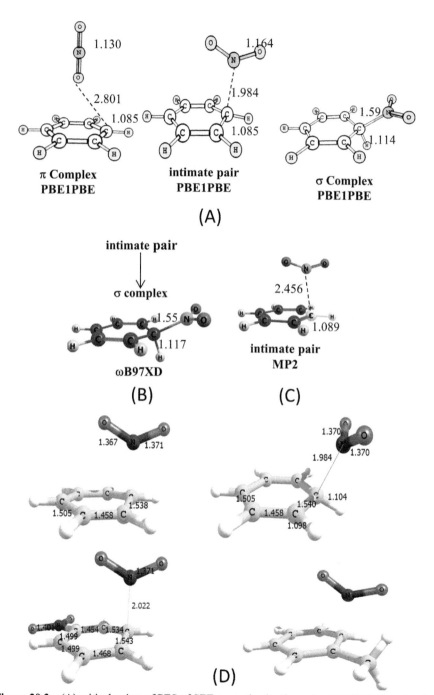

Figure 20.3 (A) critical points of PES of SET aromatic nitration using PBE1PBE functional; (B) arbitrary initial geometry of intimate pair yielding σ complex after geometry optimization using ωB97XD functional; (C) intimate pair geometry from MP2 calculation; (D) intimate pair critical points using CCSD method.

$$r_{ArNO_2} = \frac{\left(K_1 K_2 k_3 / K_{H_2SO_4}[H_2O]\right)[HNO_3]\left[H^+\right][ArH]}{1+\left(k_3/k_{-2}[H_2O]\right)[ArH]k_a} \therefore$$

$$k_a = \left\{1+\left(k_2/k_1\right)\left[1+\left[H^+\right]/K_{H_2SO_4}\right]/\left([H_2SO_4]+\left[HSO_4^-\right]\right)\right\}$$

or using QE and RDS simplifications (Peters 2017):

$$r_{ArNO_2} = k_3\left[HNO_3\right]\left[H^+\right][ArH]$$

Where the rate constant of the third step, k_3, is related to the rate determining step (chapter ten).

AROMATIC SULFONATION

The aromatic sulfonation has many industrial applications and the most important sulfonating agents are sulfur trioxide (SO_3) and fuming sulfuric acid (sulfur trioxide in sulfuric acid).

The mechanism of substitution electrophilic aromatic (SEA), for aromatic sulfonation, is commonly explained by Wheland mechanism (passing through the arenium ion). Nonetheless, it has been proved theoretically that, at least, the first step of the SEA undergoes a different mechanism.

According to B3LYP/6-31G++(d,p) level of theory, the enthalpy change in aromatic sulfonation (using SO_3 as nitrating agent) is -19.83 kcal mol^{-1} and the entropy change is -0.04 kcal mol^{-1} K^{-1}. Then, the entropy change is still low, but 10 times higher than that from aromatic nitration.

Cerfontain and coworkers have hundreds of experimental works on sulfonation of organic compounds, including aromatic molecules. Cerfontain proposed the mechanism for aromatic sulfonation using sulfur trioxide as a sulfonating agent (Cerfontain 1985). In that mechanism, the sulfur trioxide forms σ complex with benzene and then the SO_3^- moiety acts as a nucleophilic site by attacking a second molecule of SO_3 to yield a second σ complex. Afterwards, the benzenepyrosulfonic acid ($C_6H_5S_2O_6H$) is formed after deprotonation of this second σ complex. The last step is the reaction of the benzenepyrosulfonic acid with a second molecule of benzene, yielding two molecules of benzenesulfonic acid ($C_6H_5SO_3H$).

The first three steps of Cerfontain's mechanism of aromatic nitration were revisited theoretically by two researching groups.

In 2011, Schleyer's group found theoretically that the mechanism to form benzenepyrosulfonic acid occurs in a single step through concerted mechanism (Koleva et al 2011). It forms a π complex involving benzene and two molecules of sulfur trioxide, followed by the transition state represented in Fig. 20.4(A). Figure 20.4(A) also depicts one proposed mechanism (following classic Wheland mechanism) for the reaction between benzenepyrosulfonic acid and benzene to form two molecules of benzenesulfonic acid.

In 2017, using metadynamics and molecular dynamics methods in gas phase and with explicit solvent molecules, Moors et al. showed that: (i) benzene-SO_3 and benzene-S_2O_6 σ complexes are unstable at room temperature; (ii) they confirmed the transition state found for Schleyer's group (Fig. 20.4(A)) as the lowest barrier for aromatic sulfonation using sulfur trioxide; (iii) a new kinetic model was proposed in order to agree with the experimental kinetic data (Moors et al. 2017).

When considering sulfuric acid and sulfur trioxide as sulfonating agents, Morkovnik and Alkopova proposed a single step reaction in a concerted mechanism whose transition state is shown in Fig. 20.4(B) (Morkovnik and Alkopova 2013).

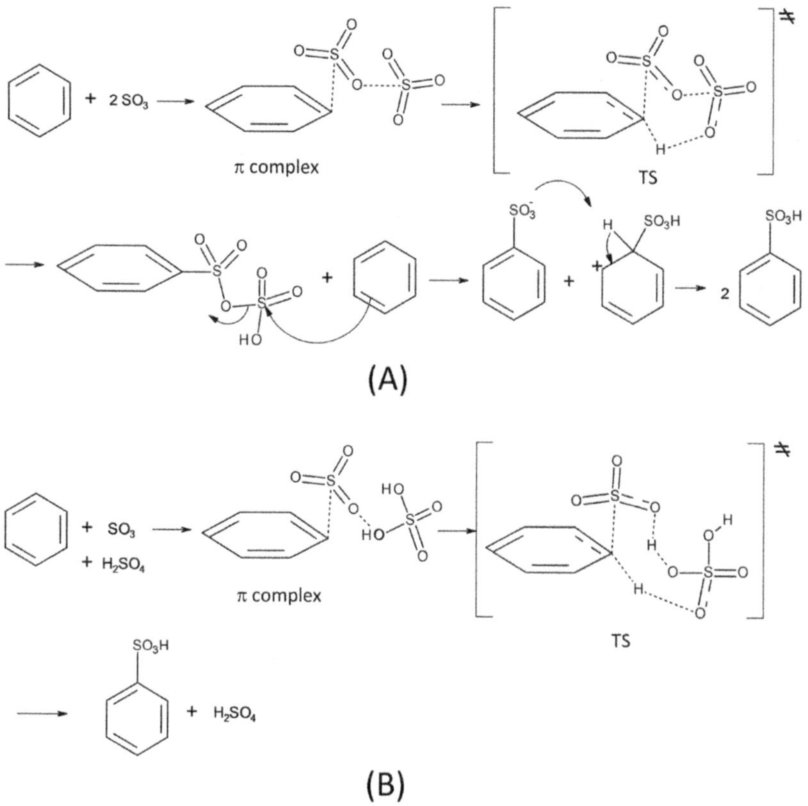

Figure 20.4 Mechanisms for the aromatic nitration using (A) sulfur trioxide, and (B) sulfur trioxide and sulfuric acid as sulfonating agents.

FRIEDEL-CRAFTS ALKYLATION

The alkylation (and acylation) or aromatic compounds have been widely applied before Friedel-Crafts. In 1877, Friedel and Crafts reacted n-pentyl chloride (amyl chloride) with aluminum metal to yield $C_{10}H_{11}Cl$ and HCl (Calloway 1935). Later, they found that aluminum chloride was the catalyst. When using benzene, the reaction gave:

$$C_6H_6 + CH_3Cl \xrightarrow{AlCl_3} C_6H_5CH_3 + HCl$$

$$C_6H_6 + C_5H_{11}Cl \xrightarrow{AlCl_3} C_6H_5C_5H_{11} + HCl$$

Afterwards, Friedel and Crafts found that other dry metallic compounds could be used as catalysts: ferrous chloride, ferric chloride, zinc chloride, sodium aluminum chloride, where aluminum chloride was the most effective so far (Calloway 1935).

Calloway defined Friedel-Crafts reaction as an activation reaction by aluminum chloride. Usually, Friedel-Crafts reactions are divided into alkylation and acylation, although alkylation seems to be more unreliable and less effective than acylation (Calloway 1935). Both reactions of benzene shown above are examples of Friedel-Crafts alkylation.

Friedel-Crafts alkylations usually form polymethylated benzenes. When the aromatic compound reacts with primary or secondary alkyl halides higher than ethyl chloride, it tends to form the highest branched compound possible. See below (Calloway 1935):

$$C_6H_6 + n-/iso-C_4H_9Cl \xrightarrow{AlCl_3} C_6H_5(tert-)C_4H_9 + HCl$$

On the other hand, other aromatic compounds, such as phenols, aryl aldehydes, aryl esters, alkybenzenes, and acenes can undergo Friedel-Crafts alkylations as well.

Besides alkyl halides, alkenes can also be used as alkylating agents. For example, ethene reacts with benzene catalyzed by aluminum chloride (Calloway 1935). The reactions with higher alkenes occur according to the Markovnikov regioselectivity. Other alkylating agents can also be found in Calloway's review paper.

$$C_6H_6 + C_2H_6 \xrightarrow{AlCl_3} C_6H_5C_2H_5 + H_2$$

Marsi and Wilen stated that Friedel-Crafts alkylation using n-alkyl halides can form unbranched alkylated aromatic compounds with considerable amounts, ranging from 28 to 40% in most cases. This result is different from former results, which had indicated small and insignificant amounts of non-branched alkylated aromatic compounds as Friedel-Crafts products. These former results of the Friedel-Crafts reactions (as well as any other reaction before the era of spectroscopic analysis of organic compounds) were based only on boiling points, indices of refraction, and/ or densities. The results presented by Marsi and Wilen were based on the analysis by infrared spectroscopy, mass spectrometry, vapor phase chromatography, and formation of solid derivatives (Marsi and Wilen 1963).

Dunathan proposed an undergraduate experiment of Friedel-Crafts alklylation of n-butyl chloride in benzene using aluminum chloride as a catalyst, which yielded different amounts of n-butylbenzene, sec-butylbenzene, and iso-butylbenzene depending on the reaction temperature (Dunathan 1964).

In Friedel-Crafts reactions using metallic compounds as catalysts and alkyl halides as alkylating agents, the halogen atom from the alkyl halide (e.g., chlorine atom) binds the metallic atom from the metallic compound. In Fig. 20.5(A), there appear several types of alkyl halides binding aluminum or iron atoms in aluminum chloride or iron bromide, respectively, forming the corresponding alkyl halide-metallic compound complex. In some cases, the rearrangement reaction occurs by means of the hydrogen, or methyl shift occurs after the formation of alkyl halide-metallic compound complex (Fig. 20.5(A)). The subsequent formation of secondary/tertiary carbocation and metal halide anion does not occur in gas phase or apolar phase, at least. Our theoretical calculations (in Fig. 20.5(B)) indicate that *tert*-butyl chloride reacts with aluminum chloride releasing 33.2 kcal mol^{-1} (although no BSSE correction was done). On the other hand, chlorine atom from tetrachloroaluminate, AlCl$_4^-$, binds carbon atom of *tert*-butyl cation, releasing 55.8 kcal mol^{-1} from ωB97XD/6-31G++(*d*,*p*) level of theory (Fig. 20.5(C)).

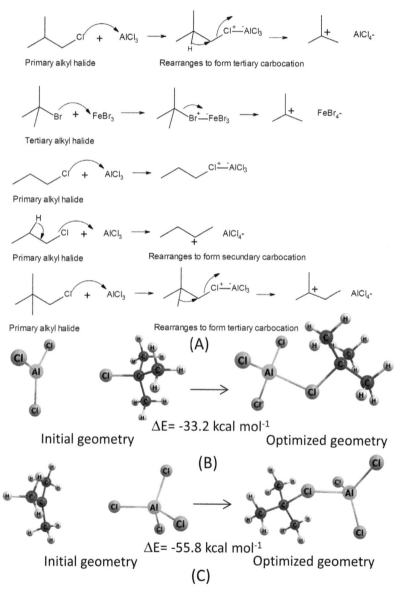

Figure 20.5 (A) Some reactions between alkyl halides and metal halides followed by rearrangement reaction or not; (B) reaction between *tert*-butyl chloride and aluminum chloride; (C) reaction between tetrachloroaluminate and *tert*-butyl cation from ωB97XD/6-31G++(*d*,*p*) level of theory.

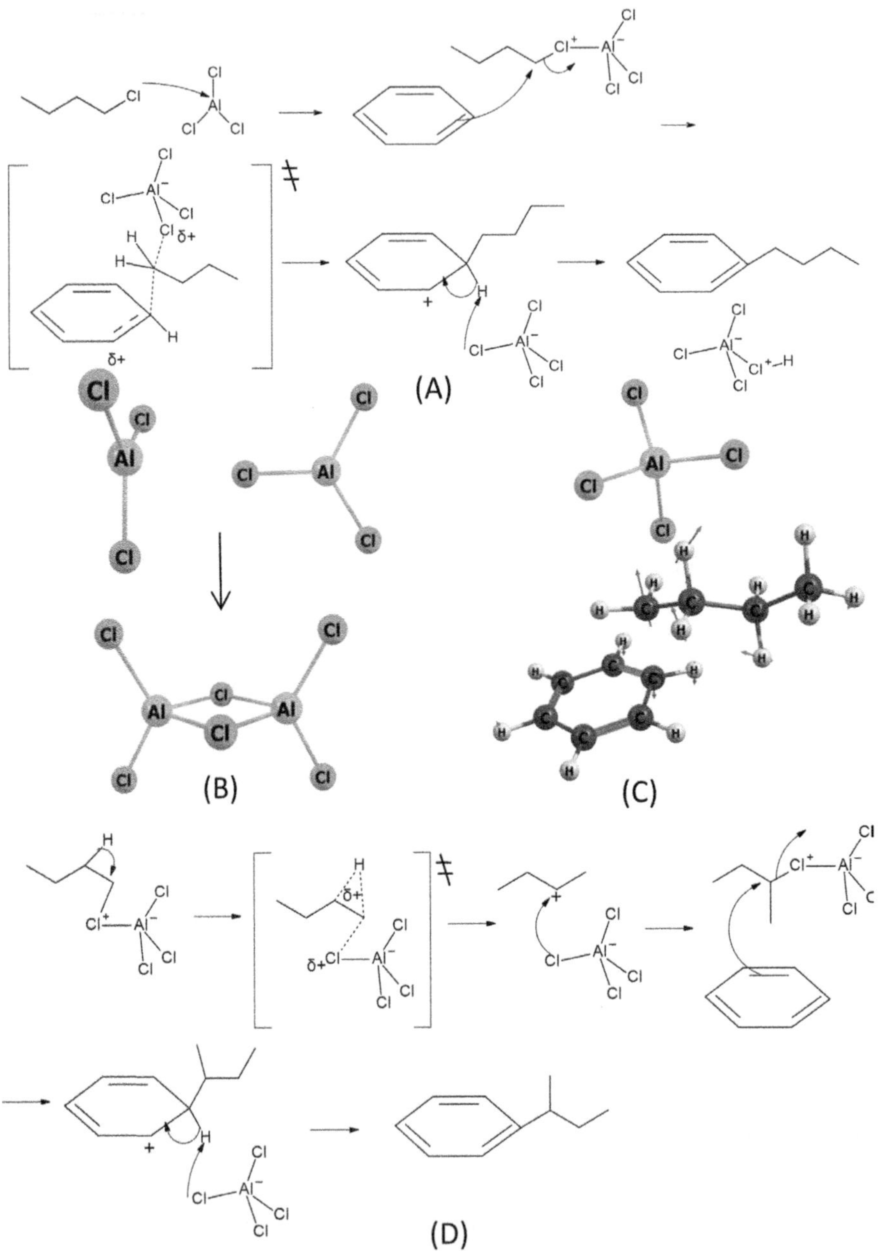

Figure 20.6 (A) Mechanism for the Friedel-Crafts alkylation between *n*-butyl chloride and benzene catalyzed by aluminum chloride yielding *n*-butylbenzene; (B) dimerization reaction of aluminum chloride; (C) activated complex for the mechanism in item (A) from ωB97XD/6-31G++(d,p) level of theory; and (D) mechanism for the Friedel-Crafts alkylation between *n*-butyl chloride and benzene catalyzed by aluminum chloride yielding *sec*-butylbenzene.

As mentioned above, the reaction between *n*-butyl chloride and benzene using aluminum chloride as a catalyst yields considerable amounts of *n*-butylbenzene, *sec*-butylbenzene, and *iso*-butylbenzene. The mechanism for the formation of the normal product (without rearrangement) in Friedel-Crafts alkylation is a displacement reaction where a pair of *p* electrons from benzene attacks the carbon atom bonded to the chlorine atom in the alkyl halide-metal halide complex, for example, butyl chloride-aluminum chloride complex. The transition state is the "π bond" being broken in benzene, a C-C σ bond being formed, and a C-Cl bond being broken (Fig. 20.6(A)). This transition state was confirmed in our own calculations using ωB97XD/6-31G++(*d*,*p*) level of theory (Fig. 20.6(C)). Afterwards, the tetrachloroaluminate removes the proton from the σ complex in a barrierless step to yield the *n*-butylbenzene (Fig. 20.6(A)).

It is important to emphasize that the usual structure of aluminum chloride is Al_2Cl_6 (as a dimer) and not the monomer $AlCl_3$. Figure 20.6(B) shows that two monomers of $AlCl_3$ give the Al_2Cl_6 dimer. However, throughout this book we will refer to $AlCl_3$ instead of Al_2Cl_6 because no sensitive change in the mechanisms involving aluminum chloride will change when using its monomer or its dimer.

As for the rearrangement reaction in the Friedel-Crafts alkylation between *n*-butyl chloride and benzene, for example, to form the sec-adduct, we propose the C2-C1 hydrogen shift in a simultaneous cleavage of C1-Cl bond in the butyl chloride-aluminum chloride complex (Fig. 20.6(D)). Afterwards, one chloride atom from tetrachloroaluminate binds the positively charged C2, which will be subsequently attacked by benzene in a displacement reaction like the mechanism in Fig. 20.6(A).

Unlike primary and secondary alkyl halides, tertiary alkyl halides do not undergo displacement reaction in Friedel-Crafts alkylation, although the alkyl halide-metal halide complex is much more stable than tertiary carbocation and tetrachloroaluminate (when the initial catalyst is aluminum chloride). Probably, there is a dynamic equilibrium with the transient structure of tertiary carbocation and tetrachloroaluminate, from which the benzene attacks the tertiary carbocation to form the σ complex. Afterwards, the tetrachloroaluminate migrate to the opposite plane of benzenoid ring from where it removes the proton of the σ complex in a barrierless step (Fig. 20.7).

Figure 20.8 shows the optimized structures of the reaction between *tert*-butyl chloride and benzene catalyzed by aluminum chloride. The π complex is 20 kcal mol^{-1} more stable than the isolated reactants. There is a huge energy barrier (although no BSSE correction was done) for the rate determining transition state. It follows the σ complex where the tetrachloroaluminate is in the opposite side of the hydrogen atom of the sp^3-hybridized carbon in the arenium ion. Afterwards, the tetrachloroaluminate moves downwards, where it removes this hydrogen (actually, the corresponding proton) without any barrier. All attempts to find the corresponding TS were unsuccessful. In addition, the minimum optimization of the expected σ complex where tetrachloroaluminate is at the same side as the hydrogen of the sp^3 carbon automatically gives the final products (Fig. 20.8).

As we already mentioned, the alkenes are also alkylating agents in Friedel-Crafts alkylation. The alkenes can be activated by aluminum chloride and oxyacid. The oxyacid protonates the alkene (e.g., cyclohexene), forming a carbocation which reacts directly with benzene, yielding the corresponding σ complex. Afterwards, a barrierless step for removal of the proton from sp^3 carbon gives the alkylated aryl compound (Fig. 20.9).

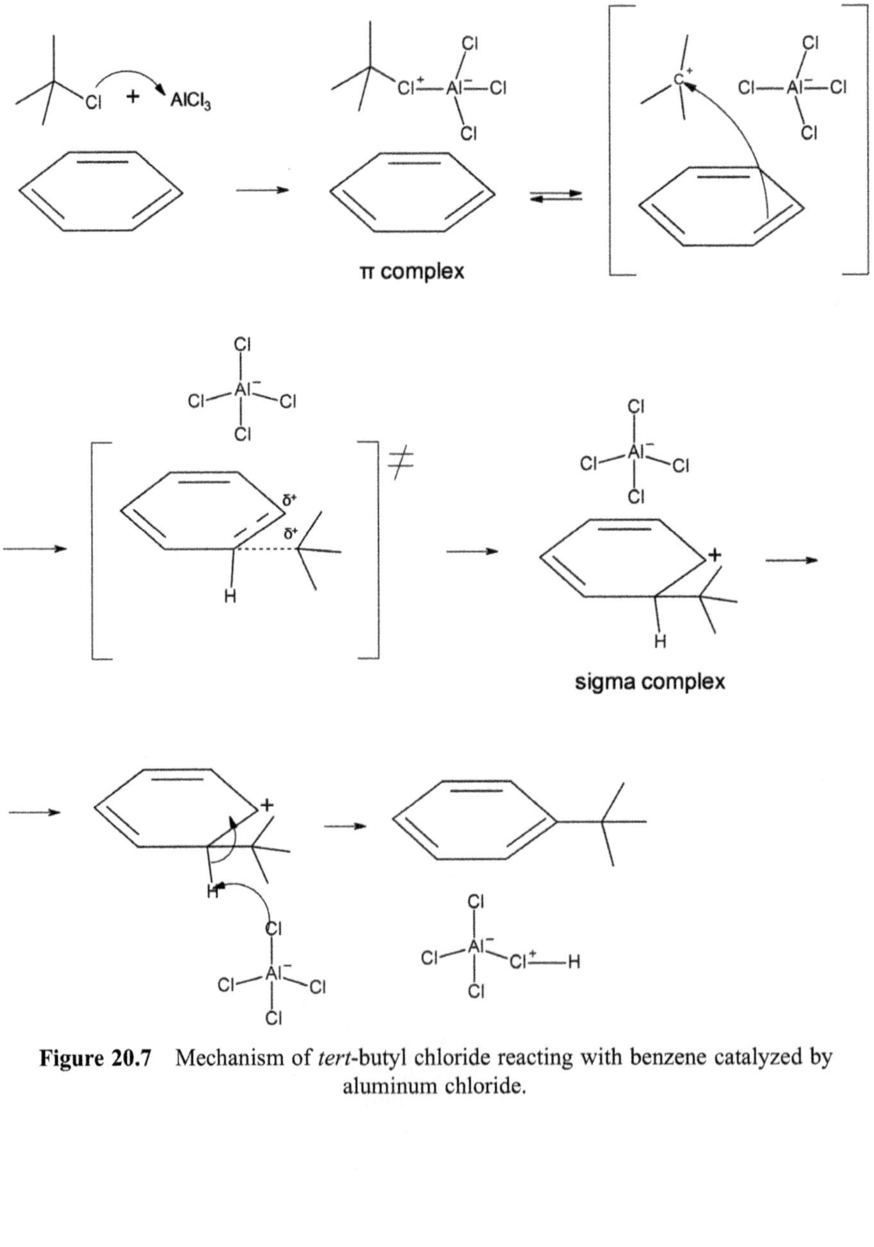

Figure 20.7 Mechanism of *tert*-butyl chloride reacting with benzene catalyzed by aluminum chloride.

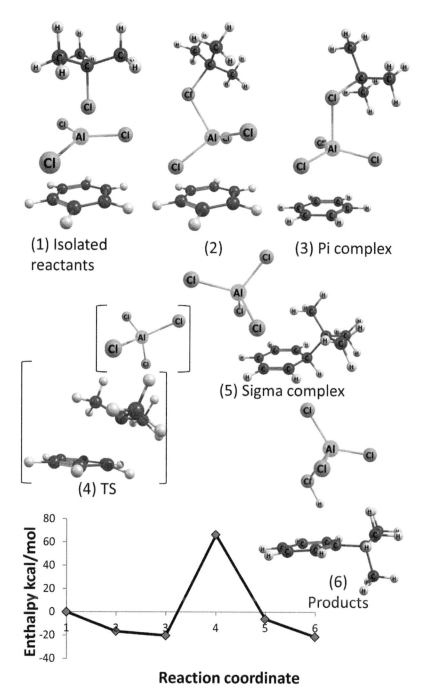

(1) Isolated reactants   (2)   (3) Pi complex

(5) Sigma complex

(4) TS

(6) Products

Figure 20.8   Optimized structures corresponding to the critical points of the potential energy surface of the Friedel-Crafts alkylation of *tert*-butyl chloride in benzene catalyzed by aluminum chloride from ωB97XD/6-31G++(*d,p*) level of theory along with its corresponding potential energy surface.

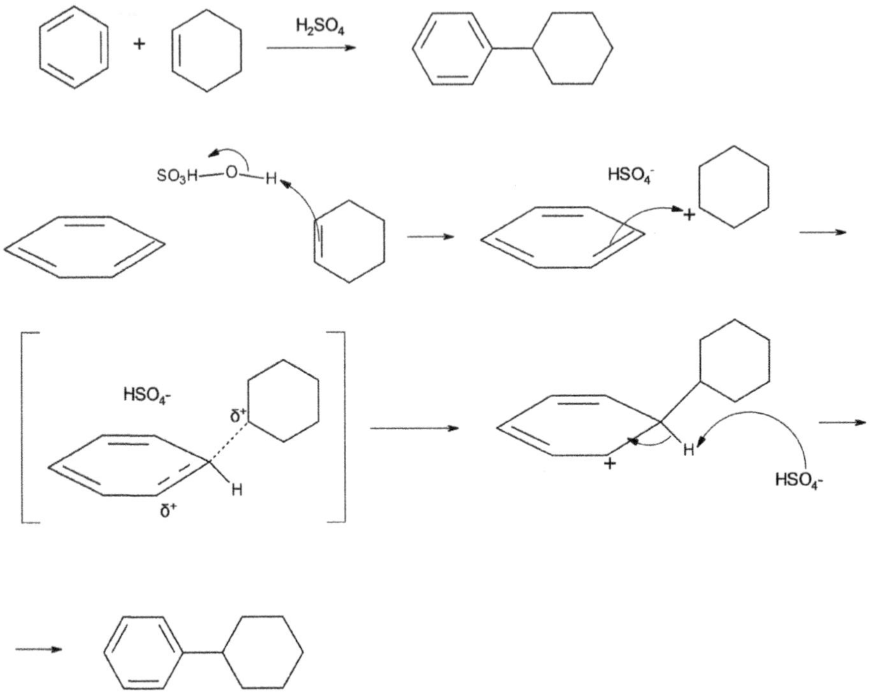

Figure 20.9 Mechanism of Friedel-Crafts alkylation where cyclohexene is the alkylating agent and sulfuric acid is its activator (catalyst).

AROMATIC HALOGENATION

The halogens, such as chlorine and bromine, can also be activated by Friedel-Crafts catalysts, such as aluminum chloride and iron (III) bromide for aromatic halogenation. Figure 20.10 shows the optimized structures of the aluminum chloride-chlorine-benzene π complex. Figure 20.11 shows that aluminum chloride binds chlorine before the π complex. Then, benzene attacks the chlorine atom from chlorine-aluminum chloride complex. Afterwards, it forms the transition state whose barrier is nearly 10 kcal mol^{-1} followed by the σ complex (where the tetrachloroaluminate is at the opposite side of the hydrogen of sp^3 carbon of the arenium ion). When the tetrachloroaluminate moves downwards, it removes the proton without any barrier to form the final product. This reaction is much more exothermic than Friedel-Crafts alkylation because H-Cl and C-Cl bonds are stronger than the Cl-Cl bond.

The reactivity of the halogens in aromatic halogenation is: $I_2 < Br_2 < Cl_2 < F_2$. In some cases, the aromatic halogenation using iodine needs HNO_3 to oxidize it to I^+ in order to react with benzene and toluene. In polar solvents such as aqueous acetic acid, there is an increase in the reactivity of aromatic halogenation in comparison with apolar solvents.

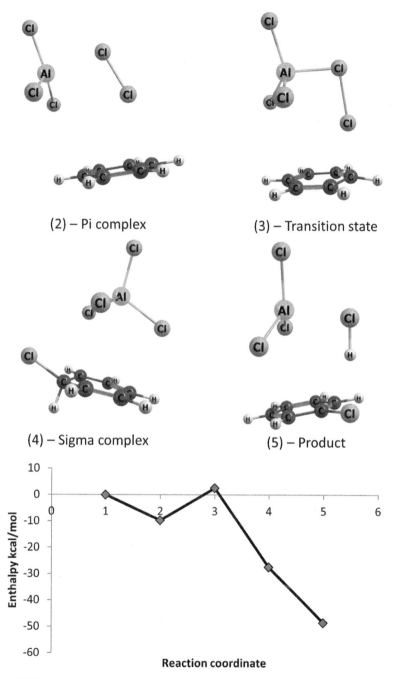

Figure 20.10 Optimized structures corresponding to the critical points of the potential energy surface of the aromatic halogenation of chlorine in benzene catalyzed by aluminum chloride from ωB97XD/6-31G++(d,p) level of theory along with its corresponding potential energy surface.

When not using aluminum chloride as a catalyst, different mechanisms arise. Liljenberg et al. investigated theoretically the halogenation of several mono-substituted benzene molecules using water as a solvent in an uncatalyzed reaction (Liljenberg et al 2018). In this work, they found both σ- and π-complexes. Nonetheless, Galabov and coworkers found distinguished mechanisms for aromatic halogenation which do not pass through σ complex. They found the concerted and the addition-elimination mechanisms for uncatalyzed and HCl-catalyzed chlorinations (Galabov et al. 2014). By using metadynamics, Lommel et al. confirmed that the 1,4-*syn* addition-elimination is the most favorable in HCl-catalyzed aromatic chlorination. They also confirmed the non-existence of σ complex (or a short-lived σ-complex) in an apolar medium (Lommel et al. 2018). Nonetheless, they found the σ-complex in polar solvents.

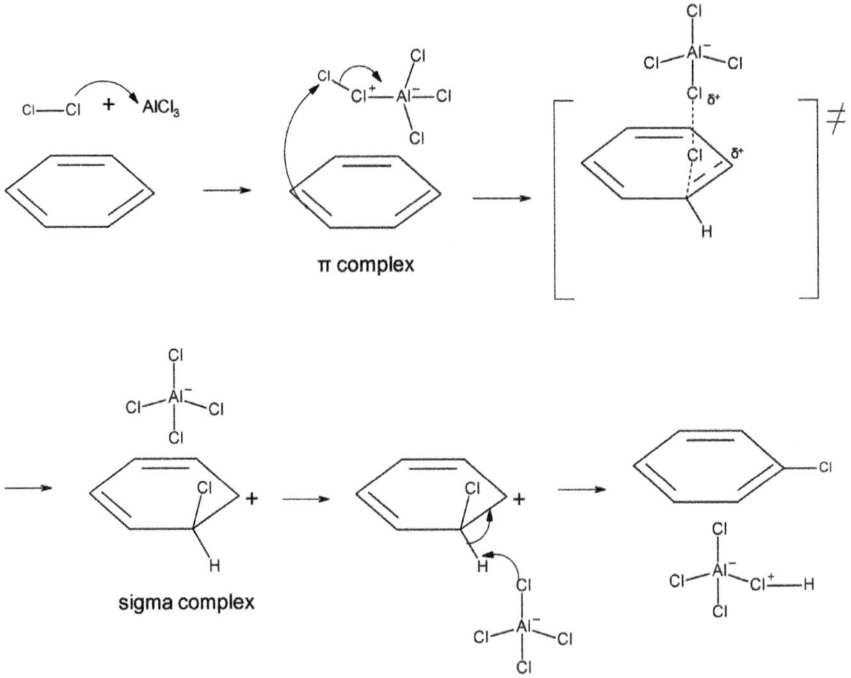

Figure 20.11   Mechanism of the aromatic halogenation of chlorine in benzene catalyzed by aluminum chloride.

FRIEDEL-CRAFTS ACYLATION

An acyl group is an alkyl substituent (R) bonded to a carbonyl carbon (-C(=O)-X), i.e., R-C(O)-X, where X might be -Cl, -OR, -OH, etc. In Friedel-Crafts acylation, this acyl group is added to the benzene (or other aromatic compound) to form $C_6H_5C(O)R$ by means of a metallic compound as a catalyst. The most used solvents are carbon disulfide and nitrobenzene. The alkyl substituent bonded to the alkyl group is unlikely to undergo rearrangement in the Friedel-Crafts acylation.

Then, the Friedel-Crafts acylation can provide alkyl aromatic compound without rearrangement after the reduction of the product of the Friedel-Crafts acylation.

The most used acylating agents are acyl chloride and acyl anhydrides. They form a ketone-metal halide complex (see Fig. 20.12(A)), which reacts with benzene or the other aromatic compound. Prior to the activated complex, the ketone-metal halide complex forms the π complex with benzene (see Fig. 20.12(B)). However, our threotical calculation has shown that benzene cannot attack this ketone-metal halide

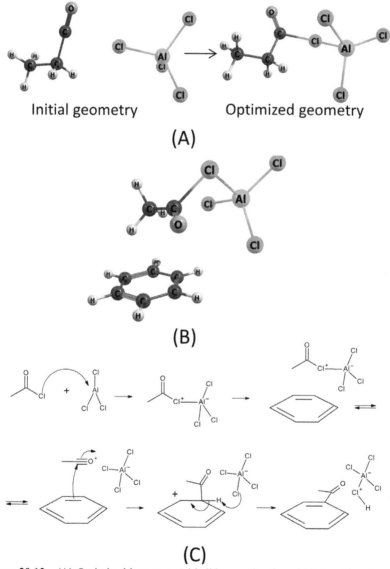

Figure 20.12 (A) Optimized ketone-metal halide complex from initial acylium ion and tetrachloroaluminate; (B) benzene- ketone-metal halide complex π complex; (C) proposed mechanism for Friedel-Crafts acylation.

complex because of the formation of a highly unstable four-charged intermediate. Most probably, there is an equilibrium between the ketone-metal halide complex and the transient structure of acylium ion (R-C≡O⁺) and tetrachloroaluminate (when the catalyst is aluminum chloride). Then, most probably, the acylium ion is attacked by benzene or the other aromatic compound to yield the corresponding σ-complex. Further investigation using metadynamics is necessary to confirm (or not) the mechanism depicted in Fig. 20.12(C).

SUBSTITUENT GROUPS

The substituent groups are divided in two different categories: (1) electron withdrawing group (EWG) or electron donating group (EDG), and (2) EWG/EDG by resonance or inductive effect. Remember that inductive effect has an electrostatic

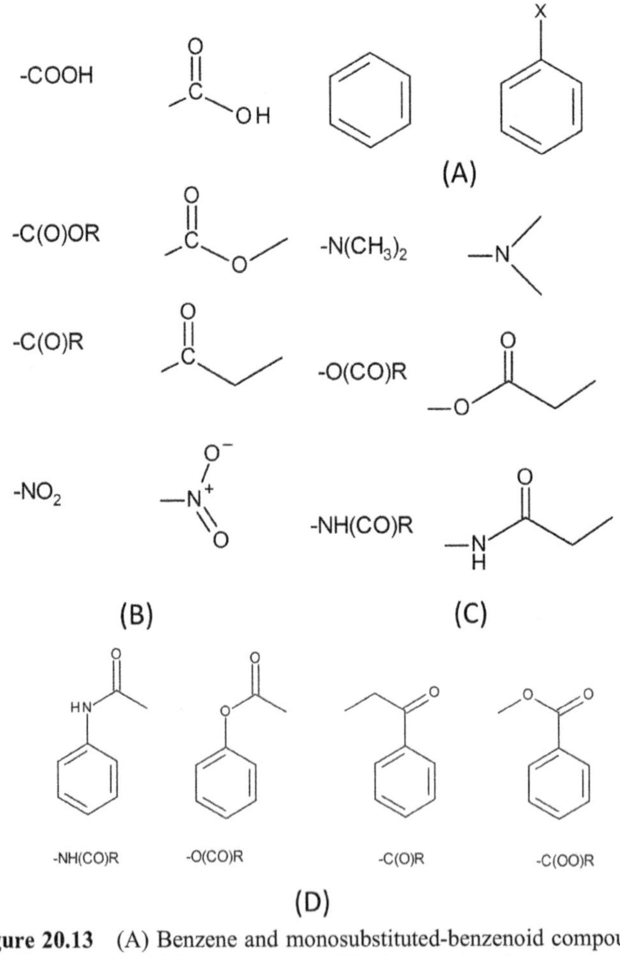

Figure 20.13 (A) Benzene and monosubstituted-benzenoid compound; (B) electron withdrawing groups; (C) electron donating groups; (D) examples of some monosubtituted aromatic molecules.

nature through bond(s) connecting the substituent group and the moiety whose local electron density or charge is attracted or repelled by the former. These categories for substituent groups are universal, and not only applied to the chemistry of aromatic compounds. A general representation of a substituent group is X. For example, Fig. 20.13(A) shows a general monosubstituted aromatic compound C_6H_5X.

The electron withdrawing groups by inductive effect are nitro group ($-NO_2$), nitrile group ($-C\equiv N$), and trifluoromethyl ($-CF_3$), while the EWGs by resonance (and by inductive effect as well) are carbonyl groups ($-C(O)-$), for instance, $-COOH$, $-C(O)OR$, $-C(O)R$ (see Fig. 20.13(B)).

The electron donating groups by inductive effect are alkyl groups ($-CRR'R''$) and sylanyl groups ($-SiRR'R''$). However, one must remember that when the environment is electron-rich, as in alkenes and carbanions, the alkyl groups are electron withdrawing groups. When the first atom of the substituent group has one or more lone pair(s), the substituent group is EDG, no matter if this atom is more electronegative than carbon, and/or it is added to a carbonyl group. They are EDGs by resonance effect, for example, $-NH_2$, $-NRR'$, $-OH$, $-OR$, halogens, $-O(CO)R$, and $NH(CO)R$ (see Fig. 20.13(C)). One exception is the fluorine atom, which can be EWG in some cases.

Figure 20.13(D) shows some mono-substituted aromatic compouinds. The first two (from right to left) have EDG and the other two have EWG.

HAMMETT EQUATION

Hammett defined a scale to measure the influence of substituents on the acidity of substituted benzoic acids and found a linear relation between their acidity and reaction rates of correlated compounds (Hammett 1935, 1937). For example, acidity constants of dissociation reactions (ionization) of p-substituted benzoic acids and rate constants of hydrolysis of p-substituted thyl benzoates (Fig. 20.14).

$$m \cdot \log \frac{K}{K_0} = \log \frac{k}{k_0}$$

Another (and actually the original) Hammett equations are:

$$\log \frac{K}{K_0} = \sigma\rho \quad \log \frac{k}{k_0} = \sigma\rho$$

$$\sigma = \frac{A}{2.303R} \quad \therefore \rho = \frac{1}{d^2T}\left(\frac{B_1}{D} + B_2\right)$$

Where σ is the substituent constant; ρ is the reaction constant; d is the distance from the substituent to the reacting group; D is the dielectric constant; A is a constant dependent upon the substituent; B_1 and B_2 are constants dependent upon the reaction.

In Hammett equations, the substituent constant, σ, shows the effect of the substituent group in free energy of ionization of substituted benzoic acids by the interaction of the substituent group with the reactive site through resonance and/or inductive effect (Hansch et al. 1991). On the other hand, the reaction constant, ρ, measures the sensitivity of a particular reaction to the effects of the substitutent group. While σ is constant for any reaction condition (although there several types of specific substitutent constant), ρ changes with the reaction conditions (e.g., reaction medium). Hammett attributed the value of a unit for the reaction constant involving the ionization of substituted benzoic acids. Hereafter, the value of each substituent constant in this reaction is obtained.

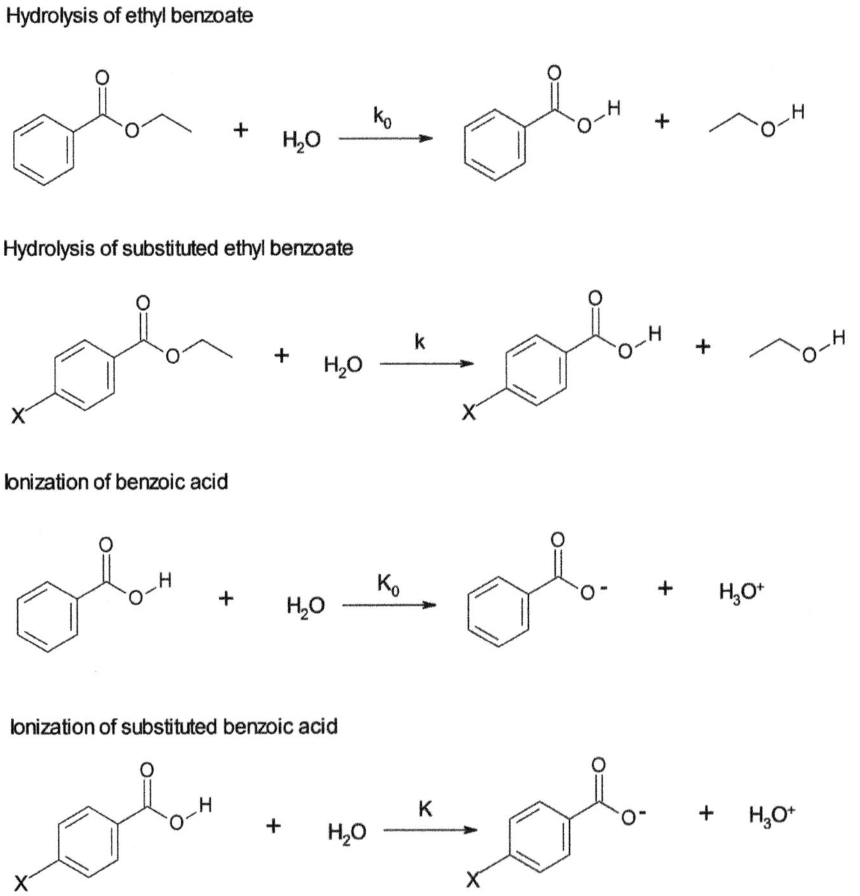

Hydrolysis of ethyl benzoate

Hydrolysis of substituted ethyl benzoate

Ionization of benzoic acid

Ionization of substituted benzoic acid

Figure 20.14 Hydrolysis reaction of (substituted) ethyl benzoate and ionization reaction of (substituted) benzoic acid.

See in Table 20.2 how the reaction constant changes with the type and proportion of solvent mixture in the ionization of substituted benzoic acids. As the ethanol (EtOH) content increases in the solvent, the reaction constant increases, because the substituent tends to increase its influence in the reaction site of the substituted benzoic acid. This is due to the less solvated conjugate base of the substituted benzoic acid (i.e., less delocalized charge of the carboxylate moiety by its smaller interaction with the solvent) as the amount of ethanol in the solvent increases.

Table 20.2   Reaction constant of substituted benzoic acid in its ionization reaction according to solvent type.

Solvent	Reaction constant
H_2O	1.000
H_2O (50%)/EtOH (50%)	1.473
H_2O (30%)/EtOH (70%)	1.738
H_2O (20%)/EtOH (80%)	1.791
H_2O (10%)/EtOH (90%)	1.869
H_2O (5%)/EtOH (95%)	1.890

As mentioned above, in water, the reaction constant is a unit, then it is possible to obtain the values of the substituent constant of several groups. Table 20.3 shows these values along with nucleophilicity of the corresponding benzenoid compounds. The comprehensive table of substituent constants can be found in Hansch and coworkers' paper (Hansch et al. 1991).

THEORETICAL NUCLEOPHILICITY AND SUBSTITUENT CONSTANT

There are several factors that influence nucleophilicity, which will be discussed in the next book (see notes at the beginning of this book). As stated in chapter ten, a nucleophile is a molecule with concentrated electron charge in a specific region called the nucleophilic site. It is possible to obtain experimental and theoretical values for the nucleophilicity, N, of nucleophiles. The experimental values will be discussed in the next book. The theoretical nucleophilicity, N, is the difference in the HOMO energies of the nucleophile and the reference compound (tretracyanoethylene). Table 20.3 shows the values of the theoretical nucleophilicity, in eV, of some aromatic compounds whose corresponding substituent group is indicated in the second column along with their substituent constant. For EWG, Table 20.3 gives the substituent constant of the substituent in meta position, σ_m, while for EDG, it is depicted as σ_p, where the substituent is in para position (see ahead).

Table 20.3   HOMO and theoretical nucleophilicity values, in eV, of some aromatic compounds along with substituent constant of the substituent group attached in each of these aromatic compounds.

Molecule	Substituent group	HOMO/eV	N/eV	$\sigma_{(p/m)}$
Tetracyanoethylene		-11.9		
Benzene	-H	-9.15	2.75	0
Anisole	$-OCH_3$ (EDG)	-8.94	2.96	$-0.27(p)$
Toluene	$-CH_3$ (EDG)	-8.79	3.11	$-0.17(p)$
Phenol	-OH (EDG)	-8.65	3.25	$-0.37(p)$
Aniline	$-NH_2$ (EDG)	-8.09	3.81	$-0.66(p)$
Chlorobenzene	-Cl (EWG)	-9.19	2.71	$0.23(p)$
Benzoic acid	$-CO_2H$ (EWG)	-9.53	2.37	$0.37(m)$
Trifluoromethylbenzene	$-CF_3$ (EWG)	-9.76	2.14	$0.43(m)$
Nitrobenzene	$-NO_2$ (EWG)	-10.09	1.81	$0.71(m)$

One can see that the negative values of the substituent constant belong to EDGs and the positive values of the substituent constant belong to EWG. *The substituent constant can quantify the electron attraction or electron releasing capacity of an EWG or EDG, respectively, of a determined substituent.* The more negative the substituent constant, the higher the electron donating capacity of the corresponding group. The more positive the substituent constant, the higher the electron withdrawing capacity of the corresponding group. From Table 20.3, based on the substituent constant, the group with the highest electron releasing capacity is -NH2 while the group with the highest electron attraction capacity is -NO$_2$.

Figure 20.15 shows the linear correlation between the nucleophility of the aromatic compound with a determined substituent group and the substituent constant of the corresponding substituent. *Then, the substituent constant indicates indirectly the nucleophility of the aromatic compound whose substituent is attached to it.* In another words, an EWG decreases the nuclophilicity of the corresponding aromatic compound, while an EDG increases the nucleophilicity of the corresponding aromatic compound. The higher the electron attraction capacity of the substituent group (i.e., the more positive the substituent constant), the lower the nucleophilicity of the corresponding aromatic compound. The higher the electron releasing capacity of the group (i.e., the more negative the substituent constant), the higher the nucleophilicity of the corresponding aromatic compound.

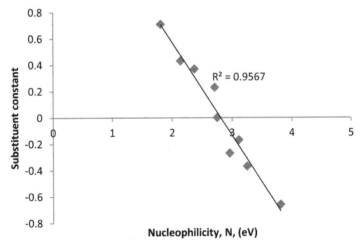

Figure 20.15 Plot of the linear relation between substituent constant and theoretical nucleophilicity.

SUBSTITUENT GROUP AND ELECTROPHILIC AROMATIC SUBSTITUTION

As mentioned above, the EDG increases the nucleophilicity of the aromatic compound in which it is attached (in comparison with benzene), while the opposite occurs with the EWG. As a consequence, *the EDG is called activating group and the EWG is called deactivating group*.

There is another feature associated with EWGs and EDGs, which is specifically related to the regiochemistry/regioselectivity of the electrophilic aromatic substitution (EAS) involving a monosubstituted aromatic compound (C_6H_5X). See below:

$$C_6H_5X+E^+ \rightarrow C_6H_4EX+H^+$$

The EDG attached to the aromatic compound is ortho/para director while the EWG attached to the aromatic compound is meta director. In another words, the EDGs lead to ortho/para regioselective EAS, while the EWGs lead to meta regioselective EAS.

There is one exception: the halogen groups are EWGs (they have positive substituent constant), but they are ortho/para directors. Then, the halogen deactivates the reaction, i.e., decreases the nucleophilicity of the corresponding aromatic compound, but it directs the product of the aromatic substitution to ortho/para positions.

The ortho position means a 1,2-positions in the benzenoid ring for X and E; the meta position means 1,3-positions for X and E; and the para position means the 1,4-positions for X and E; where X (the substituent group attached to the benzenoid ring before the reaction occurs, as a reactant) is the reference group. See Fig. 20.16(A).

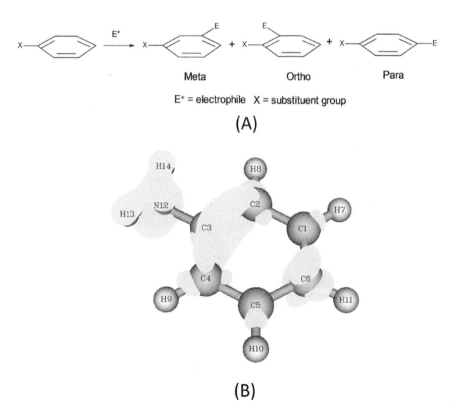

Figure 20.16 (A) General example of EAS of a monosubstituted aromatic compound; (B) Fukui surface for the electrophilic attack, $f^-(r)$, of aniline.

Let us introduce the concept of the Fukui function for the electrophilic attack, $f^-(r)$. It is an important function that indicates visually that EDG is the ortho/para director and EWG is the meta director. It is defined as:

$$f^-(r) = \rho_N(r) - \rho_{N-1}(r)$$

In Fig. 20.16(B), we show the result of the Fukui function for the electrophilic attack of aniline (where -NH$_2$ is the substituent group). It clearly indicates the region where the electrophile is most probable to attack, i.e., in ortho and para positions with respect to the -NH$_2$ group.

The regioselectivity of the monosubstituted aromatic compound can be understood from the resonance model of the arenium ion (σ complex). It is necessary to add the electrophile, E, in the three possible and distinguished positions: ortho, meta, and para with respect to the substituent group.

In the case of EWGs, let us take the nitro group (-NO$_2$) as the substituent group in the aromatic compound. This monosubstituted aromatic compound (nitrobenzene) is attacked in three distinguished positions giving three different arenium ions: 1-nitro, 2-E cyclohexadienyl cation (from ortho attack); 1-nitro, 3-E cyclohexadienyl cation (from meta attack); and 1-nitro, 4-E cyclohexadienyl cation (from para attack). All three arenium ions have three resonance structures (Fig. 20.17). The nitro group is an EWG by inductive effect. Then, removing an electron from a cation moiety is destabilizing, which accounts for the deactivating capacity of the nitro group. Moreover, only in 1-nitro, 3-E cyclohexadienyl cation (from meta attack), there is no resonance structure with a positive charge in C1 (bonded to nitro group). This

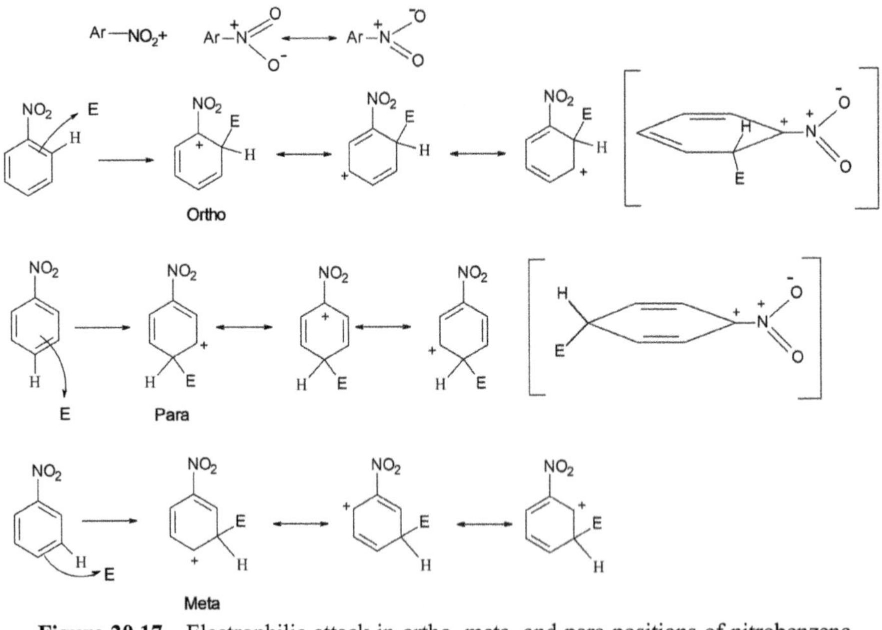

Figure 20.17   Electrophilic attack in ortho, meta, and para positions of nitrobenzene along with the corresponding resonance structures of each arenium cation.

resonance structure, in particular, is specially destabilizing due to the proximity of the positive charge in C1 and in N. This resonance structure occurs only when an electrophile attacks in ortho and para positions. This is the reason the nitro group as well as other EWGs, are meta director.

As for the EDGs, let us take the amino group (-NH$_2$) as the substituent group in the aromatic compound. This monosubstituted aromatic compound (aniline) is attacked in three distinguished positions, giving three different arenium ions: 1-amino,2-E cyclohexadienyl cation (from ortho attack); 1-amino, 3-E cyclohexadienyl cation (from meta attack); and 1-amino, 4-E cyclohexadienyl cation (from para attack). Since amino group is EDG by resonance effect, when the electrophile attacks from ortho and para positions, it yields four resonance structures, while in meta position it gives only three resonance structures (Fig. 20.18). Then, in ortho and para attacks the corresponding arenium ion is more stabilized than in meta position, which accounts for the regioselectivity of monosubstituted aromatic compounds with EDG.

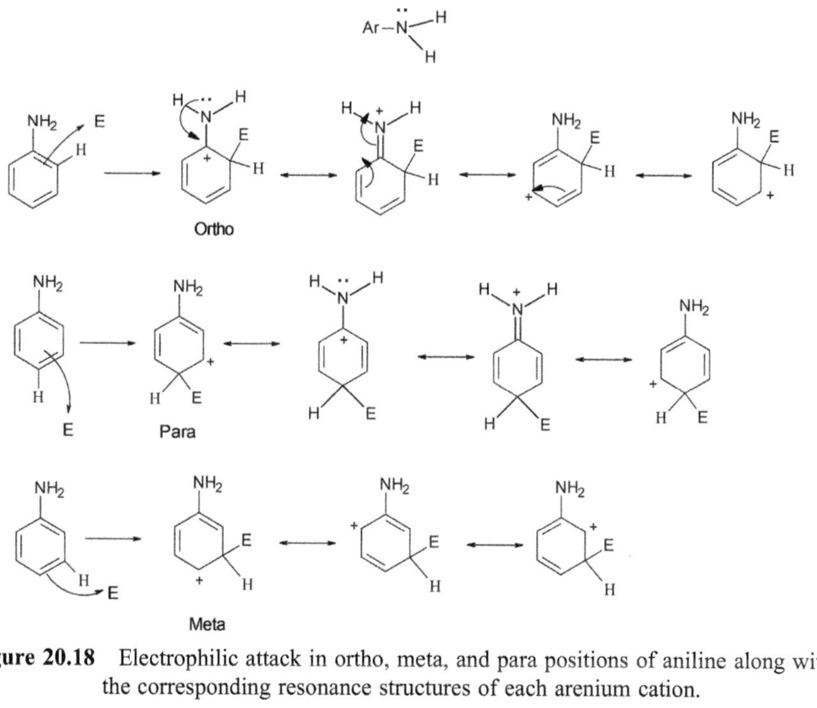

Figure 20.18   Electrophilic attack in ortho, meta, and para positions of aniline along with the corresponding resonance structures of each arenium cation.

As mentioned above, the halogen group is an exception towards the behavior of a substituent group in EAS. They have an ambiguous behavior because they are deactivating, i.e., mono-halogen aromatic compound is less nucleophilic than benzene (as the halogen has positive substituent constant), but the halogen group is ortho/para director. In another words, the halogen group is EWG deactivating but ortho/para director (instead of meta director). The halogen group attracts electron density by inductive effect (which decreases the nucleophilicity of the corresponding aryl halide or haloarene), but it donates electron through resonance effect to the electron-deficient arenium ion. Then, when the electrophile attacks the aryl halide in ortho/para positions, the corresponding arenium ion has four resonance structures while from meta attack the corresponding arenium ion has only three resonance structures (Fig. 20.19). The fourth resonance structure from ortho/para attack of the electrophile results from the electron donation from the halogen group to the electron-deficient cyclohexadienyl moiety, which provides an extra stability to the arenium ion from ortho/meta attacks.

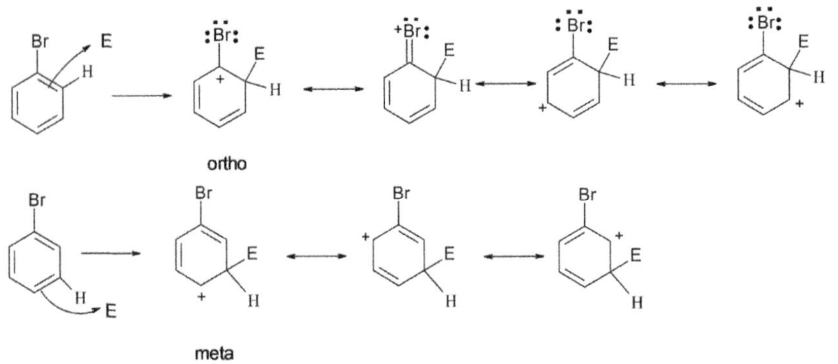

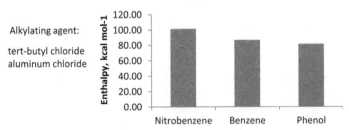

Figure 20.19 Electrophilic attack in ortho and meta positions of bromobenzene along with the corresponding resonance structures of each arenium cation.

From all the observations in this section and the last section we can infer that *the nucleophilicity of the aromatic compounds is directly dependent upon the electron density in the aromatic ring*. The higher the electron density in the aromatic ring, the higher the nucleophilicity of the corresponding aromatic compound, and vice versa. Let us take as example the Friedel-Crafts alkylation of nitrobenzene (where $-NO_2$ is an EWG), benzene (as reference) and phenol (where $-OH$ is an EDG). Figure 20.20 shows the enthalpy barrier of these reactions in gas phase

Friedel-Crafts alkylation Energy barrier

Alkylating agent:

tert-butyl chloride
aluminum chloride

Enthalpy, kcal mol-1

120.00
100.00
80.00
60.00
40.00
20.00
0.00

Nitrobenzene   Benzene   Phenol

Figure 20.20 Bar chart of enthalpy barrier, in kcal mol^{-1}, of the Friedel-Crafts alkylation, using *tert*-butyl chloride/aluminum chloride as alkylating agent, of nitrobenzene, benzene, and phenol from ωB97XD/6-31G++(d,p) level of theory.

calculated by ωB97XD/6-31G++(d,p) level of theory where the alkylating agent was *tert*-butylchloride and aluminum chloride was the catalyst. One can see that the enthalpy barrier decreases in the order nitrobenzene > benzene > phenol. This is because their nucleophilicity increases in the same order nitrobenzene < benzene < phenol owing to the fact the nitro group decreases the electron density of the aromatic ring (and the nucleophilicity of the corresponding substituted aromatic compound), while hydroxyl group increases the electron density of the aromatic ring.

As an example of the ortho/para regioselectivity of an aromatic compound having an EDG as substituent, there is the aromatic nitration of the toluene (whose substituent, -CH$_3$, is an electron donating group by inductive effect). We see only high regioselectivity of this reaction since there is only 4.4% of meta nitrotoluene or *m*-nitrotoluene (Fig. 20.21). On the other hand, chlorination of the nitrobenzene gives a high proportion of meta product (75%) and only one fourth of para and orto products (Fig. 20.21).

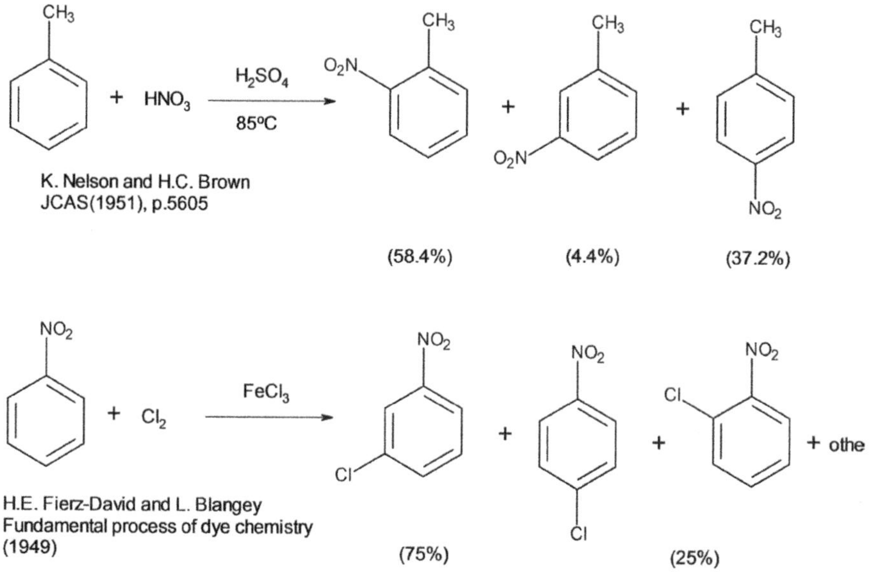

Figure 20.21 Aromatic nitration of toluene and aromatic chlorination of nitrobenzene along with their corresponding references.

SANDMEYER REACTION AND DEAMINATION

The Sandmeyer reaction is used to obtain aryl chloride from aniline, having the diazonium salt as an intermediate. In the first reaction, the aniline is converted into the corresponding diazonium salt when reacting with nitrous acid and hydrogen chloride.

$$C_6H_5NH_2 + NaNO_2 + 2HCl \rightarrow C_6H_5N_2{}^+Cl^- + 2H_2O + NaCl$$

One proposed mechanism for this reaction is given in Fig. 20.22. Nitrous acid is protonated by hydrogen chloride, yielding nitrosonium ion (NO$^+$), which is a strong electrophile and is attacked by amino group of aniline to yield diazonium salt after two prototropisms.

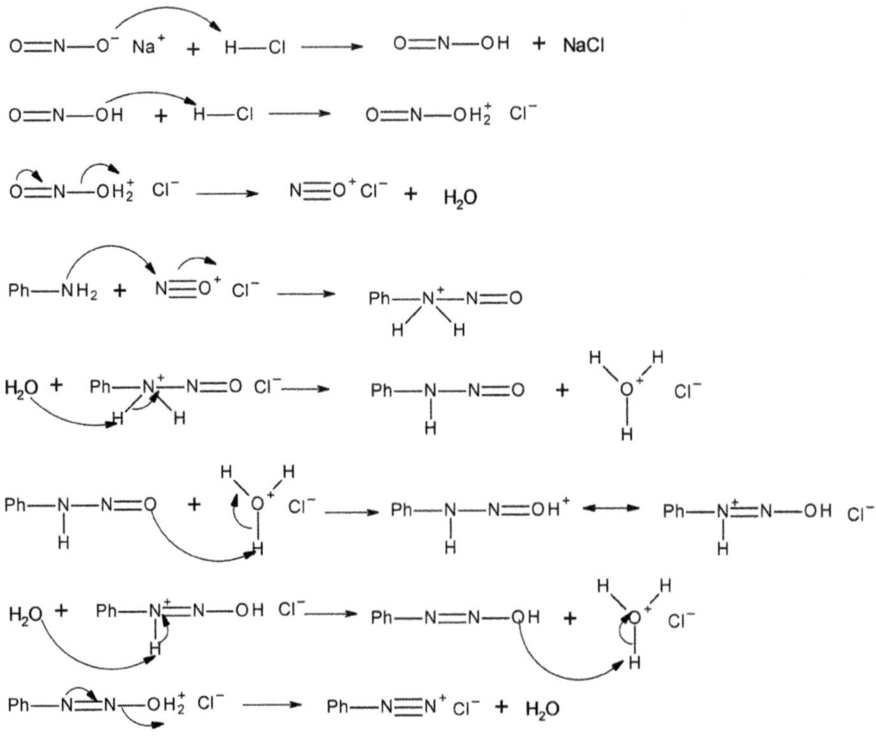

Figure 20.22 Mechanism for the formation of the diazonium salt from aniline reacting with nitrous acid and hydrogen chloride.

The second reaction involves a SET mechanism (a single electron transfer reaction) from copper atom (of cuprous chloride or copper I chloride) to nitrogen of the diazonium salt yielding aryl radical and nitrogen molecule. Afterwards, a radical reaction occurs where the aryl radical abstracts the halogen atom from cupric chloride (or copper II chloride) to yield the aryl chloride and cuprous chloride (Kochi 1957). In this reaction, the cuprous acid has unparalleled electrons and it functions as one-electron reducing agent.

$$PhN_2{}^+Cl^- + Cu^ICl \rightarrow PhN_2\bullet + Cu^{II}Cl_2$$
$$PhN_2\bullet \rightarrow Ph\bullet + N_2$$
$$Ph\bullet + Cu^{II}Cl_2 \rightarrow PhCl + Cu^ICl$$

The aryl diazonium salt can also be used for the deamination reaction with hypophosphorus acid (H$_3$PO$_2$), where one proton of hypophosphorus acid is replaced by the diazo group in the benzenoid moiety (Alexander and Burge 1948). See in Fig. 20.23 one example where deamination reaction is important.

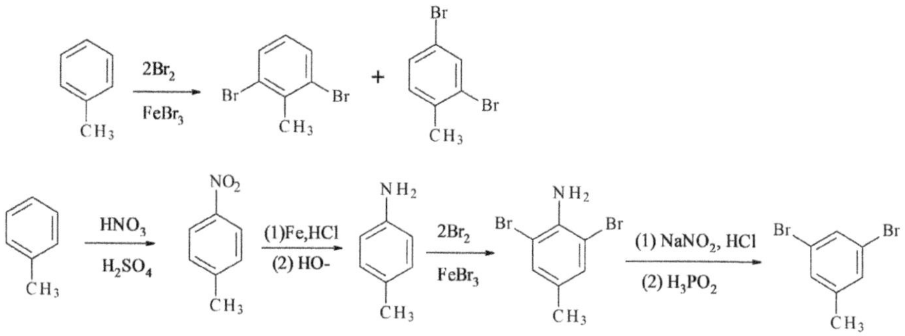

Figure 20.23   Example of the practical importance of deamination reaction.

Acknowledgment

We appreciated the permission of the high performance computer center (NPAD/ IMD) at Federal University of Rio Grande do Norte for the calculation of aromatic nitration stationary points from CCSD method.

EXERCISES

1. Give only the principal products of dibromation and tribromation of benzene. (Tip: the starting point is the bromination of bromobenzene).
2. From ortho-xylene, meta-xylene, and para-xylene, indicate which reactant gives one, two, and three distinguished chloroxylenes.
3. Explain the increasing order of the obtained meta-products of an EAS from the reactants indicated below.
   a. $PhCH_3 < PhCH_2Cl < PhCHCl_2 < PhCCl_3$
   b. $PhCH_2CH_2CF_3 < PhCH_2CF_3 < PhCF_3$
4. Write three chemical reactions to obtain the ortho-nitro-para-bromo-toluene from benzene. (Tip: there is more than one way to obtain the final product, but choose the best procedure, i.e., the most synthetically viable route).
5. Give only the principal products from aromatic sulfonation of o-nitrotoluene, m-nitrotoluene, p-nitrotoluene, o-methylanisole, m-methylanisole, and p-methylanisole.
6. Obtain the 3,5-dichloronitrobenzene from p-nitroaniline.
7. Give the reaction for:
   a. aromatic nitration of 1,3-dinitrobenzene
   b. Friedel-Crafts acylation of para-hydroxyacetophenone using acetyl chloride and aluminum chloride as acylating agent.

☰ **REFERENCES CITED** ☰

Alexander, E.R. and Burge, R.E. 1948. Studies on the mechanism of the deamination of diazonium salts with hypophosphorous acid. J. Am. Chem. Soc. 70: 876-877.

Calloway, N.O. 1935. The Friedel-crafts syntheses. Chem. Rev. 17: 327-392.

Cerfontain, H. 1985. Sulfur trioxide sulfonation of aromatic hydrocarbons. Recl. Trav. Chim. Pays-Bas 104: 153-165.

Chen, Z. and Mo, Y. 2013. Electron transfer in electrophilic aromatic nitration and nitrosation: computational evidence for the Marcus inverted region. J. Chem. Theory Comput. 9: 4428-4435.

Dunathan, H.C. 1964. The Friedel-Crafts alkylation of benzene. A first year organic laboratory experiment. J. Chem. Educ. 41: 278-279.

Galabov, B., Koleva, G., Kong, J. Schaefer III, H.F. and Schleyer, P.v.R. 2014. Addition-elimination versus direct substitution mechanisms for arene chlorination. Eur. J. Org. Chem. 6918-6924.

Gwaltney, S.R., Rosokha, S.V., Head-Gordon, M. and Kochi, J.K. 2003. Charge-transfer mechanism for electrophilic aromatic nitration and nitrosation via the convergence of (ab initio) molecular-orbital and Marcus-Hush theories with experiments. J. Am. Chem. Soc. 125: 3273-3283.

Guk, Y.V., Ilyushin, M.A., Golod, E.L. and Gidaspov, B.V. 1983. Nitronium salts in organic chemistry. Russ. Chem. Rev. 52: 284-297.

Hammett, L.P. 1935. Some relations between reaction rates and equilibrium constants. Chem. Rev. 17: 125-136.

Hammett, L.P. 1937. The effect of structure upon the reactions of organic compounds. Benzene derivatives. J. Am. Chem. Soc. 59: 96-103.

Hansch, C., Leo, A. and Taft, R.W. 1991. A survey of Hammett substituent constants and resonance and field parameters. Chem. Rev. 91: 165-195.

Hughes, E.D., Ingold, C.K. and Reed, R.I. 1950. Kinetics and mechanism of aromatic nitration. Part II. Nitration by nitronium ion, NO_2^+, derived from nitric acid. J. Chem. Soc. 0: 2400-2440.

Kochi, J.K. 1957. The mechanism of the Sandmeyer and Meerwein reactions. J. Am. Chem. Soc. 79: 2942-2948.

Kochi, J.K. 1992. Inner-sphere electron transfer in organic chemistry. Relevance to electrophilic aromatic nitration. Acc. Chem. Res. 25: 39-47.

Koleva, G., Galabov, B., Kong, J., Schaefer, H.F. and Schleyer, P.v.R. 2011. Electrophilic aromatic sulfonation with SO_3: concerted or classic S_EAr mechanism? J. Am. Chem. Soc. 133: 19094-19101.

Liljenberg, M., Stenlid, J.H. and Brinck, T. 2018. Theoretical investigation into rate-determining factors in electrophilic aromatic halogenation. J. Phys. Chem. A. 122: 3270-3279.

Lommel, R.V., Moors, S.L.C. and De Proft, F. 2018. Solvent and autocatalytic effects on the stabilization of the σ-complex during electrophilic aromatic chlorination. Chem. Eur. J. 24: 7044-7050.

Marsi, K.L. and Wilen, S.H. 1963. Friedel-crafts alkylation. J. Chem. Educ. 40: 214-215.

Moors, S.L.C., Deraet, X., Assche, G.V., Geerlings, P. and De Proft, F. 2017. Aromatic sulfonation with sulfur trioxide: mechanism and kinetic model. Chem. Sci. 8: 680-688.

Morkovnik, A.S. and Alkopova, A.R. 2013. Low-barrier concerted electrophilic aromatic substitution under the action of sulfur trioxide. Doklady Chemistry 450: 122-126.

Olah, G.A. 1965. Stable carbonium ions. IX. Methylbenzenonium hexafluoroantimonates. J. Am. Chem. Soc. 87: 1103-1108.

Parker, V.D., Kar, T. and Bethell, D. 2013. The polar mechanism for the nitration of benzene with nitronium ion: ab initio structures of intermediates and transition state. J. Org. Chem. 78: 9522-9525.

Peters, B. 2017. Reaction rate theory and rare events. Elsevier. Amsterdam.

Weiss, J. 1946. Simple electron transfer processes in systems of conjugated double bonds. Trans. Faraday Soc. 42: 116-121.

Wheland, G.W. 1942. A quantum mechanical investigation of the orientation of substituents in aromatic molecules. J. Am. Chem. Soc. 64: 900-908.

Appendices

Appendix One

Bader used the "principle of least action" and the "principle of stationary action" to define the relation between Lagrangian integral and zero surface condition. The "principle of least action" or the "principle of stationary action" is a variational principle (from the calculus of variations) applied to the action (S) of a physical system on the coordinates q_i along a path from time t_1 and t_2 where no change occurs at these two time end-points. Then, the small change in action is null, i.e., $\delta S=0$.

$$S[q(t)] = \int_{t1}^{t2} L[q(t),\dot{q}(t),t]dt \therefore \dot{q} = \frac{dq}{dt}$$

$$L = T - V = \sum_i \tfrac{1}{2}m\dot{q}_i^2 - V(q)$$

Then, δS becomes:

$$\delta S = \int_{t1}^{t2}\sum_i \left\{ (\partial L/\partial q_i)\delta q_i + (\partial L/\partial \dot{q}_i)\delta \dot{q}_i \right\} dt = 0$$

Where $\delta q(t)$ is an arbitrary change in the path at time t.

Then, Bader defined the action on an atom with the Lagrangian integral $L[\psi,t]$ which, in turn, is obtained by integration over the coordinates of all particles in the system of the Lagrangian density, $L[\psi, \nabla \psi, d\psi/dt, t]$.

$$S[\psi] = \int_{t1}^{t2} L[\psi,t] = \int_{t1}^{t2} dt \int d\tau \left\{ L[\psi,\nabla\psi,\dot{\psi},t] \right\}$$

$$L[\psi,\nabla\psi,\dot{\psi},t] = \frac{i\hbar}{2}(\psi * \dot{\psi} - \dot{\psi} * \psi)$$

$$-\frac{\hbar^2}{2m}\sum_i \nabla_i\psi * \cdot \nabla_i\psi - V\psi * \psi$$

After using time-dependent Schrödinger equation on the Lagrangian density, Bader obtained:

$$L^0 = -\frac{\hbar^2}{4m}\sum_i \nabla_i^2 (\psi * \psi)$$

Hence, the single-particle Lagrangian density, $L^0[r,t]$, in terms of the charge density becomes:

$$L^0[r,t] = \int d\tau' L^0 = -\frac{\hbar^2}{4mN}\nabla^2\rho(r,t)$$

Then, the Lagrangian integral that satisfies Schrödinger equation, $L^0[\psi,t]$, becomes:

$$L^0[\psi,t] = \int d\tau' L^0[r,t] = -\frac{\hbar^2}{4mN}\oint dS(r,t)\nabla\rho(r,t)\cdot\mathbf{n}(r,t)$$

When using the Lagrangian integral in a zero flux surface

$$L^0[\psi,t] = -\frac{\hbar^2}{4mN}\oint dS(r,t)\nabla\rho(r,t)\cdot\mathbf{n}(r,t) = 0$$

$$\nabla\rho(r,t)\cdot\mathbf{n}(r,t) = 0 \text{ at ZFS}$$

The *atomic Lagrangian integral* for a subsystem, $L[\psi,\Omega,t]$, and corresponding *atomic action integral*, $S[\psi,\Omega]$, are:

$$L[\psi,\Omega,t] = \int dr \int_\Omega d\tau' \left\{ L[\psi,\nabla\psi,\dot\psi,t] \right\}$$

$$S[\psi,\Omega] = \int_{t1}^{t2} L[\psi,t] = \int_{t1}^{t2} dt \left\{ L[\psi,\Omega,t] \right\}$$

When atomic Lagrangian integral satisfies Schröedinger equation, it becomes $L^0[\psi,\Omega,t]$. The $L^0[\psi,\Omega,t]$ is deduced in the same way $L^0[\psi,t]$ was.

$$L^0[\psi,\Omega,t] = -\frac{\hbar^2}{4mN}\int_\Omega \nabla^2\rho(r,t)$$

$$L^0[\psi,\Omega,t] = -\frac{\hbar^2}{4mN}\oint dS(r,t)\nabla\rho(r,t)\cdot\mathbf{n}(r,t)$$

For the zero-flux surface, where it applies the zero-flux condition, $\rho(r).\mathbf{n}(r)=0$, the atomic Lagrangian integral (and also the atomic action integral) will vanish.

$$L^0[\psi,\Omega,t] = -\frac{\hbar^2}{4mN}\oint dS(r,t)\nabla\rho(r,t)\cdot\mathbf{n}(r,t) = 0 \therefore \textbf{ For ZFS}$$

As a consequence, *the integration of Laplacian of the charge density applied to the entire atomic basin will vanish.* This is the local zero-flux condition.

Appendix Two

The chemical potential and its related Gibbs free energy equations are:

$$\mu_i = \mu_i^0 + RT \ln a_i$$

$$dG = V dp - S dT + \sum_{i=1}^{k} \mu_i dN_i$$

Where μ_i, a_i and N_i are the chemical potential, thermodynamic activity, and number of moles of chemical component i, respectively.

Moreover, we have:

$$dN_i = v_i d\xi$$

Where v_i is the stoichiometric coefficient of i and ξ is the extent of reaction.

By substituting previous equation into second equation, for constant pressure and temperature, we have:

$$\left(\frac{dG}{d\xi}\right)_{T,p} = \sum_{i=1}^{k} \mu_i v_i = \Delta_r G_{T,p}$$

Where $\Delta_r G_{T,p}$ is:

$$\Delta_r G_{T,p} = \chi \mu_{P1} + \delta \mu_{P2} - \alpha \mu_{R1} - \beta \mu_{R2}$$

For the generic reaction:

$$\alpha R_1 + \beta R_2 \rightleftarrows \chi P_1 + \delta P_2$$

By substituting the chemical potentials, we obtain:

$$\Delta_r G_{T,p} = \left(\chi \mu_{P1}^0 + \delta \mu_{P2}^0\right) - \left(\alpha \mu_{R1}^0 + \beta \mu_{R2}^0\right) + \left(\chi RT \ln a_{P1} + \delta RT \ln a_{P2}\right)$$
$$- \left(\alpha RT \ln a_{R1} + \beta RT \ln a_{R2}\right)$$

After some algebraic operations, the previous equation becomes:

$$\Delta_r G_{T,p} = \sum_{i=1}^{k} \mu_i^0 v_i + RT \ln \frac{\left(a_{P1}\right)^{\chi} \cdot \left(a_{P2}\right)^{\delta}}{\left(a_{R1}\right)^{\alpha} \cdot \left(a_{R2}\right)^{\beta}}$$

The reaction quotient, Q_r, is given by:

$$Q_r = \frac{\left(a_{P1}\right)^{\chi} \cdot \left(a_{P2}\right)^{\delta}}{\left(a_{R1}\right)^{\alpha} \cdot \left(a_{R2}\right)^{\beta}}$$

When equilibrium reaction is reached, we have the following relations:

$$\left(\frac{dG}{d\xi}\right)_{T,p} = \Delta_r G_{T,p} = 0 \therefore Q_r = K$$

Knowing the relation between chemical potential and Gibbs free energy as:

$$\sum_{i=1}^{k} \mu_i^0 v_i = \Delta_r G^0$$

Then, after some substitutions, we have:

$$\Delta_r G^0 = -RT \ln K$$

Appendix Three

Relation between Statistical Thermodynamics and $K^{\neq}$

The proof of the statistical thermodynamics relation with $K^{\neq}$ comes from the relation between entropy and partition function and between entropy and molar partition function:

$$S = \frac{[U - U_0]}{T} + Nk_B \ln q$$

$$S = \frac{[U - U_0]}{T} + k_B \ln Q$$

and the equation for absolute Gibbs free energy (Fogler 2011):

$$G = U - TS + PV$$

$$G = U - TS + nRT$$

By combining:

$$S = \frac{[U - U_0]}{T} + k_B \ln Q$$

$$G = U - TS + nRT$$

Then:

$$G = U_0 - k_B T \ln Q + nRT$$

Since:

$$Q = \frac{q^N}{N!}$$

Then:

$$G = U_0 + nRT - Nk_B T \ln q + k_B T \ln N!$$

Since:

$$N = nN_A$$

Where N_A is the Avogadro number and using Stirling approximation, we have:

$$G = U_0 + nRT - nN_A k_B T \ln q + k_B T (N \ln N - N)$$

Since:

$$k_B = \frac{R}{N_A} \therefore R = k_B N_A$$

$$k_B TN = k_B TN_A n = nRT$$

Then:

$$G = U_0 - nRT \ln q + nRT \ln N - nRT + nRT$$

$$G = U_0 - nRT \ln \frac{q}{N}$$

To put the values of G and U_0 given per mole we have to divide the last equation by n and using $N = nN_A$ and $q_m = q/n$ relations:

$$G = U_0 - RT \ln \frac{q}{N}$$

$$q_m = \frac{q}{n} \therefore N = nN_A$$

$$G = U_0 - RT \ln \frac{q_m}{N_A}$$

From the equilibrium $A + B \rightleftarrows [A\text{-}\text{-}\text{-}B]^{\neq}$, we have

$$G^{\neq} - G_A - G_B = \Delta G = -RT \ln K^{\neq}$$

$$G^{\neq} - G_A - G_B = \Delta U_0 - RT \ln \left(\frac{q_m^{\neq}}{q_{mA} q_{mB}} N_A \right)$$

By combining both equations and using the partition function per unit volume, q', we have:

$$K^{\neq} = \left(\frac{q'^{\neq}}{q'_A q'_B} N_A \right) \exp \left(-\frac{\Delta E_0}{RT} \right)$$

REFERENCE CITED

Fogler, H.S. 2011. Essentials of Chemical Reaction Engineering. Prentice Hall, Pearson Education, Inc. Westford.

Appendix Four

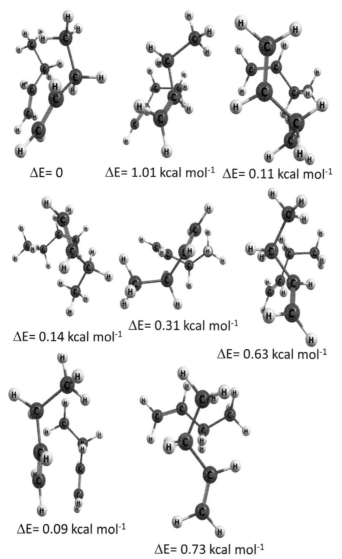

$\Delta E= 0$ $\quad\quad$ $\Delta E= 1.01$ kcal mol^{-1} $\quad$ $\Delta E= 0.11$ kcal mol^{-1}

$\Delta E= 0.14$ kcal mol^{-1}

$\Delta E= 0.31$ kcal mol^{-1}

$\Delta E= 0.63$ kcal mol^{-1}

$\Delta E= 0.09$ kcal mol^{-1}

$\Delta E= 0.73$ kcal mol^{-1}

Figure A3.1 Optimized complex conformers of but-1-ene from ωB97XD/6-311G++(d,p) level of theory.

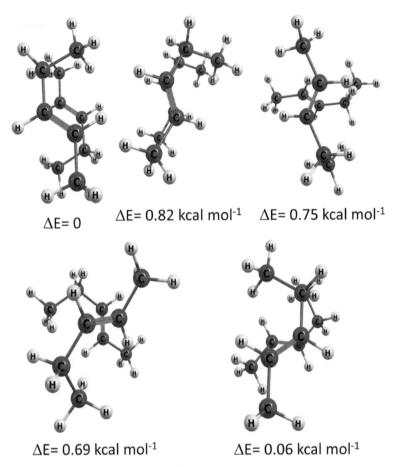

$\Delta E = 0$ $\qquad$ $\Delta E = 0.82$ kcal mol^{-1} $\qquad$ $\Delta E = 0.75$ kcal mol^{-1}

$\Delta E = 0.69$ kcal mol^{-1} $\qquad\qquad$ $\Delta E = 0.06$ kcal mol^{-1}

Figure A3.2 Optimized complex conformers of *trans*-pent-2-ene from ωB97XD/6-311G++(d,p) level of theory.

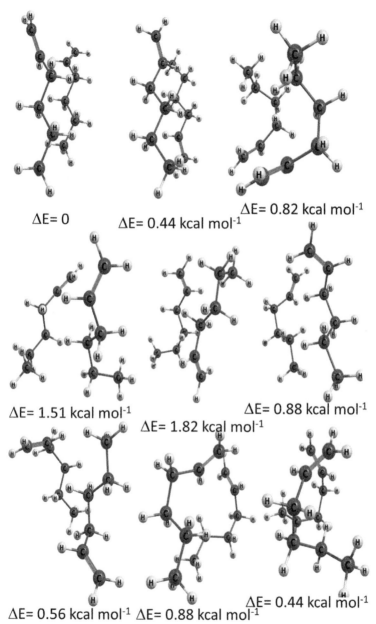

ΔE= 0   ΔE= 0.44 kcal mol⁻¹   ΔE= 0.82 kcal mol⁻¹

ΔE= 1.51 kcal mol⁻¹   ΔE= 1.82 kcal mol⁻¹   ΔE= 0.88 kcal mol⁻¹

ΔE= 0.56 kcal mol⁻¹ ΔE= 0.88 kcal mol⁻¹   ΔE= 0.44 kcal mol⁻¹

Figure A3.3   Optimized complex conformers of hex-1-ene from
ωB97XD/6-311G++(*d,p*) level of theory.

Appendix Five

Input for the GVB calculation of benzene considering only one of its Kekulé's resonance structure in VB2000 package.

#! VB(6)/6-31++G** PRINTALL

benzene

0 1

6	1.111766000	0.840315000	0.000000000
6	1.283619000	−0.542647000	−0.000003000
6	0.171905000	−1.382914000	−0.000002000
6	−1.111801000	−0.840269000	0.000000000
6	−1.283641000	0.542594000	0.000000000
6	−0.171848000	1.382921000	0.000004000
1	1.977461000	1.494635000	0.000002000
1	2.283202000	−0.965040000	0.000005000
1	0.305758000	−2.459790000	0.000014000
1	−1.977406000	−1.494709000	0.000002000
1	−2.283168000	0.965121000	−0.000008000
1	−0.305847000	2.459779000	−0.000005000

$MEMORY

90000000

$02PIVBO

6

1,2,3,4,5,6

****

Input for the GVB calculation of benzene considering all five resonance structures in VB2000 package.

#! VB(6)/6-31++G** PRINTALL

benzene

0 1

6	1.111766000	0.840315000	0.000000000
6	1.283619000	−0.542647000	−0.000003000
6	0.171905000	−1.382914000	−0.000002000
6	−1.111801000	−0.840269000	0.000000000
6	−1.283641000	0.542594000	0.000000000
6	−0.171848000	1.382921000	0.000004000
1	1.977461000	1.494635000	0.000002000

1	2.283202000	−0.965040000	0.000005000
1	0.305758000	−2.459790000	0.000014000
1	−1.977406000	−1.494709000	0.000002000
1	−2.283168000	0.965121000	−0.000008000
1	−0.305847000	2.459779000	−0.000005000

$MEMORY
90000000
$02PIVBO
6
1,2,3,4,5,6
$02VBSTR
5
1 2 3 4 5 6
2 3 4 5 6 1
1 4 2 3 5 6
2 5 3 4 1 6
3 6 1 2 4 5

****

Input for the GVB calculation of benzene considering only one of its Kekulé's resonance structure in VB2000 package.

#! SCVB(6)/6-31++G** PRINTALL

benzene

0 1

6	1.111766000	0.840315000	0.000000000
6	1.283619000	−0.542647000	−0.000003000
6	0.171905000	−1.382914000	−0.000002000
6	−1.111801000	−0.840269000	0.000000000
6	−1.283641000	0.542594000	0.000000000
6	−0.171848000	1.382921000	0.000004000
1	1.977461000	1.494635000	0.000002000
1	2.283202000	−0.965040000	0.000005000
1	0.305758000	−2.459790000	0.000014000
1	−1.977406000	−1.494709000	0.000002000
1	−2.283168000	0.965121000	−0.000008000
1	−0.305847000	2.459779000	−0.000005000

$MEMORY
90000000

Index

Color Plate Section

Chapter 2

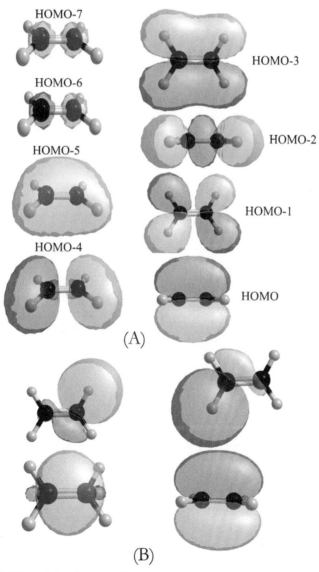

Figure 2.1 (A) Occupied molecular orbitals of ethene; (B) selected occupied NBOs of ethene.

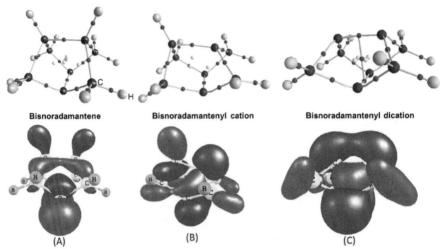

Figure 2.2 HOMO orbitals of (A) bisnoradamantane, (B) bisnoradamantenyl cation, and (C) bisnoradamantenyl dication, and their corresponding molecular graphs (see the next section). Courtesy of Springer-Verlag (see the Acknowledgment).

Chapter 11

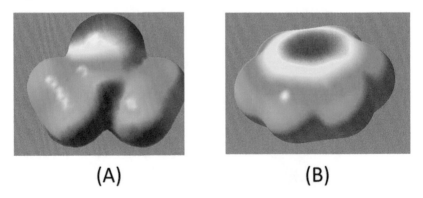

Figure 11.4 Electrostatic potential map of (A) acetone and (B) benzene.

Printed and bound by CPI Group (UK) Ltd, Croydon, CR0 4YY

24/10/2024

01778308-0015